VOLUME ONE HUNDRED AND THIRTY EIGHT

ADVANCES IN
CANCER RESEARCH

VOLUME ONE HUNDRED AND THIRTY EIGHT

Advances in
CANCER RESEARCH

Edited by

KENNETH D. TEW
Medical University of South Carolina,
Charleston, SC, United States

PAUL B. FISHER
VCU Institute of Molecular Medicine;
VCU Massey Cancer Center,
Virginia Commonwealth University,
School of Medicine,
Richmond, VA, United States

Academic Press is an imprint of Elsevier
50 Hampshire Street, 5th Floor, Cambridge, MA 02139, United States
525 B Street, Suite 1800, San Diego, CA 92101-4495, United States
The Boulevard, Langford Lane, Kidlington, Oxford OX5 1GB, United Kingdom
125 London Wall, London, EC2Y 5AS, United Kingdom

First edition 2018

Copyright © 2018 Elsevier Inc. All rights reserved.

No part of this publication may be reproduced or transmitted in any form or by any means, electronic or mechanical, including photocopying, recording, or any information storage and retrieval system, without permission in writing from the publisher. Details on how to seek permission, further information about the Publisher's permissions policies and our arrangements with organizations such as the Copyright Clearance Center and the Copyright Licensing Agency, can be found at our website: www.elsevier.com/permissions.

This book and the individual contributions contained in it are protected under copyright by the Publisher (other than as may be noted herein).

Notices
Knowledge and best practice in this field are constantly changing. As new research and experience broaden our understanding, changes in research methods, professional practices, or medical treatment may become necessary.

Practitioners and researchers must always rely on their own experience and knowledge in evaluating and using any information, methods, compounds, or experiments described herein. In using such information or methods they should be mindful of their own safety and the safety of others, including parties for whom they have a professional responsibility.

To the fullest extent of the law, neither the Publisher nor the authors, contributors, or editors, assume any liability for any injury and/or damage to persons or property as a matter of products liability, negligence or otherwise, or from any use or operation of any methods, products, instructions, or ideas contained in the material herein.

ISBN: 978-0-12-815127-3
ISSN: 0065-230X

For information on all Academic Press publications
visit our website at https://www.elsevier.com/books-and-journals

Publisher: Zoe Kruze
Acquisition Editor: Zoe Kruze
Editorial Project Manager: Fenton Coulthurst
Production Project Manager: James Selvam
Cover Designer: Greg Harris

Typeset by SPi Global, India

CONTENTS

Contributors ix

1. **Leveraging Epigenetics to Enhance the Cellular Response to Chemotherapies and Improve Tumor Immunogenicity**   1
   Liliya Tyutyunyk-Massey, Syed U. Haqqani, Reshma Mandava, Kirubel Kentiba, Mallika Dammalapati, Nga Dao, Joshua Haueis, David Gewirtz, and Joseph W. Landry

   1. Introduction   2
   2. Targeting Epigenetic Regulators to Achieve Sensitization to ICD-Inducing Chemotherapies   9
   3. Conclusions and Future Directions   23
   Acknowledgments   27
   Conflicts of Interest   27
   References   28

2. **VDAC Regulation: A Mitochondrial Target to Stop Cell Proliferation**   41
   Diana Fang and Eduardo N. Maldonado

   1. Introduction   42
   2. VDAC Channels and Mitochondrial Metabolism   48
   3. VDAC–Tubulin Interaction   50
   4. Tumor Metabolic Flexibility: Advantages of Targeting Metabolism in Chemotherapy   52
   5. VDAC–Tubulin Antagonists: A Strategy for Opening VDAC   55
   6. Concluding Remarks   61
   Acknowledgment   61
   References   61

3. **Acquired Resistance to Drugs Targeting Tyrosine Kinases**   71
   Steven A. Rosenzweig

   1. Introduction   73
   2. Inhibition of Bcr-Abl and Nonreceptor Tyrosine Kinases   74
   3. Receptor and Nonreceptor Tyrosine Kinases Activate Common Pathways   77
   4. Receptor TKIs and the EGFR Family   80
   5. Epigenetic Mechanisms of Resistance   84
   6. IGF-1R and Dependence Receptors in Drug Resistance   89

7.	Conclusions and Future Perspective	91
	Acknowledgments	92
	References	92

4. Extracellular-Regulated Kinases: Signaling From Ras to ERK Substrates to Control Biological Outcomes   99
Scott T. Eblen

1.	Introduction	100
2.	Identification of Extracellular-Regulated Kinases	101
3.	Ras to MAP Kinase Kinases	102
4.	ERK1 and ERK2	109
5.	ERK Substrates	116
6.	Concluding Remarks	129
	Acknowledgments	130
	References	130

5. Role of MDA-7/IL-24 a Multifunction Protein in Human Diseases   143
Mitchell E. Menezes, Praveen Bhoopathi, Anjan K. Pradhan, Luni Emdad, Swadesh K. Das, Chunqing Guo, Xiang-Yang Wang, Devanand Sarkar, and Paul B. Fisher

1.	Introduction	144
2.	Characteristic Features of MDA-7/IL-24	145
3.	Physiological Role of MDA-7/IL-24	150
4.	Functional Role of MDA-7/IL-24 in Cancer	152
5.	Role of MDA-7/IL-24 in Other Diseases	162
6.	Immunological Effects of MDA-7/IL-24	169
7.	Conclusions and Future Perspectives	171
	Acknowledgments	172
	Conflict of Interest	172
	References	172

6. Advances and Challenges of HDAC Inhibitors in Cancer Therapeutics   183
Jesse J. McClure, Xiaoyang Li, and C. James Chou

1.	Histone Deacetylase (HDAC) Inhibitors and Their Targets	184
2.	Classes of HDACs	185
3.	Lysine Deacylases	186
4.	Major Classes of HDAC Inhibitors	187

5.	HDAC Inhibitors Have Unique NCI 60 Screening Profiles	190
6.	HDAC Inhibitors in the Clinic	192
7.	Challenges in Solid Tumor	193
8.	Pharmacokinetic Challenges of HDAC Inhibitors	195
9.	The History of Hydrazide-Containing Compounds in Clinic and the Future of Next-Generation HDAC Inhibitors	200
10.	Conclusion	203
	References	203

7. Prospects of Gene Therapy to Treat Melanoma — 213

Mitchell E. Menezes, Sarmistha Talukdar, Stephen L. Wechman, Swadesh K. Das, Luni Emdad, Devanand Sarkar, and Paul B. Fisher

1.	Introduction	214
2.	Targets for Gene Therapy	215
3.	Gene Therapy in Melanoma	219
4.	Challenges of Gene Therapy	222
5.	Immunotherapy and Combination Therapy	227
6.	Conclusions and Future Directions	231
	Acknowledgments	232
	Conflict of Interest	232
	References	232

CONTRIBUTORS

Praveen Bhoopathi
Virginia Commonwealth University, School of Medicine, Richmond, VA, United States

C. James Chou
Medical University of South Carolina, College of Pharmacy, Charleston, SC, United States

Mallika Dammalapati
VCU Institute of Molecular Medicine, VCU Massey Cancer Center, Virginia Commonwealth University School of Medicine, Richmond, VA, United States

Nga Dao
VCU Institute of Molecular Medicine, VCU Massey Cancer Center, Virginia Commonwealth University School of Medicine, Richmond, VA, United States

Swadesh K. Das
VCU Institute of Molecular Medicine; VCU Massey Cancer Center, Virginia Commonwealth University, School of Medicine, Richmond, VA, United States

Scott T. Eblen
Medical University of South Carolina, Charleston, SC, United States

Luni Emdad
VCU Institute of Molecular Medicine; VCU Massey Cancer Center, Virginia Commonwealth University, School of Medicine, Richmond, VA, United States

Diana Fang
Medical University of South Carolina, Charleston, SC, United States

Paul B. Fisher
VCU Institute of Molecular Medicine; VCU Massey Cancer Center, Virginia Commonwealth University, School of Medicine, Richmond, VA, United States

David Gewirtz
VCU Massey Cancer Center, Virginia Commonwealth University School of Medicine, Richmond, VA, United States

Chunqing Guo
Virginia Commonwealth University, School of Medicine, Richmond, VA, United States

Syed U. Haqqani
VCU Institute of Molecular Medicine, VCU Massey Cancer Center, Virginia Commonwealth University School of Medicine, Richmond, VA, United States

Joshua Haueis
VCU Institute of Molecular Medicine, VCU Massey Cancer Center, Virginia Commonwealth University School of Medicine, Richmond, VA, United States

Kirubel Kentiba
VCU Institute of Molecular Medicine, VCU Massey Cancer Center, Virginia Commonwealth University School of Medicine, Richmond, VA, United States

Joseph W. Landry
VCU Institute of Molecular Medicine, VCU Massey Cancer Center, Virginia Commonwealth University School of Medicine, Richmond, VA, United States

Xiaoyang Li
Medical University of South Carolina, College of Pharmacy, Charleston, SC, United States

Eduardo N. Maldonado
Hollings Cancer Center, Medical University of South Carolina, Charleston, SC, United States

Reshma Mandava
VCU Institute of Molecular Medicine, VCU Massey Cancer Center, Virginia Commonwealth University School of Medicine, Richmond, VA, United States

Jesse J. McClure
Medical University of South Carolina, College of Pharmacy, Charleston, SC, United States

Mitchell E. Menezes
Virginia Commonwealth University, School of Medicine, Richmond, VA, United States

Anjan K. Pradhan
Virginia Commonwealth University, School of Medicine, Richmond, VA, United States

Steven A. Rosenzweig
Hollings Cancer Center, Medical University of South Carolina, Charleston, SC, United States

Devanand Sarkar
VCU Institute of Molecular Medicine; VCU Massey Cancer Center, Virginia Commonwealth University, School of Medicine, Richmond, VA, United States

Sarmistha Talukdar
Virginia Commonwealth University, School of Medicine, Richmond, VA, United States

Liliya Tyutyunyk-Massey
VCU Massey Cancer Center, Virginia Commonwealth University School of Medicine, Richmond, VA, United States

Xiang-Yang Wang
VCU Institute of Molecular Medicine; VCU Massey Cancer Center, Virginia Commonwealth University, School of Medicine, Richmond, VA, United States

Stephen L. Wechman
Virginia Commonwealth University, School of Medicine, Richmond, VA, United States

CHAPTER ONE

Leveraging Epigenetics to Enhance the Cellular Response to Chemotherapies and Improve Tumor Immunogenicity

Liliya Tyutyunyk-Massey[†], Syed U. Haqqani[*], Reshma Mandava[*], Kirubel Kentiba[*], Mallika Dammalapati[*], Nga Dao[*], Joshua Haueis[*], David Gewirtz[†], Joseph W. Landry[*,1]

[*]VCU Institute of Molecular Medicine, VCU Massey Cancer Center, Virginia Commonwealth University School of Medicine, Richmond, VA, United States
[†]VCU Massey Cancer Center, Virginia Commonwealth University School of Medicine, Richmond, VA, United States
[1]Corresponding author: e-mail address: joseph.landry@vcuhealth.org

Contents

1. Introduction 2
   1.1 Chemotherapy and the Induced Immune Response 2
   1.2 The Value of Chemosensitization 5
   1.3 Targeting Epigenetics to Achieve Chemosensitization 6
2. Targeting Epigenetic Regulators to Achieve Sensitization to ICD-Inducing Chemotherapies 9
   2.1 Writers 9
   2.2 Erasers 16
   2.3 Readers 20
   2.4 miRNAs 21
3. Conclusions and Future Directions 23
Acknowledgments 27
Conflicts of Interest 27
References 28

Abstract

Cancer chemotherapeutic drugs have greatly advanced our ability to successfully treat a variety of human malignancies. The different forms of stress produced by these agents in cancer cells result in both cell autonomous and cell nonautonomous effects. Desirable cell autonomous effects include reduced proliferative potential, cellular senescence, and cell death. More recently recognized cell nonautonomous effects, usually in the form of stimulating an antitumor immune response, have significant roles in therapeutic efficiency for a select number of chemotherapies. Unfortunately, the success of these therapeutics is not universal as not all tumors respond to treatment, and those that do respond will frequently relapse into therapy-resistant disease. Numerous strategies have been developed to sensitize tumors toward chemotherapies as a means to either improve initial responses, or serve as a secondary treatment strategy for therapy-resistant disease. Recently, targeting epigenetic regulators has emerged as a viable method of sensitizing tumors to the effects of chemotherapies, many of which are cytotoxic. In this review, we summarize these strategies and propose a path for future progress.

1. INTRODUCTION

1.1 Chemotherapy and the Induced Immune Response

The advent of chemotherapies that selectively targets cancer cells revolutionized the treatment of cancer (Weinberg, 2007). These compounds express selective toxicity to cancer cells which inhibits cancer cells growth or, more preferably, inducing cancer cell death. Chemotherapies induce cell stress in a variety of forms, most prominantly DNA damage (Helleday, Petermann, Lundin, Hodgson, & Sharma, 2008), endoplasmic reticulum (ER) stress (Wang, Yang, & Zhang, 2010), and mitotic disruption (Jackson, Patrick, Dar, & Huang, 2007). The cell autonomous response to this stress varies and includes cellular senescence (Shay & Roninson, 2004), autophagy (Gewirtz, 2014), apoptosis (McConkey, 2007), and necrosis (Amaravadi & Thompson, 2007).

Therapy-induced senescence is a state of growth arrest accompanied by both morphological, biochemical, and genetic changes in the tumor cell. These changes most often include enlarged and flattened morphology, upregulation of cell cycle control proteins such as Rb, p16, and p21 and increased expression of the lysosomal senescence-associated β-galactosidase (SA-β-Gal) (Lee et al., 2006; Rodier & Campisi, 2011; Toussaint, Medrano, & von Zglinicki, 2000). Senescent cells are further characterized by a unique secretome termed the senescence-associated secretory phenotype (SASP) (Coppe, Desprez, Krtolica, & Campisi, 2010). Senescence may be viewed as a desirable outcome of chemotherapy when it is considered to represent a permanent loss of replicative capacity by the tumor cell (Roninson, 2003; Roninson, Broude, & Chang, 2001). Conversely, senescence can also be considered as a mechanism of resistance to treatment when cell death is avoided and if the cells retain the capacity to ultimately recover proliferative function. In this context, an analysis of senescent cells revealed unregulated expression of antiapoptotic proteins such as Bcl-2 and Bcl-XL (Yosef et al., 2016).

Senescent growth arrest is classified as a cell autonomous effect that results from DNA damage or other forms of stress. The SASP, on the other hand, exerts cell nonautonomous effects on both the originating cancer cell itself and the surrounding environment (Coppe et al., 2008; Krtolica, Parrinello, Lockett, Desprez, & Campisi, 2001). Over 40 chemokines, cytokines, and metalloproteases comprise the secretome of senescent cells (Kuilman & Peeper, 2009). In some instances, components of the SASP have been shown to promote and/or maintain the state of senescence (Acosta et al., 2008; Chien et al., 2011). Evidence has also been presented that the secreted signaling molecules mobilize an immune response against the senescent cells which results in their clearance (Krizhanovsky et al., 2008; Rao & Jackson, 2016; Xue et al., 2007). However, many studies indicate that the SASP is primarily deleterious as its components have been found to promote chronic inflammation, cause oncogenic transformation of neighboring tissue, and facilitate neoplastic escape from senescence (Barajas-Gomez et al., 2017; Ortiz-Montero, Londono-Vallejo, & Vernot, 2017; Zhu, Armstrong, Tchkonia, & Kirkland, 2014). Recent studies suggest that years after therapy cancer cells can reemerge from senescence and drive disease relapse (Bellovin, Das, & Felsher, 2013).

Like senescence, therapy-induced autophagy can allow tumor cells to evade cell death (Belounis et al., 2016; Mani et al., 2015; Wu, Jiang, Ding, Shao, & Liu, 2015). However, depending on its nature, autophagy may not prevent apoptosis and contribute to initiation of immune surveillance of tumor cells (Martins et al., 2012). Autophagy in response to chemotherapy can be protective, potentially conferring resistance to chemotherapy, or nonprotective, in which case it does not influence the outcome of treatment (Gewirtz, 2013; Levy et al., 2014; Li et al., 2012). In some contexts, therapy-induced autophagy can be immunostimulatory largely by triggering a danger response from the immune system through the secretion of danger-associated molecular pattern molecules (DAMPs) (Michaud et al., 2011). DAMPs are a class of immune stimulatory molecules which induce sterile inflammation in the tumor (Krysko et al., 2012). These molecules have key functions in the immune systems recognition of nonself under the danger model of immune recognition (Matzinger, 1994). Therapy-induced DAMPs include cell surface calreticulin (Gardai et al., 2005;

Obeid et al., 2007), heat-shock proteins (Melcher et al., 1998; Spisek et al., 2007), released HMGB1 (Rovere-Querini et al., 2004), and ATP (Elliott et al., 2009; Garg et al., 2012). The efficacy with which DAMP release occurs largely depends on the type of chemotherapy and the degree of autophagy which is induced (Obeid et al., 2007).

Similar to autophagy, the ER stress response, that is also termed the unfolded protein response, is a housekeeping mechanism designed to clear damaged primarily misfolded proteins from the cell (Garg, Maes, van Vliet, & Agostinis, 2015). Some chemotherapeutic drugs, perhaps most prominently the anthracyclines, are capable of inducing significant DNA damage and ER stress (Gewirtz, 1999), which also results in the secretion of DAMPs. Under genotoxic stress, the ER stress pathway may become overwhelmed and switch in function from protecting the cell to promoting programmed cell death (Garg et al., 2015). To initiate programmed cell death, the ER stress response regulates cell autonomous caspase-dependent apoptosis (Bhat et al., 2017). When DAMP secretion is coupled with apoptosis, it can provide tumor-specific antigens (TSA) to the immune system in conjunction with immune stimulating DAMPs, effectively vaccinating against the tumor in situ (Michaud et al., 2011). This process has been termed immunogenic cell death (ICD) (Krysko et al., 2012).

Through targeted studies and hypothesis-driven approaches, anthracyclines and oxaliplatin were first observed to induce ICD (Casares et al., 2005; Tesniere et al., 2010). To more systematically identify compounds which induce DAMP secretion, and possibly undergo ICD, several screens were conducted of FDA approved drugs including both anticancer chemotherapies (Menger et al., 2012) and the NCI mechanistic diversity set (Sukkurwala et al., 2014). These screens tested for a compound's ability to induce classic DAMP responses including cell surface calreticulin exposure and ATP/HMGB1 release. Results from these studies confirmed that several anthracyclines (doxorubicin, mitoxantrone, and daunorubicin) can induce DAMP secretion (Menger et al., 2012). In addition, microtubule-targeting drugs including vincristine and vinorelbine were identified (Menger et al., 2012). Interestingly, the Ca^{++} mobilizing compounds digitoxin and digoxin were also shown to improve the immunogenicity of tumor cells treated with chemotherapies which do not induce ICD (Menger et al., 2012). In addition, the compound septacidin was identified from the NCI diversity set (Sukkurwala et al., 2014). Septacidin has anticancer activity to a limited number of cell lines (Aszalos, Lemanski, Berk, & Dutcher, 1965). To improve these activities, a variety of analogs have been

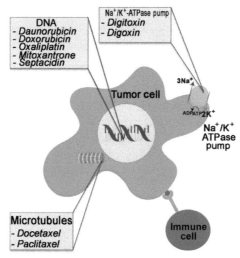

Fig. 1 Chemotherapies which induce immunogenic cell death. Cartoon of tumor cell showing the subcellular location of function for chemotherapies which have been confirmed to result in immunogenic cell death. The anthracyclines daunorubicin, doxorubicin, and mitoxantrone have activity in both the nucleus and in generating endoplasmic reticulum stress. The platinum agent oxaliplatin damages DNA, and septacidin is a nucleoside analog which inhibits DNA synthesis. Docetaxel and paclitaxel primarily function to manipulate microtubule polymerization. In contrast digitoxin and digoxin are cardiac glycosides which inhibit the Na^+/K^+ ATPase to increase intracellular Ca^{++} levels. Each of these compounds, either as a monotherapy or in combination, induces immunogenic cell death. Immunogenic cell death results in the secretion of DAMPs which attract and improve the activity of effector immune cells.

developed but not extensively characterized (Acton, Ryan, & Luetzow, 1977). Several of these DAMP stimulating compounds were later confirmed to induce bona fide ICD using a gold standard vaccination assay (Menger et al., 2012; Sukkurwala et al., 2014). From these analysis, and the previous literature, a list of nine ICD-inducing compounds has been identified which includes daunorubicin, docetaxel, doxorubicin, mitoxantrone, oxaliplatin, paclitaxel, digitoxin, digoxin, and septacidin (Fig. 1) (Kepp, Senovilla, & Kroemer, 2014). Because of their potential to stimulate both cell autonomous and cell nonautonomous responses, there is a strong interest in improving the efficiency of these compounds.

1.2 The Value of Chemosensitization

Chemotherapies are frequently but not always successful in tumor growth control. In addition, tumors which do initially respond to therapy frequently relapse into therapy-resistant disease. To improve the initial tumor response rate, and to resensitize tumors once they have relapsed, a variety of chemosensitization strategies have been employed which seek to inhibit acquired pathways of chemotherapy resistance (Mansoori, Mohammadi, Davudian, Shirjang, & Baradaran, 2017). Several pathways have been successfully targeted to this end including apoptosis (Cree, Knight, Di Nicolantonio, Sharma, & Gulliford, 2002), DNA damage repair (Luo & Leverson, 2005), and drug metabolism (Huang et al., 2004; Lob, Konigsrainer, Rammensee, Opelz, & Terness, 2009). In some cases, blocking autophagy by pharmacologic inhibitors such as chloroquine and hydroxychloroquine resulted in sensitization of tumor cells and increased cell death without increasing the dose of the chemotherapeutic drug and associated toxicity (Gao, Xu, & Qiu, 2017; Qin et al., 2016). Successful targeting and elimination of senescent tumor cells have also been reported (Chang, Wang, et al., 2016; Zhu, Tchkonia, et al., 2015). Although senolytic agents such as small-molecule Bcl-2 and Bcl-XL inhibitors do not induce cell death as an independent treatment, these agents appear to potentiate the effectiveness of chemotherapies that induce cellular senescence (Shoemaker et al., 2006). The ability to pharmacologically target gene expression has recently been achievable by targeting epigenetic regulatory mechanisms. This strategy can be successful because in many instances acquisition of cancer hallmarks, including acquired resistance, is the result of abnormal gene expression resulting from epigenetic changes.

1.3 Targeting Epigenetics to Achieve Chemosensitization

In many instances, the development of cancer (Hanahan & Weinberg, 2011) is a result of abnormal stable gene expression at the level of chromatin or the transcript (Esteller & Pandolfi, 2017). As such it seems plausible that the inherent or acquired resistance to chemotherapies, particularly chemotherapies which induce ICD and require epigenetically regulated processes like apoptosis (Casares et al., 2005), would also be suppressed by abnormal gene expression. Much of epigenetics functions to alter chromatin structure with an emphasis on the nucleosome. The nucleosome is composed of a histone octamer, which in turn is composed of two copies each of histone H2A, H2B, H3, and H4, wrapped by ~146 bp of DNA (Luger, Mader, Richmond, Sargent, & Richmond, 1997). Nucleosomes can prevent access to the DNA sequence from transacting factors such as transcription factors and polymerases. These functions regulate nuclear processes including gene expression, DNA replication, and DNA damage repair (Polach & Widom, 1995). The histone proteins and the DNA sequence in nucleosomes are heavily posttranslationally modified to regulate its chromatin structure (Bannister & Kouzarides, 2011). Protein complexes which catalyze these reactions are grouped into three main functional categories including "writers," "erasers," and "readers" (Fig. 2) (Allis et al., 2007). Writers are transferases which catalyze the addition of a number of posttranslational modifications. Prominent writers include histone acetyltransferases, histone methyltransferases, and histone kinases which catalyze the acetylation and methylation of lysine residues, and phosphorylation of serine and threonine residues, respectively. These posttranslational additions are removed by erasers to return the histone or DNA to the unmodified state. Every known histone and DNA modification has a writer and an eraser, and the opposing functions of these enzymes reversibly modify chromatin. Erasers which would counter the writers presented earlier include histone deacetylases, histone demethylases, and histone phosphatases. Posttranslational modifications are recognized by readers which represent a number of well-characterized domains (Patel & Wang, 2013). One reader domain which has attracted interest is the bromodomains which recognize acetylated histones (Marchand & Caflisch, 2015). While writers,

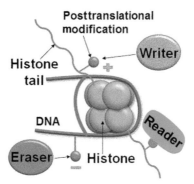

Fig. 2 Classification of epigenetic regulatory complexes. Cartoon of a nucleosome with posttranslationally modified histone N-terminal tails and DNA. These posttranslational modifications are added to unmodified proteins and DNA by "writers," removed by "erasers" and serve as ligands for proteins with functional "reader" domains. Proteins or subunits of multiprotein epigenetic complexes may have more than one, and frequently several, of these functions.

erasers and readers encompass many functions of epigenetic complexes there are also chromatin remodeling complexes, which do have reader domains, but also ATP-dependent nucleosome remodeling activity (Becker & Workman, 2013). These activities serve to regulate transacting factors (e.g., transcription factors and polymerases) access to the DNA to regulate nuclear processes most prominently transcription. Two properties of epigenetic regulators make them a very attractive drug target. First, epigenetic regulation is catalyzed by an opposing set of writers and readers and therefore its effects are reversible. Second, because writers and readers are enzymes, with potentially multiple active sites, they are potently druggable (Dawson & Kouzarides, 2012; Ribich, Harvey, & Copeland, 2017).

Noncoding RNAs are important regulators of gene expression and cancer biology which can stably regulate gene expression after cellular division (Esteller & Pandolfi, 2017). In addition to their direct effects on transcript abundance, they can also target epigenetic chromatin modifiers to have wider effects on chromatin structure (Guil & Esteller, 2009). Because of these functions, noncoding RNAs are considered by some to have properties of an epigenetic regulator and should be included alongside other mechanisms of chromatin modifications (Guil & Esteller, 2009). For the purpose of this review, we will include studies which have exploited noncoding RNAs as a means to sensitize cancer cells to chemotherapies.

Because epigenetic and noncoding RNAs are important regulators of tumor cell gene expression, it would naturally follow that they would have essential roles in promoting resistance to chemotherapies. These effects would likely apply to all chemotherapies, including those which induce ICD. Manipulating epigenetic regulators as a means to improve the effectiveness of ICD-inducing chemotherapies is of great current interest because of their ability to stimulate the antitumor immune response (Pol et al., 2015). These effects have standalone benefits, but would be further beneficial if combined with a variety of immunotherapies, most relevant of which would be immune checkpoint blockades (Pardoll, 2012). To precipitate these studies, we have summarized the available literature in which targeting epigenetic regulators sensitizes cancer cells to the nine confirmed ICD-inducing chemotherapies (daunorubicin, docetaxel, doxorubicin, mitoxantrone, oxaliplatin, paclitaxel, digitoxin, digoxin, and septacidin), provided an analysis of this literature, and suggested future directions for moving the field forward (Fig. 3).

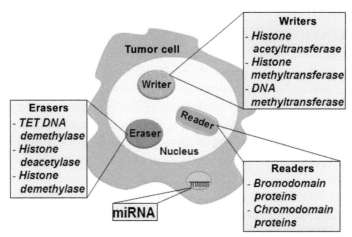

Fig. 3 Classes of epigenetic complexes which sensitize cancer cells to ICD-inducing chemotherapies. Cartoon of cancer cell showing classes of "writers," "erasers," "readers," and miRNAs which when inhibited can sensitize cancer cells to ICD-inducing chemotherapies.

2. TARGETING EPIGENETIC REGULATORS TO ACHIEVE SENSITIZATION TO ICD-INDUCING CHEMOTHERAPIES

2.1 Writers

2.1.1 Histone Acetyltransferases

HAT inhibitors can sensitize neuroblastomas to chemotherapy. A pan HAT inhibitor PU139, which has activity to Gcn5, PCAF, CBP, and p300, can sensitize neuroblastoma SK-N-SH cells to increase caspase-independent apoptosis, as validated by zVad-fmk inhibition (Gajer et al., 2015). Using xenograft models, PU139 sensitizes the same neuroblastoma cell line to the cytotoxic effects of doxorubicin. However, the immune systems effects were not evaluated.

2.1.2 Histone Methyltransferases

Genetic depletion of the H3K9 histone methyltransferase EZH2 induces senescence in the human gastric cancer cell line SGC-7901 (Bai et al., 2014). EZH2 depletion inhibited cell proliferation, arrested the cell cycle, and promoted doxorubicin-induced cell senescence. These effects correlated with elevated cell cycle regulators p21 and p16, with the p21 activation being p53 dependent. It was subsequently shown that EZH2 depletion promoted apoptosis in p53 mutant cells and that the cooperative relationship with doxorubicin required a mutant p53 (Bai et al., 2014). Similar roles for EZH2-regulating cell cycle regulators and, in turn, sensitizing to doxorubicin were observed in pancreatic cancer. In this study, EZH2 was found to be overexpressed in the nuclei of human primary pancreatic cancers compared to normal tissues. Additional significance of EZH2 nuclear enrichment was observed in poorly differentiated pancreatic adenocarcinomas. Genetic depletion of EZH2 resulted in the reexpression of $p27^{Kip1}$, decreasing pancreatic cell proliferation. These results coincide with sensitization of pancreatic cells to doxorubicin, as measured by an increase in apoptosis (Ougolkov, Bilim, & Billadeau, 2008).

In addition to its expression in gastric cancer, EZH2 is overexpressed in metastatic prostate cancer cells. In these same cells it suppresses the expression of miR-205 and miR-31, two proapoptotic miRNAs. Genetic knockdown (KD) of EZH2 increases the expression of these miRs which decreases the antiapoptotic proteins Bcl-w and E2F6. This, in turn, results in the sensitization of prostate cells to docetaxel-induced apoptosis (Zhang, Padi, Tindall, & Guo, 2014).

2.1.3 DNA Methyltransferases

DNA methyltransferase inhibitors (DNMTi) have been aggressively investigated as a means to chemosensitize breast cancers. Using the triple-negative breast cancer (TNBC) model MDA-MB-231, it was shown that the DNMTi decitabine induces tumor necrosis factors related apoptosis-inducing ligand (TRAIL) and inhibits AKT (Xu, Zhou, Tainsky, & Wu, 2007). Combined treatment of decitabine and doxorubicin significantly increased apoptotic cell death activating both death receptor and mitochondrial apoptotic pathways. Activation of both pathways was not observed for doxorubicin alone, which activated only the mitochondrial pathway, and neither pathway was activated with decitabine treatment alone. Genetic shRNA KD of the TRAIL pathway abolished the enhanced effects of decitabine + doxorubicin including apoptosis and caspase activation, demonstrating that the TRAIL pathway is critical for enhanced apoptosis.

Docetaxel-resistant MCF7 and MDA-MB-231 cell lines were studied to determine the importance of DNA methylation to chemoresistance (Kastl, Brown, & Schofield, 2010). At a global level, the MCF7-resistant cell lines had increases in DNA methylation, whereas the MDA-MB-231 cell line did not. Decreased DNMT activity and decreased expression of DNMT1 and DNMT3b were associated with docetaxel resistance in both cell lines.

Decitabine treatment decreased global methylation, activity of DNMT, and expression of DNMT1, DNMT3a, and DNMT3b in MDA-MB-231-resistant cells. However, decitabine-treated MCF7-resistant cells showed increased DNMT1, DNMT3a, and DNMT3b expression. Decitabine treatment increased resistance in MCF7-resistant cell lines, MCF7-sensitive cell lines, and MDA-MB-231-sensitive cell lines, but not in MDA-MB-231-resistant cell lines (Kastl et al., 2010). Similar effects were observed when a low dose of decitabine was delivered via nanoparticles to the MDA-MB-231 mammary spheres. Systemic delivery of decitabine nanoparticles reduced DNMT1 and DNMT3b while also inducing caspase-9 expression. When the decitabine nanoparticles were combined with doxorubicin nanoparticles, a significant reduction in cancer stem cells (ALDH^{hi} cells), increased apoptotic cells and enhanced tumor-suppressive effects were observed (Li et al., 2015).

Utilizing a slightly different strategy, doxorubicin-resistant breast cancer cells were sequentially treated with either decitabine, or the histone deacetylase inhibitor (HDACi) SAHA, followed by doxorubicin (Vijayaraghavalu, Dermawan, Cheriyath, & Labhasetwar, 2013). This sequential treatment resulted in ∼90% of cancer cells entering into G2/M cell cycle arrest. This correlated with a depletion of DNMT1, which possibly resulted from p21 upregulation. Interestingly, the sequential treatment strategy was more effective for growth control than simultaneous treatment, and decitabine was more effective than SAHA. Microarray results indicated the upregulation of tumor suppressor genes (TSGs) and downregulation of protumor genes suggesting that the resulting changes in gene expression contributed to chemosensitization.

In addition to monotherapy, DNMTi has been combined with HDACi to sensitize breast cancer cells to conventional therapies. Breast cancer patients treated with the DNA demethylating agent hydralazine + HDACi valproate in combination with the cytotoxic chemotherapies doxorubicin and cyclophosphamide resulted in a 31% complete response rate (Arce et al., 2006). In primary tumors, there was a significant reduction in global CpG methylation content suggesting DNMT inhibition, as well as histone deacetylation suggesting effective HDAC inhibition. Both of these outcomes correlate with gene reactivation. Microarray experiments further indicated that the majority of changes in gene expression are the result of expression upregulation consistent with DNMT and HDAC acting as repressors of gene expression.

In one of the only studies to focus on the antitumor immune response, azacitidine was used to sensitize CT26 colon cancer and EM6 breast cancer cells to the effects of photodynamic therapy (Wachowska et al., 2014). Enhanced antitumor activity resulting from azacitidine was dependent on CD8 T cells, as measured by mAb depletion and adoptive cell transfer studies. These activities were proposed to occur through the enhanced expression of the tumor antigen P1A as a result of the azacitidine treatments.

In addition to breast cancer cells, the inhibition of epigenetic regulators has been significantly investigated as a means to sensitize colon cancer cells to chemotherapy. Pretreatment of the colon cancer cell lines HCT116 and DLD1 with decitabine and azacitidine sensitized the cells to doxorubicin and mitoxantrone; however, this did not occur in the case of HT-29 cells (Pawlak et al., 2016). Sensitization included a reduction in cell viability, clonogenicity, and an increase in programmed cell death. In the case of etoposide, a chemotherapy which does not induce strong ICD (Obeid et al., 2007), sensitization was long lasting and occurred with merely a single DNMTi dose.

In other colorectal cancer cell lines, the DNMTis decitabine, azacitidine, and zebularine were tested for sensitization to several chemotherapies which included oxaliplatin (Ikehata et al., 2014). Sensitization, as measured by cell viability, was observed with DNMTi, but interestingly no sensitization was seen for the HDACis TSA, SAHA, and valproate. Broadening the significance of these results, DNMTi sensitization to oxaliplatin was observed in a variety of other colon cancer cell lines including HT29, SW48, and HCT116.

The ability of DNMTi to reverse gene suppression important for sensitization to platinum drugs was tested in the clinic on patients with colon and rectal cancer (Tsimberidou et al., 2015). In this study, an oxaliplatin + azacitidine combination was found to be safe. However, only in 5.4% of patients completion of all six cycles of therapy resulted in stable disease. Interestingly, the combination resulted in a 3–18-fold increase in tumor platinum concentration suggesting increased drug accumulation in the tumor cells. Hypomethylation of DNA in the tumor was observed in four out of seven patients. In the responding patients, the nuclear and cytoplasmic copper transporter 1 (CTR1) scores decreased, suggesting a decreased ability to transport platinum compounds into the tumor cells (Howell, Safaei, Larson, & Sailor, 2010). Higher platinum levels correlated with progression-free survival but did not correlate with CTR scores. In combination these results suggest that transporters other than CTR1 could be responsible for increased platinum levels in the tumor. In a separate study colon cancer patients whose tumors harbored hypermethylation of promoter CpG islands (CIMP) were treated with a combination of azacitidine and oxaliplatin. CIMP is associated with chemotherapy resistance and the pathogenesis of CIMP-high colon cancers (Kondo & Issa, 2004). The combination therapy was well tolerated with high rates of stable disease, but there was a lack of objective responses (Overman et al., 2016).

To determine the significance of DNA methylation to oxaliplatin resistance, a DNA methylation array was used to identify differences in DNA methylation between oxaliplatin-sensitive and resistant colorectal cancer cells (Moutinho et al., 2014). Serum deprivation response factor-related gene product that binds to c-kinase (SRBC) were identified using this approach and validated using locus-specific DNA methylation and gene expression assays. SRBC is a BRCA1 interacting factor, when overexpressed or depleted it gives rise to sensitivity or resistance to oxaliplatin, respectively. SRBC epigenetic inactivation through DNA methylation occurred in 29.8% of patient's primary tumors. In a validation cohort of patients treated with oxaliplatin, SRBC hypermethylation was associated with shorter progression free survival.

In an interesting study, the epigenetic silencing of TSGs by DNA methylation in colon cancer cells was reversed by novel drugs, which mobilize intracellular Ca^{++} (Raynal et al., 2016). Eleven drugs were identified in a screen to reexpress a DNA methylated and silenced GFP reporter gene. In addition to the GFP reporter, these drugs also induced endogenous TSGs in several cell lines, suggesting bona fide DNA demethylation activity. These newly identified drugs prominently featured cardiac glycosides and included digitoxin and digoxin, two ICD-inducing chemotherapies. These drugs did not alter DNA methylation levels, but rather changed the intracellular distribution of DNA methyl-binding protein MECP2 by translocating it to the cytoplasm. Blocking Ca^{++}/calmodulin-dependent protein kinases (CamK) activity abolished gene reactivation and cancer cell killing, suggesting that mobilization of intracellular Ca^{++} stores is an essential component of their action.

Using a similar assay, the same authors identified decitabine as an epigenetic regulator which sensitizes colon cancer cells to platinum compounds (Qin et al., 2015). Interestingly, none of the 16 anticancer compounds tested had activity on the expression of the GFP marker alone. However, decitabine showed significant synergy with platinum compounds in activating GFP, which included oxaliplatin. This synergy was also observed at endogenous hypermethylated TSGs. Platinum compounds did not cause DNA demethylation, but rather suppressed an essential component of heterochromatin protein 1 (Eissenberg & Elgin, 2014), increased H3K4me3 and H3K9ac and reduced binding of methyl-binding proteins MeCP2 and methyl-binding domain 2 (MBD2). These studies suggest that the combination of DNMTi and oxaliplatin could synergize in colon cancer cells to improve their anticancer activities.

Because DNMTis are used to treat hematological malignancies, it seems natural that they would be used in chemosensitization strategies. Metronomic dosing of epigenetic inhibitors has been gaining acceptance as a means to alter the epigenome of tumors without the toxicity observed by the patient at higher doses (Munzone & Colleoni, 2015). It was recently shown that metronomic exposure to the DNMTi azacitidine can sensitize chemotherapy-resistant diffuse large B-cell lymphoma (DLBCL) to doxorubicin without major toxicity to the patient (Clozel et al., 2013). Nine genes (*ETV6*, *SLC25A24*, *VAV3*, *SMAD1*, *IRF4*, *SNTB1*, *OPN3*, *CYP4V2*, and *TRIM44*) were hypermethylated with DNMTi treatment and found to be relevant to chemosensitization. Subsequent studies showed that *SMAD1* is a critical contributor to DLBCL progression and is required for chemosensitization. In a phase 1 study, azacitidine priming flowed by a standard R-CHOP chemotherapy (rituximab + cyclophosphamide, doxorubicin, vincristine, and prednisone) in high risk patients resulted in a high rate of complete remission. Biopsies confirmed that SMAD1 is demethylated with azacitidine treatment and is increased in expression, which is consistent with the model that its reexpression is required for sensitization of refractory DLBCL.

Similarly, a phase I study was performed testing the feasibility of priming acute myeloid leukemia (AML) patients with decitabine before standard induction chemotherapy (Scandura et al., 2011). Both acute 1 h infusion and continuous infusion over 3, 5, or 7 days before a standard induction with cytarabine or daunorubicin were tested. The decitabine treatments induced DNA hypomethylation in all dosing schedules. In this study, 90% of the patients responded to therapy, with 57% entering into complete remission. This report confirms that epigenetic priming with DNMTi followed by intensive chemotherapy improves response rates for hematological malignances.

In comparison, a phase 1 and 2 study was performed using decitabine priming in 52 adults followed by mitoxantrone, etoposide, and cytarabine with relapse refractory AML and other high grade myeloid neoplasms (Halpern et al., 2017). Among the 46 patients treated 33% achieved complete remission. This complete remission rate is in line with historical averages for these groups, suggesting that decitabine priming is not substantially better than cytarabine-based regimens used for relapse/refractory AML.

More mechanistically, chemotherapy resistance in acute lymphoblastic leukemia (ALL) is promoted through interactions with the bone marrow microenvironment. Using azacitidine and the HDACi panobinostat can overcome the protective effects of osteoblasts (major components of the bone marrow microenvironment) and promote the chemosensitization to daunorubicin (Quagliano, Gopalakrishnapillai, & Barwe, 2017). These results can be replicated ex vivo in a number of xenograft lines from both B-cell ALL and T-cell ALL patients.

A combination of decitabine and the HDACi vorinostat was administered to patients with relapse/refractory acute ALL 1–4 days followed by a cocktail of chemotherapies including doxorubicin and vincristine (Burke et al., 2014). Four of 13 patents achieved complete remission without platelet recovery. Following decitabine treatment significant genome-wide demethylation was observed.

Osteosarcoma is the most common bone cancer. Osteosarcoma cell lines were exposed to azacitidine, SAHA, and 3-deazaneplanocin A prior to exposure to doxorubicin (Pettke et al., 2016). The effects of doxorubicin on the cell lines were mixed with some cell lines (HOS and MG-63) showing agonistic effects and others (MNNG or ZK-58) showing antagonistic effects on growth. The nature of the reductions in cell growth was not investigated.

DNA methylation patterns for the drug resistant cervical cancer cell line SiHa and its oxaliplatin-resistant S3 cells were first assessed by methylation-specific microarray (Chen et al., 2015). Differences in the DNA methylation status genome wide, and at specific loci, were detected in the oxaliplatin-resistant S3 cells over the oxaliplatin-sensitive SiHa cells. Next, the drug-resistant cells were treated with the DNMTi decitabine to determine if DNA demethylation could reverse oxaliplatin resistance. Treatment of S3 with decitabine restored sensitivity of S3 to cisplatin, paclitaxel, and oxaliplatin to the same level as SiHa. Methylation of the gene Casp8AP2 was sufficient to increase drug resistance in different cells.

Guadecitabine, a DNMTi prodrug, was used on hepatocellular carcinoma (HCC) cell lines and xenograft models to determine its antitumor activity as a single agent or in combination with oxaliplatin (Kuang, El-Khoueiry, Taverna, Ljungman, & Neamati, 2015). Treatment with low doses of guadecitabine significantly synergized with oxaliplatin yielding enhanced cytotoxicity. The combination significantly delayed tumor growth in mice compared to oxaliplatin alone. Measuring transcription by RNA-Seq showed that the combination treatment inhibited expression of genes involved in the WNT/EGF/IGF pathways, which could have relevance to chemosensitization.

Hormone refractory prostate cancer is prone to develop chemoresistance (Heidegger et al., 2013). Azacitidine was used in two models of aggressive prostate cancer to test if it can sensitize the cancer cells to docetaxel and cisplatin (Festuccia et al., 2009). Azacitidine exposure shows increased G0/G1 cell cycle arrest and apoptosis in the 22vr1 cell line, and increased G2/M phase arrest with an absence of cell killing in the PC3 cells. In mouse models, azacitidine reduced tumor proliferation and induced apoptosis in both 22vr1 and PC3 xenografts increasing p16/INK4A, Bax, Bak, p21/WAF1, and p27/KIP1 and inhibiting AKT activity and the expression of cyclin D1, Bcl2, and Bcl-XL. In vitro, azacitidine treatments resulted in increased caspase-3 and PARP cleavage, consistent with enhanced apoptosis and showed synergistic effects with docetaxel and cisplatin.

Esophageal squamous cell carcinoma (ESCC) has increased rates of methylation of the checkpoint with forkhead and ring finger domains (CHFR) gene as the cancer progresses (Yun et al., 2015). CHFR is an E3 ubiquitin–protein ligase which functions in a checkpoint regulator during mitosis (Sanbhnani & Yeong, 2012). When treated with docetaxel or paclitaxel, cell viability was lower in CHFR-methylated cancer cells than unmethylated cancer cells. Docetaxel and paclitaxel treatment of ESCC with methylated CHFR resulted in all cells being arrested in the G1/G0 phase of the cell cycle. After azacitidine treatment, there was an increase in the fraction of CHFR-methylated cells in the S and G2/M phases of the cell cycle. Thus methylation of CHFR sensitizes ESCC cells to taxanes and azacitidine and may sensitize the chemotherapy-resistant refractory tumors by inducing cell cycle phase redistribution.

2.2 Erasers
2.2.1 Histone Deacetylases
Histone deacetylases have prominent roles in promoting tumor biology by repressing TSGs in cancer (Falkenberg & Johnstone, 2014). In addition, they can also repress the expression of genes which sensitize tumor cells to chemotherapies. As such, combining HDAC inhibitors with chemotherapies, including those which induce ICD, has been aggressively pursued. Using both breast and prostate cancer cell models HDAC6 has been shown to

promote resistance to doxorubicin in cancer cells but not normal foreskin fibroblasts (Namdar, Perez, Ngo, & Marks, 2010). Both pharmacologic (tubacin) and genetic (shRNA) HDAC6 inhibition enhance cell death induced by doxorubicin as measured by PARP cleavage and its inhibition by zVAD-fmk. HDAC6 inhibition induces γH2AX and activation of checkpoint kinase Chk2, and the ER stress transcription factor CHOP. The combination of DNA damage and ER stress could possibly lead to ICD.

Combined treatment of TNBC xenografts with entinostat, all trans-retinoic acid, and doxorubicin (EAD therapy) resulted in significant tumor regression and restored RARβ expression from epigenetic silencing (Merino et al., 2016). Entinostat and doxorubicin treatment inhibited topoisomerase II and relieved RARβ repression. EAD combination therapy was the most effective, compared to individual therapies, at inducing differentiation of breast tumor initiating cells in vivo. These results suggest that entinostat potentiates doxorubicin cytotoxicity and retinoids drive differentiation to regress TNBC.

The novel HDACi AN446, a valproyl ester-valpramide of acyclovir, was found to be 2–5-fold more potent as an anticancer agent than butyric acid (Tarasenko et al., 2014). AN446 improved the toxicity of doxorubicin to cancer cells but also reduced doxorubicin toxicity to noncancer cells. The enhancements of doxorubicin toxicity to cancer cells were synergistic in the mouse 4T1 TNBC model and human U251 glioma model. In U251 glioblastoma xenografts, both AN446 and AN7 reduced the body weight loss, protected against doxorubicin induced heart failure, and increased the anticancer efficiency of doxorubicin.

TSA and paclitaxel have synergistic effects decreasing proliferation of gastric cancer cells lines OCUM-8 and MKN-74 (Zhang, Yashiro, Ren, & Hirakawa, 2006). TSA increased the expression of p21, p53, DAPK-1, and DAPK-2 in both cell lines which might be associated with the synergistic effects.

Sensitivity to doxorubicin can be induced through decreased expression of catalase (CAT). Catalase inactivates reactive oxygen which reduces the toxic effects of doxorubicin. CAT downregulation in the doxorubicin-resistant AML-ll line AML-2/DX-100 was recovered after exposure to the broad HDAC inhibition TSA but not the DNMTi azacitidine (Lee, Moon, & Choi, 2012). The reason for this is that the CpG island at the CAT gene promoter is not methylated, but is significantly deacetylated compared to doxorubicin-sensitive AML-2/DX-100. Some caution is warranted with this combination, however, because multidrug resistance

1 transporter (MDR1) expression is increased with exposure to azacitidine and TSA (Lee, Park, Min, Kim, & Choi, 2008). When DNMTi and HDACi are used in combination they can synergize in some cancer cell lines to reexpress genes promoting a chemoresistant phenotype.

Osteosarcoma is an incurable and fatal disease which is treated with doxorubicin as a standard of care therapy (Bielack, Hecker-Nolting, Blattmann, & Kager, 2016). Using human and canine cell line models, the HDACi valproic acid enhances bulk histone acetylation levels and increases the nuclear accumulation of doxorubicin. The combination induces significant growth inhibition and potentiation of apoptosis compared to each monotherapy in the canine osteosarcoma xenograft model (Wittenburg, Bisson, Rose, Korch, & Thamm, 2011).

A significant problem in chemotherapy resistance is cross-resistance. Hence, when cancer cells become resistant to one DNA-damaging agent they can develop cross-resistance to other DNA-damaging agents. HDACs by nature close chromatin, making access to the underlying DNA more difficult, possibly providing resistance to DNA-damaging agents. This hypothesis was tested in several cell lines resistant to cisplatin. HDAC inhibitors increased the sensitivity of cisplatin-resistant ovarian cancer cell lines to a variety of DNA-damaging agents, including doxorubicin (Ozaki et al., 2008). The cellular response includes enhanced γH2AX, reactive oxygen species generation, and apoptosis. In contrast, the HDAC inhibitors did not increase the sensitivity of the cell lines to metabolic antagonists or microtubule-targeting agents.

HCC is the fifth most common cancer worldwide; therefore, there is a strong need to develop novel therapeutic approaches to treatment. The Raf-1 kinase inhibitor protein (RKIP) is a tumor suppressor and an overall inhibitor of NF-κB signaling and is frequently downregulated in HCC. Exposure to the HDACi TSA induced cell growth inhibition and proapoptotic effects of doxorubicin in several HCC cell lines. TSA caused histone hyperacetylation and the upregulation of gene expression but not RKIP expression (Poma et al., 2012). Therefore the mechanism behind the enhanced effects of HDACi on chemosensitization to doxorubicin in HCC are unknown.

Rhabdomyosarcoma is the most common soft-tissue sarcoma in childhood and has a very poor prognosis. HDACi SAHA sensitizes the rhabdomyosarcoma RMS cell line to several anticancer agents including doxorubicin (Heinicke & Fulda, 2014). The functional interactions between doxorubicin and HDACi are synergistic as calculated by a combination index (Chou, 2010). SAHA changes the ratio of expressed pro- and antiapoptotic genes including activation of Bax and Bak, resulting in caspase activation and shifting the balance to apoptosis. The importance of apoptosis was confirmed using zVAD-fmk, which significantly inhibited the effectiveness of the SAHA/doxorubicin combination.

Glioblastoma cell lines are induced into apoptosis by the HDACi MS275 in combination with doxorubicin (Bangert, Hacker, Cristofanon, Debatin, & Fulda, 2011). As with other HDACi, MS275 and doxorubicin act synergistically in glioblastoma. In contrast, MS275 inhibited the apoptotic effects of vincristine and paclitaxel possibly because MS275 arrests cells in the G1 phase, when cells are insensitive to microtubule inhibitors.

In uterine cancers combining the HDACi sodium butyrate with doxorubicin inhibits proliferation and increases caspase-dependent apoptosis (Yu et al., 2014). Accompanying these changes is a decrease in human telomerase reverse transcriptase (hTERT) expression. The defects in proliferation and increased apoptosis were reduced by increased hTERT expression and KD of hTERT sensitizes cancer cells to doxorubicin. Sodium butyrate significantly sensitizes primary uterine cancer cells to the effects of doxorubicin. It is proposed that hTERT downregulation is a means by which sodium butyrate sensitizes cancer cells to doxorubicin.

2.2.2 Histone Demethylases

The histone 3 K27 demethylase KDM6B is overexpressed in a germinal center B-cell subtype of DLBCL, and higher KDM6B expression levels associate with worse survival from the disease. Inhibition of KDM6B in DLBCL using the small-molecule inhibitor GSK-J4 induced apoptosis in several cell line models of DLBCL which includes SU-DHL-6, OCI-Ly1, Toledo, OCI-Ly8, and SU-DHL-8. GSK-J4 treatment resulted in downregulation of B-cell receptor (BCR) signaling and a downstream BCR target gene, BCL6. BCL6 is a transcription repressor which acts as an oncogene in germinal center-derived lymphomas such as DLBCL (Basso & Dalla-Favera, 2012). Relevant to this review, GSK-J4 treatments resulted in sensitization to doxorubicin as measured by increased apoptosis (Mathur et al., 2017).

2.2.3 TET Family of DNA Demethylases

The ten eleven translocation (TET) enzymes catalyze the demethylation of CpG-methylated DNA (Kohli & Zhang, 2013). The TET enzymes are commonly mutated in hematological malignancies including diffuse large

B-cell (DLBC) and peripheral T-cell lymphomas resulting in aberrant DNA methylation (Couronne, Bastard, & Bernard, 2012; Quivoron et al., 2011). Ascorbic acid (AA) is a required cofactor for the TET family of enzymes and when supplied pharmacologically can increase the TET enzyme activity resulting in DNA demethylation (Yin et al., 2013; Young, Zuchner, & Wang, 2015). Using a model of DLBC, pharmacologic addition of AA increased gene expression of TSGs. AA treatment resulted in DNA demethylation and increased sensitivity to doxorubicin, as measured by cell viability in vitro (Shenoy et al., 2017).

2.3 Readers
2.3.1 Bromodomains
Bromodomains are epigenetic reader domains which recognize proteins with acetylated lysine residues, prominently histones (Marchand & Caflisch, 2015). Several small-molecule inhibitors have been made which serve to inhibit the function of epigenetic complexes with bromodomains (Gallenkamp, Gelato, Haendler, & Weinmann, 2014). The prominent histone acetyltransferases CBP/p300 are bromodomain-containing complexes that have roles in a variety of solid and hematological cancers (Attar & Kurdistani, 2017). The small-molecule inhibitor I-CBP112 which targets the p300/CBP bromodomains inhibits its histone acetyltransferase activity (Zucconi et al., 2016). Use of I-CBP112 in vitro results in reduced colony formation and induced cellular differentiation, impairing the self-renewal capability of leukemia cells. These defects also translate to in vivo studies where I-CBP112 inhibits tumor growth. In addition to these functions, I-CBP112 improves the cytotoxic activity of doxorubicin and the BRD4 inhibitor JQ1 (Picaud et al., 2015). Genome-wide gene expression analysis shows that I-CBP112 selectively alters gene expression with enrichment for genes regulating the immune response.

In addition to the p300/CBP bromodomains, the bromodomain-containing protein BRD8 has been targeted as a means to achieve chemosensitization. BRD8 is a component of the TRRAP/TIP60 histone acetyltransferase complex (Judes et al., 2015). BRD8 expression is increased in aggressive colon tumors compared to normal colon tissue, suggesting functions in colon cancer biology (Yamada & Rao, 2009). Knockdown of BRD8 sensitizes colon cancer cells to paclitaxel, resulting in G2/M cell cycle arrest, reduced clonogenic survival, and increased cell death (Yamada & Rao, 2009). There are >60 human bromodomain-containing proteins, that have important functions in human cancers, and can be targeted by small-molecule inhibitors (Gallenkamp et al., 2014). As such, there is clear potential for targeting these epigenetic regulators to achieve chemosensitization.

2.3.2 Chromodomain Helicase DNA-Binding Proteins

The chromatin remodeling complex NURD is unique among epigenetic regulatory complexes because it has reader functions through its chromodomains, eraser activity through HDAC subunits and ATP-dependent chromatin remodeling activity from its ATPase domain (Lai & Wade, 2011). The CHD4 ATPase subunit has prominent functions in the repair of DNA damage with documented roles in resolving DNA lesions once created, and promoting signaling pathways activated after DNA damage (Larsen et al., 2010). Recently CHD4 depleted AML cell lines were shown to be sensitized to the DNA-damaging effects of daunorubicin resulting in increased caspase cleavage and apoptosis (Sperlazza et al., 2015). Roles for NURD in regulating sensitivity to ICD-inducing chemotherapies also extend to the NURD metastasis-associated 1 (MTA1) subunit. In a similar study, depletion of MTA1 sensitized pancreatic cells to the cytotoxic effects of docetaxel (Yu, Su, Zhao, Wang, & Li, 2013).

2.4 miRNAs

miRNA expression profiles were compared between TNBC tumors and normal breast tissues to identify differently expressed miRNAs (Ouyang et al., 2014). From this analysis, 11 specific miRs, 5 upregulated and 6 downregulated, were discovered each of which regulates cancer-associated pathways. These studies identified for the first time miR-451a and miR-130a-3p as being downregulated in breast cancer, and where their upregulation significantly increased MDA-MB-231 sensitivity to doxorubicin. These results identified several miRNAs which could be manipulated to sensitize TNBC cells to doxorubicin.

In similar studies, miRNA profiling was performed on the estrogen and progesterone receptor positive MCF7 and doxorubicin-resistant MCF7 cells to identify miRNAs differently expressed with chemoresistance (Kovalchuk et al., 2008). Constitutive deregulation of the miRNA transcriptome was observed including deregulated miRNA processing enzymes Dicer and Argonaut 2. Focused studies on miR-451 show that it regulates the expression of MDR1 and that its overexpression results in increased sensitivity of drug-resistant MCF7 cells to doxorubicin.

In the same MCF7 breast tumor cell line, a pathway enrichment approach was applied to discover that miR-222 regulates the PTEN/Akt/FOXO1 pathway to regulate sensitivity of breast cancer cells to doxorubicin (Shen et al., 2017). MiR-222 overexpression in doxorubicin-sensitive MCF7 cells correlated with decreased FOXO1 and PTEN expression and increased phospho-AKT abundance. Inversely, in doxorubicin-resistant MCF7 cells inhibition of miR-222 increased PTEN expression and decreased phospho-AKT, leading to increased cell death. Use of the AKT pathway inhibitor LY294002 confirmed that AKT signaling is responsible for the doxorubicin resistance of MCF7 cells.

miR-16 regulates the expression of wild-type p53-induced phosphatase (WIP1); an oncogene overexpressed in several cancers (Zhang et al., 2010). WIP1 dephosphorylates several proteins important for DNA damage repair and deactivates DNA damage checkpoints. MiR-16 is downregulated in mammary cancers and when overexpressed in the MCF7 breast cancer cell lines, results in the suppression of growth, decreased self-renewal and increases sensitivity to doxorubicin.

Initial studies discovered that miR-128 is reduced in chemoresistant stem cell like breast tumor initiating cells (Zhu et al., 2011). Subsequent studies showed that reduced miR-128 expression accompanied the overexpression of chemoresistance proteins BMI-1 and ABCC5, each of which is a target of miR-128. Overexpression of miR-128 in these cell lines reduced BMI-1 and ABCC5 protein expression in BT-ICs, decreasing cell viability, and increasing DNA damage and apoptosis after treatment with doxorubicin.

In addition to doxorubicin, miRs regulate the sensitivity of breast cancer cells to paclitaxel. Increased expression of Dicer mediated by adenovirus-type 5 E1A enhances paclitaxel sensitization and reduces cancer stem cell properties (Chang, Chen, et al., 2016; Chang, Wang, et al., 2016). Overexpressing Dicer increased levels of miR-494 which targets the AXL kinase and which results in paclitaxel sensitivity. More broadly, these studies identify the nature of E1A-mediated chemosensitization to paclitaxel, which involves the suppression of breast cancer stem cell like cells.

The activation of an epithelial-to-mesenchymal transition (EMT) in pancreatic cancer and its association with cancer stem cell properties is a key determinant of therapy resistance (Meidhof et al., 2015). MiR-203 was identified in an attempt to discover regulators of the EMT activator ZEB1. Subsequent studies showed that the HDAC1 inhibitor mocetinostat, which is more effective than other HDACi including SAHA, interferes with ZEB1 function, restores miR-203 expression to repress stem cell properties and sensitizes tumor cells to paclitaxel and docetaxel.

3. CONCLUSIONS AND FUTURE DIRECTIONS

In this review, we summarize the available literature where epigenetics was targeted to sensitize cancer cells to the effects of ICD-inducing chemotherapies. These studies heavily focused on apoptosis, a required pathway for ICD. The full extent of pathways which could be targeted by epigenetic drugs to enhance ICD is unknown, largely because it is also unknown why some chemotherapies induce ICD and other do not. Until this is resolved, there will be shortfalls in our abilities to design effective combinational strategies. For example, related compounds such as cisplatin and oxaliplatin similarly induce DNA crosslinks which induce DNA damage (Dilruba & Kalayda, 2016), yet oxaliplatin can induce ICD whereas cisplatin cannot (Tesniere et al., 2010). Selective activity is likely due to differences in chemical structure (Dilruba & Kalayda, 2016), which would give oxaliplatin its ICD-inducing abilities. It is unlikely that DNA damage itself is required for ICD, because many DNA-damaging chemotherapies do not induce ICD, with cisplatin as an example (Tesniere et al., 2010).

ICD-inducing activities could result from induced autophagy, ER stress, and caspase-dependent apoptosis. As described earlier, DNMTi and HDACi in part generally sensitize cancer cells to chemotherapies by reactivating silenced apoptosis mechanisms (see references earlier). These activities would also benefit ICD-inducing chemotherapies. Epigenetic regulation of ER stress is less well known, but has been documented. In one study, HDACi treatment of pancreatic cells resulted in the upregulation of ER stress factors including BiP, ATF4, CHOP, and caspase 4, which promoted ER stress (Klieser et al., 2015). In a second study, the histone H3K9 methyltransferase SUV39H1 inhibitor chaetocin upregulated the ER stress response in lung cancer cells as measured by the markers ATF3 and CHOP, which in turn contributed to death receptor 5 induced apoptosis (Liu, Guo, & Su, 2015). Therefore, in addition to reactivating apoptosis, HDACi and histone methyltransferase inhibitor (HMTi) could prime tumor cells for ER stress.

In addition to apoptosis and ER stress pathways, autophagy has been known for some time to be regulated by epigenetics in cancer cells. Under nutrient-rich conditions, the expression of autophagy gene promoters (LC3B, WIPI1, and DOR) is repressed by G9a-mediated H3K9 methylation (Artal-Martinez de Narvajas et al., 2013). In addition to G9a, EZH2 has autophagy-suppressing activities (Wei et al., 2015). In this capacity, EZH2 methylates and suppresses several negative regulators of the mammalian target of the rapamycin (mTOR) pathway. These functions activate mTOR which suppresses autophagy. Both G9a and EZH2 are commonly overexpressed in a variety of cancers and have procancer functions (Casciello, Windloch, Gannon, & Lee, 2015; Kim & Roberts, 2016), one of which could be the suppression of autophagy. Since autophagy could be important for ICD (Michaud et al., 2011), it is plausible that the autophagy-suppressing activity of G9a and EZH2 activity could thereby suppress therapy-induced ICD. It would then follow that inhibiting G9a or EZH2 could enhance the effects of ICD-inducing chemotherapies by making chemotherapy-treated cells more prone to undergo autophagy.

One key element of the cell nonautonomous response to ICD-inducing chemotherapies is the release and efficient cross-presentation of tumor antigens. Cross-presented tumor antigens prime and expand tumor-reactive T-cell clones which, in turn, are partially responsible for tumor growth control (Bloy et al., 2017). One well-described effect of epigenetic therapies is the reexpression of TSAs. A variety of DNMTi and HDACi can upregulate several classes of tumor antigens and cancer testis antigens (CTAs) which include the melanoma-associated antigens (MAGE) (Goodyear et al., 2010). Both MAGE and CTAs, important sources of tumor cell antigenicity, are downregulated by tumors (Rivoltini et al., 1998) and have been exploited as a means to achieve immunotherapy through vaccination (Coulie et al., 2002). The ability of epigenetic regulators to enhance tumor cell antigenicity due to MAGE and CTA upregulation would further improve the cell nonautonomous effects of ICD-inducing chemotherapies. This would occur in conjunction with potential improvements in sensitization to apoptosis, ER stress, and autophagy as articulated earlier.

In addition to enhancing tumor antigenicity, and the accompanying CD8 T-cell antitumor response, inhibiting epigenetic regulators can also improve NK cell-mediated antitumor activity. These activities occur when expression of NK cell receptor ligands is upregulated, which would normally happen in stressed tumor cells. Upregulation of stress ligands on tumors cells is commonly repressed as a result of immune editing (Mittal, Gubin, Schreiber, & Smyth, 2014; O'Sullivan, Dunn, Lacoursiere, Schreiber, & Bui, 2011). Prominent examples include HDACi-induced reexpression of NK cell receptor ligands in HCC (Shin, Kim, Lee, Park, & Lee, 2017;

Zhang, Wang, Zhou, Zhang, & Tian, 2009), pancreatic (Shi et al., 2014), colon (Zhu, Denman, et al., 2015; Zhu, Tchkonia, et al., 2015), melanoma (Wu, Tao, Hou, Meng, & Shi, 2012), and hematologic malignancies (Diermayr et al., 2008; Kato et al., 2007). In addition to HDACi, DNMTi also have these functions in ovarian (Hoogstad-van Evert et al., 2017) and hematologic malignancies (Baragano Raneros et al., 2015; Cany et al., 2017; Raneros et al., 2017; Rohner, Langenkamp, Siegler, Kalberer, & Wodnar-Filipowicz, 2007). While ICD is largely mediated and dependent on CD8 T-cell interactions with antigen-presenting cells, cytotoxic chemotherapies, irrespective of their ability to induce ICD (Khallouf et al., 2012; Siew et al., 2015), are known to upregulate NK cell ligands, which could be further enhanced by epigenetic therapy. In addition, the cell autonomous benefits of epigenetic therapies on tumor cell immunogenicity will occur in context with their effects on other immune cells. In some instances, epigenetic drugs can regulate immune cells to improve their antitumor response. This can occur from suppressing immune suppressor cells such as Treg (Shen et al., 2012) and myeloid-derived suppressor cells (Kim et al., 2014). In contrast, there is evidence that epigenetic therapies can inhibit the activity of effector cells, which suppresses the antitumor response (Terranova-Barberio, Thomas, & Munster, 2016). The ultimate outcome of epigenetic therapies on immune response will likely involve a balance of these activities.

A majority of the studies cited in this review do not study how the immune system responds to the combinatorial therapies. Because ICD by definition results in an enhanced antitumor immune response, it will be critical to perform many of the key studies in the context of a functional immune system. Popular models to this end would be syngeneic mouse transplantable tumor models, which have been heavily used and well characterized over years of research. Ideally, studies should also include one or more genetically engineered spontaneous models (Ngiow, Loi, Thomas, & Smyth, 2016). Efficacy to both spontaneous (low mutation burden) and transplantable (high mutation burden) models must be investigated to assess the requirement of tumor mutation burden to the success of therapy-induced ICD. Testing the relevance of tumor mutation burden for the success of ICD, and its epigenetic manipulation, is important because previous reports have shown a lack of ICD when spontaneous models (Ciampricotti, Hau, Doornebal, Jonkers, & de Visser, 2012) or poorly immunogenic transplantable models are used (Starobinets et al., 2016). It is plausible that targeting epigenetic regulations, in combination with ICD-inducing chemotherapies, could improve the response to these low mutation burden or poorly immunogenic models.

The current strategies targeting epigenetics as a means to achieve therapy rely heavily on DNMTi and HDACi. This is likely due, in part, because pharmacologic grade inhibitors to these regulators have been used in the clinic for decades (Nebbioso, Carafa, Benedetti, & Altucci, 2012), with next-generation inhibitors constantly being developed (Fang et al., 2014). These agents have been pursued not only for convenience, but also because both DNMTs and HDACs serve a central and partially overlapping role in the repression of genes with tumor suppressor activity. These genes include positive regulators of apoptosis, cellular differentiation, and DNA damage repair. These categories all have central roles in tumor cells acquisition of drug resistance, including those which induce ICD (see references earlier). Looking to the horizon, new classes of HDACs could be targeted to achieve chemosensitization. These include the SIR2 family of HDACs, which have been inhibited in preliminary studies to sensitize human breast, prostate, skin, and ovarian cancers as well as leukemia to chemotherapy (Kim et al., 2015; Sociali et al., 2015; Xiong, Li, Tang, & Chen, 2011).

Work over the last several decades has discovered many new epigenetic regulators with important functions in cancer biology. Some of these functions can be exploited to improve chemosensitization as a means to enhance ICD. For example, the HATs are an important group of epigenetic regulators for which functional small-molecule regulatory molecules are being designed (Wapenaar & Dekker, 2016). As described earlier, the CBP/p300 (Dutto, Scalera, & Prosperi, 2017) and Tip60 (Zhang, Wu, & Luan, 2017) family of HATs has critical functions in regulating DNA damage repair and apoptosis in cancer cells, two proven pathways which can be targeted to achieve chemosensitization. For either of these families there is preliminary evidence that they do regulate resistance to chemotherapy, with outcomes including apoptosis in pancreatic (Ono, Basson, & Ito, 2016), colon (Du et al., 2017), and breast (Sen et al., 2011) cancer. In addition to HATs, the histone H3K9 methyltransferase G9a has shown promise as a means to enhance sensitivity to chemotherapy to pancreatic (Pan et al., 2016), colon (Agarwal & Jackson, 2016), and head and neck cancers (Liu et al., 2017). In addition, the H3K9 demethylases KDM3B (Dalvi et al., 2017) and PHF8 (Wang et al., 2016) also have roles in sensitization to chemotherapy. While these initial studies on HATs, HMTs, and HDMs are potentially of clinical relevance, they did involve the use of chemotherapies which induce ICD and therefore could have broader therapeutic effectiveness.

One often overlooked and, thus, underexplored group of epigenetic regulators are the chromatin remodeling complexes. These complexes are classified by the presence of an ATPase from one of four families; switching defective/sucrose nonfermenting (SWI/SNF), imitation switch (ISWI), chromodomain helicase DNA-binding (CHD), or inositol requiring 80 (INO80) (Becker & Workman, 2013). Chromatin remodeling complexes are frequently mutated and deregulated in human cancers (Becker & Workman, 2013). In fact, it has been estimated that one or more subunits of the SWI/SNF family of chromatin remodeling complexes is mutated in ~20% of all cancers, and in ~50% of some specific human cancers (Kadoch et al., 2013). Therefore, it is not surprising that the expression of the SWI/SNF family of complexes in several cancer cells including pancreatic (Liu et al., 2014), lung (Fillmore et al., 2015; Kothandapani, Gopalakrishnan, Kahali, Reisman, & Patrick, 2012), and colon (Kwon et al., 2015) cancer regulates the sensitivity to chemo or radiation therapy. These studies are relevant to our own work, as targeting the nucleosome remodeling factor (NURF) (Alkhatib & Landry, 2011) could serve as a means to sensitize tumors to chemotherapies. We recently showed that NURF depleted TNBC cells are more sensitive to the cell nonautonomous effects of gemcitabine using immunocompetent mice (Mayes et al., 2016). These effects could be related to the improved antigenicity and immunogenicity we observe in NURF depleted tumor cells (Mayes et al., 2016, 2017). It is currently under investigation if NURF depleted tumors are sensitized to other chemotherapies, particularly those which induce ICD.

ACKNOWLEDGMENTS

The authors would like to apologize for omitting any publications relevant to this review because of space constraints. The authors thank Luni Emdad for taking the time to critically review the manuscript. Financial support for this review was provided by the Virginia Commonwealth University School of Medicine and the Massey Cancer Center. Work in Dr. Gewirtz's laboratory was supported by the Office of the Assistant Secretary of Defense for Health Affairs through the Breast Cancer Research Program (grant no. W81XWH-14-1-0088).

CONFLICTS OF INTEREST

The authors do not declare any conflicts of interest.

REFERENCES

Acosta, J. C., O'Loghlen, A., Banito, A., Guijarro, M. V., Augert, A., Raguz, S., et al. (2008). Chemokine signaling via the CXCR2 receptor reinforces senescence. *Cell, 133*(6), 1006–1018.
Acton, E. M., Ryan, K. J., & Luetzow, A. E. (1977). Antitumor septacidin analogues. *Journal of Medicinal Chemistry, 20*(11), 1362–1371.
Agarwal, P., & Jackson, S. P. (2016). G9a inhibition potentiates the anti-tumour activity of DNA double-strand break inducing agents by impairing DNA repair independent of p53 status. *Cancer Letters, 380*(2), 467–475.
Alkhatib, S. G., & Landry, J. W. (2011). The nucleosome remodeling factor. *FEBS Letters, 585*(20), 3197–3207.
Allis, C. D., Berger, S. L., Cote, J., Dent, S., Jenuwien, T., Kouzarides, T., et al. (2007). New nomenclature for chromatin-modifying enzymes. *Cell, 131*(4), 633–636.
Amaravadi, R. K., & Thompson, C. B. (2007). The roles of therapy-induced autophagy and necrosis in cancer treatment. *Clinical Cancer Research, 13*(24), 7271–7279.
Arce, C., Perez-Plasencia, C., Gonzalez-Fierro, A., de la Cruz-Hernandez, E., Revilla-Vazquez, A., Chavez-Blanco, A., et al. (2006). A proof-of-principle study of epigenetic therapy added to neoadjuvant doxorubicin cyclophosphamide for locally advanced breast cancer. *PLoS One, 1*, e98.
Artal-Martinez de Narvajas, A., Gomez, T. S., Zhang, J. S., Mann, A. O., Taoda, Y., Gorman, J. A., et al. (2013). Epigenetic regulation of autophagy by the methyltransferase G9a. *Molecular and Cellular Biology, 33*(20), 3983–3993.
Aszalos, A., Lemanski, P., Berk, B., & Dutcher, J. D. (1965). Septacidin analogues. *Antimicrobial Agents and Chemotherapy (Bethesda), 5*, 845–849.
Attar, N., & Kurdistani, S. K. (2017). Exploitation of EP300 and CREBBP lysine acetyltransferases by cancer. *Cold Spring Harbor Perspectives in Medicine, 7*(3), a026534.
Bai, J., Chen, J., Ma, M., Cai, M., Xu, F., Wang, G., et al. (2014). Inhibiting enhancer of zeste homolog 2 promotes cellular senescence in gastric cancer cells SGC-7901 by activation of p21 and p16. *DNA and Cell Biology, 33*(6), 337–344.
Bai, J., Ma, M., Cai, M., Xu, F., Chen, J., Wang, G., et al. (2014). Inhibition enhancer of zeste homologue 2 promotes senescence and apoptosis induced by doxorubicin in p53 mutant gastric cancer cells. *Cell Proliferation, 47*(3), 211–218.
Bangert, A., Hacker, S., Cristofanon, S., Debatin, K. M., & Fulda, S. (2011). Chemosensitization of glioblastoma cells by the histone deacetylase inhibitor MS275. *Anti-Cancer Drugs, 22*(6), 494–499.
Bannister, A. J., & Kouzarides, T. (2011). Regulation of chromatin by histone modifications. *Cell Research, 21*(3), 381–395.
Baragano Raneros, A., Martin-Palanco, V., Fernandez, A. F., Rodriguez, R. M., Fraga, M. F., Lopez-Larrea, C., et al. (2015). Methylation of NKG2D ligands contributes to immune system evasion in acute myeloid leukemia. *Genes and Immunity, 16*(1), 71–82.
Barajas-Gomez, B. A., Rosas-Carrasco, O., Morales-Rosales, S. L., Pedraza Vazquez, G., Gonzalez-Puertos, V. Y., Juarez-Cedillo, T., et al. (2017). Relationship of inflammatory profile of elderly patients serum and senescence-associated secretory phenotype with human breast cancer cells proliferation: Role of IL6/IL8 ratio. *Cytokine, 91*, 13–29.
Basso, K., & Dalla-Favera, R. (2012). Roles of BCL6 in normal and transformed germinal center B cells. *Immunological Reviews, 247*(1), 172–183.
Becker, P. B., & Workman, J. L. (2013). Nucleosome remodeling and epigenetics. *Cold Spring Harbor Perspectives in Biology, 5*(9), a017905.
Bellovin, D. I., Das, B., & Felsher, D. W. (2013). Tumor dormancy, oncogene addiction, cellular senescence, and self-renewal programs. *Advances in Experimental Medicine and Biology, 734*, 91–107.

Belounis, A., Nyalendo, C., Le Gall, R., Imbriglio, T. V., Mahma, M., Teira, P., et al. (2016). Autophagy is associated with chemoresistance in neuroblastoma. *BMC Cancer*, *16*(1), 891.

Bhat, T. A., Chaudhary, A. K., Kumar, S., O'Malley, J., Inigo, J. R., Kumar, R., et al. (2017). Endoplasmic reticulum-mediated unfolded protein response and mitochondrial apoptosis in cancer. *Biochimica et Biophysica Acta*, *1867*(1), 58–66.

Bielack, S. S., Hecker-Nolting, S., Blattmann, C., & Kager, L. (2016). Advances in the management of osteosarcoma. *F1000Res*, *5*, 2767.

Bloy, N., Garcia, P., Laumont, C. M., Pitt, J. M., Sistigu, A., Stoll, G., et al. (2017). Immunogenic stress and death of cancer cells: Contribution of antigenicity vs adjuvanticity to immunosurveillance. *Immunological Reviews*, *280*(1), 165–174.

Burke, M. J., Lamba, J. K., Pounds, S., Cao, X., Ghodke-Puranik, Y., Lindgren, B. R., et al. (2014). A therapeutic trial of decitabine and vorinostat in combination with chemotherapy for relapsed/refractory acute lymphoblastic leukemia. *American Journal of Hematology*, *89*(9), 889–895.

Cany, J., Roeven, M. W. H., Hoogstad-van Evert, J. S., Hobo, W., Maas, F., Franco Fernandez, R., et al. (2017). Decitabine enhances targeting of AML cells by CD34(+) progenitor-derived NK cells in NOD/SCID/IL2Rg(null) mice. *Blood*, *131*, 202–214.

Casares, N., Pequignot, M. O., Tesniere, A., Ghiringhelli, F., Roux, S., Chaput, N., et al. (2005). Caspase-dependent immunogenicity of doxorubicin-induced tumor cell death. *The Journal of Experimental Medicine*, *202*(12), 1691–1701.

Casciello, F., Windloch, K., Gannon, F., & Lee, J. S. (2015). Functional role of G9a histone methyltransferase in cancer. *Frontiers in Immunology*, *6*, 487.

Chang, T. Y., Chen, H. A., Chiu, C. F., Chang, Y. W., Kuo, T. C., Tseng, P. C., et al. (2016). Dicer elicits paclitaxel chemosensitization and suppresses cancer stemness in breast cancer by repressing AXL. *Cancer Research*, *76*(13), 3916–3928.

Chang, J., Wang, Y., Shao, L., Laberge, R. M., Demaria, M., Campisi, J., et al. (2016). Clearance of senescent cells by ABT263 rejuvenates aged hematopoietic stem cells in mice. *Nature Medicine*, *22*(1), 78–83.

Chen, C. C., Lee, K. D., Pai, M. Y., Chu, P. Y., Hsu, C. C., Chiu, C. C., et al. (2015). Changes in DNA methylation are associated with the development of drug resistance in cervical cancer cells. *Cancer Cell International*, *15*, 98.

Chien, Y., Scuoppo, C., Wang, X., Fang, X., Balgley, B., Bolden, J. E., et al. (2011). Control of the senescence-associated secretory phenotype by NF-kappaB promotes senescence and enhances chemosensitivity. *Genes & Development*, *25*(20), 2125–2136.

Chou, T. C. (2010). Drug combination studies and their synergy quantification using the Chou–Talalay method. *Cancer Research*, *70*(2), 440–446.

Ciampricotti, M., Hau, C. S., Doornebal, C. W., Jonkers, J., & de Visser, K. E. (2012). Chemotherapy response of spontaneous mammary tumors is independent of the adaptive immune system. *Nature Medicine*, *18*(3), 344–346, author reply 346.

Clozel, T., Yang, S., Elstrom, R. L., Tam, W., Martin, P., Kormaksson, M., et al. (2013). Mechanism-based epigenetic chemosensitization therapy of diffuse large B-cell lymphoma. *Cancer Discovery*, *3*(9), 1002–1019.

Coppe, J. P., Desprez, P. Y., Krtolica, A., & Campisi, J. (2010). The senescence-associated secretory phenotype: The dark side of tumor suppression. *Annual Review of Pathology*, *5*, 99–118.

Coppe, J. P., Patil, C. K., Rodier, F., Sun, Y., Munoz, D. P., Goldstein, J., et al. (2008). Senescence-associated secretory phenotypes reveal cell-nonautonomous functions of oncogenic RAS and the p53 tumor suppressor. *PLoS Biology*, *6*(12), 2853–2868.

Coulie, P. G., Karanikas, V., Lurquin, C., Colau, D., Connerotte, T., Hanagiri, T., et al. (2002). Cytolytic T-cell responses of cancer patients vaccinated with a MAGE antigen. *Immunological Reviews*, *188*, 33–42.

Couronne, L., Bastard, C., & Bernard, O. A. (2012). TET2 and DNMT3A mutations in human T-cell lymphoma. *The New England Journal of Medicine, 366*(1), 95–96.

Cree, I. A., Knight, L., Di Nicolantonio, F., Sharma, S., & Gulliford, T. (2002). Chemosensitization of solid tumor cells by alteration of their susceptibility to apoptosis. *Current Opinion in Investigational Drugs, 3*(4), 641–647.

Dalvi, M. P., Wang, L., Zhong, R., Kollipara, R. K., Park, H., Bayo, J., et al. (2017). Taxane-platin-resistant lung cancers co-develop hypersensitivity to jumonjic demethylase inhibitors. *Cell Reports, 19*(8), 1669–1684.

Dawson, M. A., & Kouzarides, T. (2012). Cancer epigenetics: From mechanism to therapy. *Cell, 150*(1), 12–27.

Diermayr, S., Himmelreich, H., Durovic, B., Mathys-Schneeberger, A., Siegler, U., Langenkamp, U., et al. (2008). NKG2D ligand expression in AML increases in response to HDAC inhibitor valproic acid and contributes to allorecognition by NK-cell lines with single KIR-HLA class I specificities. *Blood, 111*(3), 1428–1436.

Dilruba, S., & Kalayda, G. V. (2016). Platinum-based drugs: Past, present and future. *Cancer Chemotherapy and Pharmacology, 77*(6), 1103–1124.

Du, C., Huang, D., Peng, Y., Yao, Y., Zhao, Y., Yang, Y., et al. (2017). 5-Fluorouracil targets histone acetyltransferases p300/CBP in the treatment of colorectal cancer. *Cancer Letters, 400*, 183–193.

Dutto, I., Scalera, C., & Prosperi, E. (2017). CREBBP and p300 lysine acetyl transferases in the DNA damage response. *Cellular and Molecular Life Sciences*.

Eissenberg, J. C., & Elgin, S. C. (2014). HP1a: A structural chromosomal protein regulating transcription. *Trends in Genetics, 30*(3), 103–110.

Elliott, M. R., Chekeni, F. B., Trampont, P. C., Lazarowski, E. R., Kadl, A., Walk, S. F., et al. (2009). Nucleotides released by apoptotic cells act as a find-me signal to promote phagocytic clearance. *Nature, 461*(7261), 282–286.

Esteller, M., & Pandolfi, P. P. (2017). The epitranscriptome of noncoding RNAs in cancer. *Cancer Discovery, 7*(4), 359–368.

Falkenberg, K. J., & Johnstone, R. W. (2014). Histone deacetylases and their inhibitors in cancer, neurological diseases and immune disorders. *Nature Reviews. Drug Discovery, 13*(9), 673–691.

Fang, F., Munck, J., Tang, J., Taverna, P., Wang, Y., Miller, D. F., et al. (2014). The novel, small-molecule DNA methylation inhibitor SGI-110 as an ovarian cancer chemosensitizer. *Clinical Cancer Research, 20*(24), 6504–6516.

Festuccia, C., Gravina, G. L., D'Alessandro, A. M., Muzi, P., Millimaggi, D., Dolo, V., et al. (2009). Azacitidine improves antitumor effects of docetaxel and cisplatin in aggressive prostate cancer models. *Endocrine-Related Cancer, 16*(2), 401–413.

Fillmore, C. M., Xu, C., Desai, P. T., Berry, J. M., Rowbotham, S. P., Lin, Y. J., et al. (2015). EZH2 inhibition sensitizes BRG1 and EGFR mutant lung tumours to TopoII inhibitors. *Nature, 520*(7546), 239–242.

Gajer, J. M., Furdas, S. D., Grunder, A., Gothwal, M., Heinicke, U., Keller, K., et al. (2015). Histone acetyltransferase inhibitors block neuroblastoma cell growth in vivo. *Oncogene, 4*, e137.

Gallenkamp, D., Gelato, K. A., Haendler, B., & Weinmann, H. (2014). Bromodomains and their pharmacological inhibitors. *ChemMedChem, 9*(3), 438–464.

Gao, M., Xu, Y., & Qiu, L. (2017). Sensitization of multidrug-resistant malignant cells by liposomes co-encapsulating doxorubicin and chloroquine through autophagic inhibition. *Journal of Liposome Research, 27*(2), 151–160.

Gardai, S. J., McPhillips, K. A., Frasch, S. C., Janssen, W. J., Starefeldt, A., Murphy-Ullrich, J. E., et al. (2005). Cell-surface calreticulin initiates clearance of viable or apoptotic cells through trans-activation of LRP on the phagocyte. *Cell, 123*(2), 321–334.

Garg, A. D., Krysko, D. V., Verfaillie, T., Kaczmarek, A., Ferreira, G. B., Marysael, T., et al. (2012). A novel pathway combining calreticulin exposure and ATP secretion in immunogenic cancer cell death. *The EMBO Journal, 31*(5), 1062–1079.

Garg, A. D., Maes, H., van Vliet, A. R., & Agostinis, P. (2015). Targeting the hallmarks of cancer with therapy-induced endoplasmic reticulum (ER) stress. *Molecular & Cellular Oncology, 2*(1), e975089.

Gewirtz, D. A. (1999). A critical evaluation of the mechanisms of action proposed for the antitumor effects of the anthracycline antibiotics adriamycin and daunorubicin. *Biochemical Pharmacology, 57*(7), 727–741.

Gewirtz, D. A. (2013). Cytoprotective and nonprotective autophagy in cancer therapy. *Autophagy, 9*(9), 1263–1265.

Gewirtz, D. A. (2014). The four faces of autophagy: Implications for cancer therapy. *Cancer Research, 74*(3), 647–651.

Goodyear, O., Agathanggelou, A., Novitzky-Basso, I., Siddique, S., McSkeane, T., Ryan, G., et al. (2010). Induction of a CD8+ T-cell response to the MAGE cancer testis antigen by combined treatment with azacitidine and sodium valproate in patients with acute myeloid leukemia and myelodysplasia. *Blood, 116*(11), 1908–1918.

Guil, S., & Esteller, M. (2009). DNA methylomes, histone codes and miRNAs: Tying it all together. *The International Journal of Biochemistry & Cell Biology, 41*(1), 87–95.

Halpern, A. B., Othus, M., Huebner, E. M., Buckley, S. A., Pogosova-Agadjanyan, E. L., Orlowski, K. F., et al. (2017). Mitoxantrone, etoposide and cytarabine following epigenetic priming with decitabine in adults with relapsed/refractory acute myeloid leukemia or other high-grade myeloid neoplasms: A phase 1/2 study. *Leukemia, 31*(12), 2560–2567.

Hanahan, D., & Weinberg, R. A. (2011). Hallmarks of cancer: The next generation. *Cell, 144*(5), 646–674.

Heidegger, I., Massoner, P., Eder, I. E., Pircher, A., Pichler, R., Aigner, F., et al. (2013). Novel therapeutic approaches for the treatment of castration-resistant prostate cancer. *The Journal of Steroid Biochemistry and Molecular Biology, 138*, 248–256.

Heinicke, U., & Fulda, S. (2014). Chemosensitization of rhabdomyosarcoma cells by the histone deacetylase inhibitor SAHA. *Cancer Letters, 351*(1), 50–58.

Helleday, T., Petermann, E., Lundin, C., Hodgson, B., & Sharma, R. A. (2008). DNA repair pathways as targets for cancer therapy. *Nature Reviews. Cancer, 8*(3), 193–204.

Hoogstad-van Evert, J. S., Cany, J., van den Brand, D., Oudenampsen, M., Brock, R., Torensma, R., et al. (2017). Umbilical cord blood CD34(+) progenitor-derived NK cells efficiently kill ovarian cancer spheroids and intraperitoneal tumors in NOD/SCID/IL2Rg(null) mice. *Oncoimmunology, 6*(8), e1320630.

Howell, S. B., Safaei, R., Larson, C. A., & Sailor, M. J. (2010). Copper transporters and the cellular pharmacology of the platinum-containing cancer drugs. *Molecular Pharmacology, 77*(6), 887–894.

Huang, Y., Anderle, P., Bussey, K. J., Barbacioru, C., Shankavaram, U., Dai, Z., et al. (2004). Membrane transporters and channels: Role of the transportome in cancer chemosensitivity and chemoresistance. *Cancer Research, 64*(12), 4294–4301.

Ikehata, M., Ogawa, M., Yamada, Y., Tanaka, S., Ueda, K., & Iwakawa, S. (2014). Different effects of epigenetic modifiers on the cytotoxicity induced by 5-fluorouracil, irinotecan or oxaliplatin in colon cancer cells. *Biological & Pharmaceutical Bulletin, 37*(1), 67–73.

Jackson, J. R., Patrick, D. R., Dar, M. M., & Huang, P. S. (2007). Targeted anti-mitotic therapies: Can we improve on tubulin agents? *Nature Reviews. Cancer, 7*(2), 107–117.

Judes, G., Rifai, K., Ngollo, M., Daures, M., Bignon, Y. J., Penault-Llorca, F., et al. (2015). A bivalent role of TIP60 histone acetyl transferase in human cancer. *Epigenomics, 7*(8), 1351–1363.

Kadoch, C., Hargreaves, D. C., Hodges, C., Elias, L., Ho, L., Ranish, J., et al. (2013). Proteomic and bioinformatic analysis of mammalian SWI/SNF complexes identifies extensive roles in human malignancy. *Nature Genetics, 45*(6), 592–601.

Kastl, L., Brown, I., & Schofield, A. C. (2010). Altered DNA methylation is associated with docetaxel resistance in human breast cancer cells. *International Journal of Oncology, 36*(5), 1235–1241.

Kato, N., Tanaka, J., Sugita, J., Toubai, T., Miura, Y., Ibata, M., et al. (2007). Regulation of the expression of MHC class I-related chain A, B (MICA, MICB) via chromatin remodeling and its impact on the susceptibility of leukemic cells to the cytotoxicity of NKG2D-expressing cells. *Leukemia, 21*(10), 2103–2108.

Kepp, O., Senovilla, L., & Kroemer, G. (2014). Immunogenic cell death inducers as anticancer agents. *Oncotarget, 5*(14), 5190–5191.

Khallouf, H., Marten, A., Serba, S., Teichgraber, V., Buchler, M. W., Jager, D., et al. (2012). 5-Fluorouracil and interferon-alpha immunochemotherapy enhances immunogenicity of murine pancreatic cancer through upregulation of NKG2D ligands and MHC class I. *Journal of Immunotherapy, 35*(3), 245–253.

Kim, H. B., Lee, S. H., Um, J. H., Kim, M. J., Hyun, S. K., Gong, E. J., et al. (2015). Sensitization of chemo-resistant human chronic myeloid leukemia stem-like cells to Hsp90 inhibitor by SIRT1 inhibition. *International Journal of Biological Sciences, 11*(8), 923–934.

Kim, K. H., & Roberts, C. W. (2016). Targeting EZH2 in cancer. *Nature Medicine, 22*(2), 128–134.

Kim, K., Skora, A. D., Li, Z., Liu, Q., Tam, A. J., Blosser, R. L., et al. (2014). Eradication of metastatic mouse cancers resistant to immune checkpoint blockade by suppression of myeloid-derived cells. *Proceedings of the National Academy of Sciences of the United States of America, 111*(32), 11774–11779.

Klieser, E., Illig, R., Stattner, S., Primavesi, F., Jager, T., Swierczynski, S., et al. (2015). Endoplasmic reticulum stress in pancreatic neuroendocrine tumors is linked to clinicopathological parameters and possible epigenetic regulations. *Anticancer Research, 35*(11), 6127–6136.

Kohli, R. M., & Zhang, Y. (2013). TET enzymes, TDG and the dynamics of DNA demethylation. *Nature, 502*(7472), 472–479.

Kondo, Y., & Issa, J. P. (2004). Epigenetic changes in colorectal cancer. *Cancer Metastasis Reviews, 23*(1-2), 29–39.

Kothandapani, A., Gopalakrishnan, K., Kahali, B., Reisman, D., & Patrick, S. M. (2012). Downregulation of SWI/SNF chromatin remodeling factor subunits modulates cisplatin cytotoxicity. *Experimental Cell Research, 318*(16), 1973–1986.

Kovalchuk, O., Filkowski, J., Meservy, J., Ilnytskyy, Y., Tryndyak, V. P., Chekhun, V. F., et al. (2008). Involvement of microRNA-451 in resistance of the MCF-7 breast cancer cells to chemotherapeutic drug doxorubicin. *Molecular Cancer Therapeutics, 7*(7), 2152–2159.

Krizhanovsky, V., Xue, W., Zender, L., Yon, M., Hernando, E., & Lowe, S. W. (2008). Implications of cellular senescence in tissue damage response, tumor suppression, and stem cell biology. *Cold Spring Harbor Symposia on Quantitative Biology, 73*, 513–522.

Krtolica, A., Parrinello, S., Lockett, S., Desprez, P. Y., & Campisi, J. (2001). Senescent fibroblasts promote epithelial cell growth and tumorigenesis: A link between cancer and aging. *Proceedings of the National Academy of Sciences of the United States of America, 98*(21), 12072–12077.

Krysko, D. V., Garg, A. D., Kaczmarek, A., Krysko, O., Agostinis, P., & Vandenabeele, P. (2012). Immunogenic cell death and DAMPs in cancer therapy. *Nature Reviews. Cancer, 12*(12), 860–875.

Kuang, Y., El-Khoueiry, A., Taverna, P., Ljungman, M., & Neamati, N. (2015). Guadecitabine (SGI-110) priming sensitizes hepatocellular carcinoma cells to oxaliplatin. *Molecular Oncology, 9*(9), 1799–1814.

Kuilman, T., & Peeper, D. S. (2009). Senescence-messaging secretome: SMS-ing cellular stress. *Nature Reviews. Cancer, 9*(2), 81–94.

Kwon, S. J., Lee, S. K., Na, J., Lee, S. A., Lee, H. S., Park, J. H., et al. (2015). Targeting BRG1 chromatin remodeler via its bromodomain for enhanced tumor cell radiosensitivity in vitro and in vivo. *Molecular Cancer Therapeutics, 14*(2), 597–607.

Lai, A. Y., & Wade, P. A. (2011). Cancer biology and NuRD: A multifaceted chromatin remodelling complex. *Nature Reviews. Cancer, 11*(8), 588–596.

Larsen, D. H., Poinsignon, C., Gudjonsson, T., Dinant, C., Payne, M. R., Hari, F. J., et al. (2010). The chromatin-remodeling factor CHD4 coordinates signaling and repair after DNA damage. *The Journal of Cell Biology, 190*(5), 731–740.

Lee, B. Y., Han, J. A., Im, J. S., Morrone, A., Johung, K., Goodwin, E. C., et al. (2006). Senescence-associated beta-galactosidase is lysosomal beta-galactosidase. *Aging Cell, 5*(2), 187–195.

Lee, T. B., Moon, Y. S., & Choi, C. H. (2012). Histone H4 deacetylation down-regulates catalase gene expression in doxorubicin-resistant AML subline. *Cell Biology and Toxicology, 28*(1), 11–18.

Lee, T. B., Park, J. H., Min, Y. D., Kim, K. J., & Choi, C. H. (2008). Epigenetic mechanisms involved in differential MDR1 mRNA expression between gastric and colon cancer cell lines and rationales for clinical chemotherapy. *BMC Gastroenterology, 8*, 33.

Levy, J. M., Thompson, J. C., Griesinger, A. M., Amani, V., Donson, A. M., Birks, D. K., et al. (2014). Autophagy inhibition improves chemosensitivity in BRAF(V600E) brain tumors. *Cancer Discovery, 4*(7), 773–780.

Li, S. Y., Sun, R., Wang, H. X., Shen, S., Liu, Y., Du, X. J., et al. (2015). Combination therapy with epigenetic-targeted and chemotherapeutic drugs delivered by nanoparticles to enhance the chemotherapy response and overcome resistance by breast cancer stem cells. *Journal of Controlled Release, 205*, 7–14.

Li, D. D., Sun, T., Wu, X. Q., Chen, S. P., Deng, R., Jiang, S., et al. (2012). The inhibition of autophagy sensitises colon cancer cells with wild-type p53 but not mutant p53 to topotecan treatment. *PLoS One, 7*(9), e45058.

Liu, X., Guo, S., & Su, L. (2015). Chaetocin induces endoplasmic reticulum stress response and leads to death receptor 5-dependent apoptosis in human non-small cell lung cancer cells. *Apoptosis, 20*(11), 1499–1507.

Liu, C. W., Hua, K. T., Li, K. C., Kao, H. F., Hong, R. L., Ko, J. Y., et al. (2017). Histone methyltransferase G9a drives chemotherapy resistance by regulating the glutamate-cysteine ligase catalytic subunit in head and neck squamous cell carcinoma. *Molecular Cancer Therapeutics, 16*(7), 1421–1434.

Liu, X., Tian, X., Wang, F., Ma, Y., Kornmann, M., & Yang, Y. (2014). BRG1 promotes chemoresistance of pancreatic cancer cells through crosstalking with Akt signalling. *European Journal of Cancer, 50*(13), 2251–2262.

Lob, S., Konigsrainer, A., Rammensee, H. G., Opelz, G., & Terness, P. (2009). Inhibitors of indoleamine-2,3-dioxygenase for cancer therapy: Can we see the wood for the trees? *Nature Reviews. Cancer, 9*(6), 445–452.

Luger, K., Mader, A. W., Richmond, R. K., Sargent, D. F., & Richmond, T. J. (1997). Crystal structure of the nucleosome core particle at 2.8 A resolution. *Nature, 389*(6648), 251–260.

Luo, Y., & Leverson, J. D. (2005). New opportunities in chemosensitization and radiosensitization: Modulating the DNA-damage response. *Expert Review of Anticancer Therapy, 5*(2), 333–342.

Mani, J., Vallo, S., Rakel, S., Antonietti, P., Gessler, F., Blaheta, R., et al. (2015). Chemoresistance is associated with increased cytoprotective autophagy and diminished apoptosis in bladder cancer cells treated with the BH3 mimetic (-)-Gossypol (AT-101). *BMC Cancer, 15*, 224.

Mansoori, B., Mohammadi, A., Davudian, S., Shirjang, S., & Baradaran, B. (2017). The different mechanisms of cancer drug resistance: A brief review. *Advances in Pharmacology Bulletin, 7*(3), 339–348.

Marchand, J. R., & Caflisch, A. (2015). Binding mode of acetylated histones to bromodomains: Variations on a common motif. *ChemMedChem, 10*(8), 1327–1333.

Martins, I., Michaud, M., Sukkurwala, A. Q., Adjemian, S., Ma, Y., Shen, S., et al. (2012). Premortem autophagy determines the immunogenicity of chemotherapy-induced cancer cell death. *Autophagy, 8*(3), 413–415.

Mathur, R., Sehgal, L., Havranek, O., Kohrer, S., Khashab, T., Jain, N., et al. (2017). Inhibition of demethylase KDM6B sensitizes diffuse large B-cell lymphoma to chemotherapeutic drugs. *Haematologica, 102*(2), 373–380.

Matzinger, P. (1994). Tolerance, danger, and the extended family. *Annual Review of Immunology, 12*, 991–1045.

Mayes, K., Alkhatib, S. G., Peterson, K., Alhazmi, A., Song, C., Chan, V., et al. (2016). BPTF depletion enhances T-cell-mediated antitumor immunity. *Cancer Research, 76*(21), 6183–6192.

Mayes, K., Elsayed, Z., Alhazmi, A., Waters, M., Alkhatib, S. G., Roberts, M., et al. (2017). BPTF inhibits NK cell activity and the abundance of natural cytotoxicity receptor co-ligands. *Oncotarget, 8*(38), 64344–64357.

McConkey, D. J. (2007). Therapy-induced apoptosis in primary tumors. *Advances in Experimental Medicine and Biology, 608*, 31–51.

Meidhof, S., Brabletz, S., Lehmann, W., Preca, B. T., Mock, K., Ruh, M., et al. (2015). ZEB1-associated drug resistance in cancer cells is reversed by the class I HDAC inhibitor mocetinostat. *EMBO Molecular Medicine, 7*(6), 831–847.

Melcher, A., Todryk, S., Hardwick, N., Ford, M., Jacobson, M., & Vile, R. G. (1998). Tumor immunogenicity is determined by the mechanism of cell death via induction of heat shock protein expression. *Nature Medicine, 4*(5), 581–587.

Menger, L., Vacchelli, E., Adjemian, S., Martins, I., Ma, Y., Shen, S., et al. (2012). Cardiac glycosides exert anticancer effects by inducing immunogenic cell death. *Science Translational Medicine, 4*(143), 143ra199.

Merino, V. F., Nguyen, N., Jin, K., Sadik, H., Cho, S., Korangath, P., et al. (2016). Combined treatment with epigenetic, differentiating, and chemotherapeutic agents cooperatively targets tumor-initiating cells in triple-negative breast cancer. *Cancer Research, 76*(7), 2013–2024.

Michaud, M., Martins, I., Sukkurwala, A. Q., Adjemian, S., Ma, Y., Pellegatti, P., et al. (2011). Autophagy-dependent anticancer immune responses induced by chemotherapeutic agents in mice. *Science, 334*(6062), 1573–1577.

Mittal, D., Gubin, M. M., Schreiber, R. D., & Smyth, M. J. (2014). New insights into cancer immunoediting and its three component phases—Elimination, equilibrium and escape. *Current Opinion in Immunology, 27*, 16–25.

Moutinho, C., Martinez-Cardus, A., Santos, C., Navarro-Perez, V., Martinez-Balibrea, E., Musulen, E., et al. (2014). Epigenetic inactivation of the BRCA1 interactor SRBC and resistance to oxaliplatin in colorectal cancer. *Journal of the National Cancer Institute, 106*(1), djt322.

Munzone, E., & Colleoni, M. (2015). Clinical overview of metronomic chemotherapy in breast cancer. *Nature Reviews. Clinical Oncology, 12*(11), 631–644.

Namdar, M., Perez, G., Ngo, L., & Marks, P. A. (2010). Selective inhibition of histone deacetylase 6 (HDAC6) induces DNA damage and sensitizes transformed cells to anticancer agents. *Proceedings of the National Academy of Sciences of the United States of America, 107*(46), 20003–20008.

Nebbioso, A., Carafa, V., Benedetti, R., & Altucci, L. (2012). Trials with 'epigenetic' drugs: An update. *Molecular Oncology, 6*(6), 657–682.

Ngiow, S. F., Loi, S., Thomas, D., & Smyth, M. J. (2016). Mouse models of tumor immunotherapy. *Advances in Immunology, 130*, 1–24.

Obeid, M., Tesniere, A., Ghiringhelli, F., Fimia, G. M., Apetoh, L., Perfettini, J. L., et al. (2007). Calreticulin exposure dictates the immunogenicity of cancer cell death. *Nature Medicine, 13*(1), 54–61.

Ono, H., Basson, M. D., & Ito, H. (2016). P300 inhibition enhances gemcitabine-induced apoptosis of pancreatic cancer. *Oncotarget, 7*(32), 51301–51310.

Ortiz-Montero, P., Londono-Vallejo, A., & Vernot, J. P. (2017). Senescence-associated IL-6 and IL-8 cytokines induce a self- and cross-reinforced senescence/inflammatory milieu strengthening tumorigenic capabilities in the MCF-7 breast cancer cell line. *Cell Communication and Signaling, 15*(1), 17.

O'Sullivan, T., Dunn, G. P., Lacoursiere, D. Y., Schreiber, R. D., & Bui, J. D. (2011). Cancer immunoediting of the NK group 2D ligand H60a. *Journal of Immunology, 187*(7), 3538–3545.

Ougolkov, A. V., Bilim, V. N., & Billadeau, D. D. (2008). Regulation of pancreatic tumor cell proliferation and chemoresistance by the histone methyltransferase enhancer of zeste homologue 2. *Clinical Cancer Research, 14*(21), 6790–6796.

Ouyang, M., Li, Y., Ye, S., Ma, J., Lu, L., Lv, W., et al. (2014). MicroRNA profiling implies new markers of chemoresistance of triple-negative breast cancer. *PLoS One, 9*(5), e96228.

Overman, M. J., Morris, V., Moinova, H., Manyam, G., Ensor, J., Lee, M. S., et al. (2016). Phase I/II study of azacitidine and capecitabine/oxaliplatin (CAPOX) in refractory CIMP-high metastatic colorectal cancer: Evaluation of circulating methylated vimentin. *Oncotarget, 7*(41), 67495–67506.

Ozaki, K., Kishikawa, F., Tanaka, M., Sakamoto, T., Tanimura, S., & Kohno, M. (2008). Histone deacetylase inhibitors enhance the chemosensitivity of tumor cells with cross-resistance to a wide range of DNA-damaging drugs. *Cancer Science, 99*(2), 376–384.

Pan, M. R., Hsu, M. C., Luo, C. W., Chen, L. T., Shan, Y. S., & Hung, W. C. (2016). The histone methyltransferase G9a as a therapeutic target to override gemcitabine resistance in pancreatic cancer. *Oncotarget, 7*(38), 61136–61151.

Pardoll, D. M. (2012). The blockade of immune checkpoints in cancer immunotherapy. *Nature Reviews. Cancer, 12*(4), 252–264.

Patel, D. J., & Wang, Z. (2013). Readout of epigenetic modifications. *Annual Review of Biochemistry, 82*, 81–118.

Pawlak, A., Ziolo, E., Fiedorowicz, A., Fidyt, K., Strzadala, L., & Kalas, W. (2016). Long-lasting reduction in clonogenic potential of colorectal cancer cells by sequential treatments with 5-azanucleosides and topoisomerase inhibitors. *BMC Cancer, 16*(1), 893.

Pettke, A., Hotfilder, M., Clemens, D., Klco-Brosius, S., Schaefer, C., Potratz, J., et al. (2016). Suberanilohydroxamic acid (vorinostat) synergistically enhances the cytotoxicity of doxorubicin and cisplatin in osteosarcoma cell lines. *Anti-Cancer Drugs, 27*(10), 1001–1010.

Picaud, S., Fedorov, O., Thanasopoulou, A., Leonards, K., Jones, K., Meier, J., et al. (2015). Generation of a selective small molecule inhibitor of the CBP/p300 bromodomain for leukemia therapy. *Cancer Research, 75*(23), 5106–5119.

Pol, J., Vacchelli, E., Aranda, F., Castoldi, F., Eggermont, A., Cremer, I., et al. (2015). Trial Watch: Immunogenic cell death inducers for anticancer chemotherapy. *Oncoimmunology, 4*(4), e1008866.

Polach, K. J., & Widom, J. (1995). Mechanism of protein access to specific DNA sequences in chromatin: A dynamic equilibrium model for gene regulation. *Journal of Molecular Biology, 254*(2), 130–149.

Poma, P., Labbozzetta, M., Vivona, N., Porcasi, R., D'Alessandro, N., & Notarbartolo, M. (2012). Analysis of possible mechanisms accounting for raf-1 kinase inhibitor protein downregulation in hepatocellular carcinoma. *OMICS, 16*(11), 579–588.

Qin, T., Si, J., Raynal, N. J., Wang, X., Gharibyan, V., Ahmed, S., et al. (2015). Epigenetic synergy between decitabine and platinum derivatives. *Clinical Epigenetics, 7*, 97.

Qin, L., Xu, T., Xia, L., Wang, X., Zhang, X., Zhu, Z., et al. (2016). Chloroquine enhances the efficacy of cisplatin by suppressing autophagy in human adrenocortical carcinoma treatment. *Drug Design, Development and Therapy, 10*, 1035–1045.

Quagliano, A., Gopalakrishnapillai, A., & Barwe, S. P. (2017). Epigenetic drug combination overcomes osteoblast-induced chemoprotection in pediatric acute lymphoid leukemia. *Leukemia Research, 56*, 36–43.

Quivoron, C., Couronne, L., Della Valle, V., Lopez, C. K., Plo, I., Wagner-Ballon, O., et al. (2011). TET2 inactivation results in pleiotropic hematopoietic abnormalities in mouse and is a recurrent event during human lymphomagenesis. *Cancer Cell, 20*(1), 25–38.

Raneros, A. B., Puras, A. M., Rodriguez, R. M., Colado, E., Bernal, T., Anguita, E., et al. (2017). Increasing TIMP3 expression by hypomethylating agents diminishes soluble MICA, MICB and ULBP2 shedding in acute myeloid leukemia, facilitating NK cell-mediated immune recognition. *Oncotarget, 8*(19), 31959–31976.

Rao, S. G., & Jackson, J. G. (2016). SASP: Tumor suppressor or promoter? yes! *Trends Cancer, 2*(11), 676–687.

Raynal, N. J., Lee, J. T., Wang, Y., Beaudry, A., Madireddi, P., Garriga, J., et al. (2016). Targeting calcium signaling induces epigenetic reactivation of tumor suppressor genes in cancer. *Cancer Research, 76*(6), 1494–1505.

Ribich, S., Harvey, D., & Copeland, R. A. (2017). Drug discovery and chemical biology of cancer epigenetics. *Cellulose Chemistry and Biology, 24*(9), 1120–1147.

Rivoltini, L., Loftus, D. J., Squarcina, P., Castelli, C., Rini, F., Arienti, F., et al. (1998). Recognition of melanoma-derived antigens by CTL: Possible mechanisms involved in down-regulating anti-tumor T-cell reactivity. *Critical Reviews in Immunology, 18*(1-2), 55–63.

Rodier, F., & Campisi, J. (2011). Four faces of cellular senescence. *The Journal of Cell Biology, 192*(4), 547–556.

Rohner, A., Langenkamp, U., Siegler, U., Kalberer, C. P., & Wodnar-Filipowicz, A. (2007). Differentiation-promoting drugs up-regulate NKG2D ligand expression and enhance the susceptibility of acute myeloid leukemia cells to natural killer cell-mediated lysis. *Leukemia Research, 31*(10), 1393–1402.

Roninson, I. B. (2003). Tumor cell senescence in cancer treatment. *Cancer Research, 63*(11), 2705–2715.

Roninson, I. B., Broude, E. V., & Chang, B. D. (2001). If not apoptosis, then what? Treatment-induced senescence and mitotic catastrophe in tumor cells. *Drug Resistance Updates, 4*(5), 303–313.

Rovere-Querini, P., Capobianco, A., Scaffidi, P., Valentinis, B., Catalanotti, F., Giazzon, M., et al. (2004). HMGB1 is an endogenous immune adjuvant released by necrotic cells. *EMBO Reports, 5*(8), 825–830.

Sanbhnani, S., & Yeong, F. M. (2012). CHFR: A key checkpoint component implicated in a wide range of cancers. *Cellular and Molecular Life Sciences, 69*(10), 1669–1687.

Scandura, J. M., Roboz, G. J., Moh, M., Morawa, E., Brenet, F., Bose, J. R., et al. (2011). Phase 1 study of epigenetic priming with decitabine prior to standard induction chemotherapy for patients with AML. *Blood, 118*(6), 1472–1480.

Sen, G. S., Mohanty, S., Hossain, D. M., Bhattacharyya, S., Banerjee, S., Chakraborty, J., et al. (2011). Curcumin enhances the efficacy of chemotherapy by tailoring p65NFkappaB-p300 cross-talk in favor of p53-p300 in breast cancer. *The Journal of Biological Chemistry, 286*(49), 42232–42247.

Shay, J. W., & Roninson, I. B. (2004). Hallmarks of senescence in carcinogenesis and cancer therapy. *Oncogene, 23*(16), 2919–2933.

Shen, L., Ciesielski, M., Ramakrishnan, S., Miles, K. M., Ellis, L., Sotomayor, P., et al. (2012). Class I histone deacetylase inhibitor entinostat suppresses regulatory T cells and enhances immunotherapies in renal and prostate cancer models. *PLoS One, 7*(1), e30815.

Shen, H., Wang, D., Li, L., Yang, S., Chen, X., Zhou, S., et al. (2017). MiR-222 promotes drug-resistance of breast cancer cells to adriamycin via modulation of PTEN/Akt/FOXO1 pathway. *Gene, 596*, 110–118.

Shenoy, N., Bhagat, T., Nieves, E., Stenson, M., Lawson, J., Choudhary, G. S., et al. (2017). Upregulation of TET activity with ascorbic acid induces epigenetic modulation of lymphoma cells. *Blood Cancer Journal*, 7(7), e587.

Shi, P., Yin, T., Zhou, F., Cui, P., Gou, S., & Wang, C. (2014). Valproic acid sensitizes pancreatic cancer cells to natural killer cell-mediated lysis by upregulating MICA and MICB via the PI3K/Akt signaling pathway. *BMC Cancer*, 14, 370.

Shin, S., Kim, M., Lee, S. J., Park, K. S., & Lee, C. H. (2017). Trichostatin a sensitizes hepatocellular carcinoma cells to enhanced NK cell-mediated killing by regulating immune-related genes. *Cancer Genomics Proteomics*, 14(5), 349–362.

Shoemaker, A. R., Oleksijew, A., Bauch, J., Belli, B. A., Borre, T., Bruncko, M., et al. (2006). A small-molecule inhibitor of Bcl-XL potentiates the activity of cytotoxic drugs in vitro and in vivo. *Cancer Research*, 66(17), 8731–8739.

Siew, Y. Y., Neo, S. Y., Yew, H. C., Lim, S. W., Ng, Y. C., Lew, S. M., et al. (2015). Oxaliplatin regulates expression of stress ligands in ovarian cancer cells and modulates their susceptibility to natural killer cell-mediated cytotoxicity. *International Immunology*, 27(12), 621–632.

Sociali, G., Galeno, L., Parenti, M. D., Grozio, A., Bauer, I., Passalacqua, M., et al. (2015). Quinazolinedione SIRT6 inhibitors sensitize cancer cells to chemotherapeutics. *European Journal of Medicinal Chemistry*, 102, 530–539.

Sperlazza, J., Rahmani, M., Beckta, J., Aust, M., Hawkins, E., Wang, S. Z., et al. (2015). Depletion of the chromatin remodeler CHD4 sensitizes AML blasts to genotoxic agents and reduces tumor formation. *Blood*, 126(12), 1462–1472.

Spisek, R., Charalambous, A., Mazumder, A., Vesole, D. H., Jagannath, S., & Dhodapkar, M. V. (2007). Bortezomib enhances dendritic cell (DC)-mediated induction of immunity to human myeloma via exposure of cell surface heat shock protein 90 on dying tumor cells: Therapeutic implications. *Blood*, 109(11), 4839–4845.

Starobinets, H., Ye, J., Broz, M., Barry, K., Goldsmith, J., Marsh, T., et al. (2016). Antitumor adaptive immunity remains intact following inhibition of autophagy and antimalarial treatment. *The Journal of Clinical Investigation*, 126(12), 4417–4429.

Sukkurwala, A. Q., Adjemian, S., Senovilla, L., Michaud, M., Spaggiari, S., Vacchelli, E., et al. (2014). Screening of novel immunogenic cell death inducers within the NCI mechanistic diversity set. *Oncoimmunology*, 3, e28473.

Tarasenko, N., Cutts, S. M., Phillips, D. R., Berkovitch-Luria, G., Bardugo-Nissim, E., Weitman, M., et al. (2014). A novel valproic acid prodrug as an anticancer agent that enhances doxorubicin anticancer activity and protects normal cells against its toxicity in vitro and in vivo. *Biochemical Pharmacology*, 88(2), 158–168.

Terranova-Barberio, M., Thomas, S., & Munster, P. N. (2016). Epigenetic modifiers in immunotherapy: A focus on checkpoint inhibitors. *Immunotherapy*, 8(6), 705–719.

Tesniere, A., Schlemmer, F., Boige, V., Kepp, O., Martins, I., Ghiringhelli, F., et al. (2010). Immunogenic death of colon cancer cells treated with oxaliplatin. *Oncogene*, 29(4), 482–491.

Toussaint, O., Medrano, E. E., & von Zglinicki, T. (2000). Cellular and molecular mechanisms of stress-induced premature senescence (SIPS) of human diploid fibroblasts and melanocytes. *Experimental Gerontology*, 35(8), 927–945.

Tsimberidou, A. M., Said, R., Culotta, K., Wistuba, I., Jelinek, J., Fu, S., et al. (2015). Phase I study of azacitidine and oxaliplatin in patients with advanced cancers that have relapsed or are refractory to any platinum therapy. *Clinical Epigenetics*, 7, 29.

Vijayaraghavalu, S., Dermawan, J. K., Cheriyath, V., & Labhasetwar, V. (2013). Highly synergistic effect of sequential treatment with epigenetic and anticancer drugs to overcome drug resistance in breast cancer cells is mediated via activation of p21 gene expression leading to G2/M cycle arrest. *Molecular Pharmaceutics*, 10(1), 337–352.

Wachowska, M., Gabrysiak, M., Muchowicz, A., Bednarek, W., Barankiewicz, J., Rygiel, T., et al. (2014). 5-Aza-2'-deoxycytidine potentiates antitumour immune response induced by photodynamic therapy. *European Journal of Cancer, 50*(7), 1370–1381.

Wang, Q., Ma, S., Song, N., Li, X., Liu, L., Yang, S., et al. (2016). Stabilization of histone demethylase PHF8 by USP7 promotes breast carcinogenesis. *The Journal of Clinical Investigation, 126*(6), 2205–2220.

Wang, G., Yang, Z. Q., & Zhang, K. (2010). Endoplasmic reticulum stress response in cancer: Molecular mechanism and therapeutic potential. *American Journal of Translational Research, 2*(1), 65–74.

Wapenaar, H., & Dekker, F. J. (2016). Histone acetyltransferases: Challenges in targeting bi-substrate enzymes. *Clinical Epigenetics, 8*, 59.

Wei, F. Z., Cao, Z., Wang, X., Wang, H., Cai, M. Y., Li, T., et al. (2015). Epigenetic regulation of autophagy by the methyltransferase EZH2 through an MTOR-dependent pathway. *Autophagy, 11*(12), 2309–2322.

Weinberg, R. A. (2007). *The biology of cancer*. New York: Garland Science.

Wittenburg, L. A., Bisson, L., Rose, B. J., Korch, C., & Thamm, D. H. (2011). The histone deacetylase inhibitor valproic acid sensitizes human and canine osteosarcoma to doxorubicin. *Cancer Chemotherapy and Pharmacology, 67*(1), 83–92.

Wu, H. M., Jiang, Z. F., Ding, P. S., Shao, L. J., & Liu, R. Y. (2015). Hypoxia-induced autophagy mediates cisplatin resistance in lung cancer cells. *Scientific Reports, 5*, 12291.

Wu, X., Tao, Y., Hou, J., Meng, X., & Shi, J. (2012). Valproic acid upregulates NKG2D ligand expression through an ERK-dependent mechanism and potentially enhances NK cell-mediated lysis of myeloma. *Neoplasia, 14*(12), 1178–1189.

Xiong, P., Li, Y. X., Tang, Y. T., & Chen, H. G. (2011). Proteomic analyses of Sirt1-mediated cisplatin resistance in OSCC cell line. *The Protein Journal, 30*(7), 499–508.

Xu, J., Zhou, J. Y., Tainsky, M. A., & Wu, G. S. (2007). Evidence that tumor necrosis factor-related apoptosis-inducing ligand induction by 5-Aza-2'-deoxycytidine sensitizes human breast cancer cells to adriamycin. *Cancer Research, 67*(3), 1203–1211.

Xue, W., Zender, L., Miething, C., Dickins, R. A., Hernando, E., Krizhanovsky, V., et al. (2007). Senescence and tumour clearance is triggered by p53 restoration in murine liver carcinomas. *Nature, 445*(7128), 656–660.

Yamada, H. Y., & Rao, C. V. (2009). BRD8 is a potential chemosensitizing target for spindle poisons in colorectal cancer therapy. *International Journal of Oncology, 35*(5), 1101–1109.

Yin, R., Mao, S. Q., Zhao, B., Chong, Z., Yang, Y., Zhao, C., et al. (2013). Ascorbic acid enhances Tet-mediated 5-methylcytosine oxidation and promotes DNA demethylation in mammals. *Journal of the American Chemical Society, 135*(28), 10396–10403.

Yosef, R., Pilpel, N., Tokarsky-Amiel, R., Biran, A., Ovadya, Y., Cohen, S., et al. (2016). Directed elimination of senescent cells by inhibition of BCL-W and BCL-XL. *Nature Communications, 7*, 11190.

Young, J. I., Zuchner, S., & Wang, G. (2015). Regulation of the epigenome by vitamin C. *Annual Review of Nutrition, 35*, 545–564.

Yu, M., Kong, H., Zhao, Y., Sun, X., Zheng, Z., Yang, C., et al. (2014). Enhancement of adriamycin cytotoxicity by sodium butyrate involves hTERT downmodulation-mediated apoptosis in human uterine cancer cells. *Molecular Carcinogenesis, 53*(7), 505–513.

Yu, L., Su, Y. S., Zhao, J., Wang, H., & Li, W. (2013). Repression of NR4A1 by a chromatin modifier promotes docetaxel resistance in PC-3 human prostate cancer cells. *FEBS Letters, 587*(16), 2542–2551.

Yun, T., Liu, Y., Gao, D., Linghu, E., Brock, M. V., Yin, D., et al. (2015). Methylation of CHFR sensitizes esophageal squamous cell cancer to docetaxel and paclitaxel. *Genes & Cancer, 6*(1–2), 38–48.

Zhang, Q., Padi, S. K., Tindall, D. J., & Guo, B. (2014). Polycomb protein EZH2 suppresses apoptosis by silencing the proapoptotic miR-31. *Cell Death & Disease*, *5* . e1486.

Zhang, X., Wan, G., Mlotshwa, S., Vance, V., Berger, F. G., Chen, H., et al. (2010). Oncogenic Wip1 phosphatase is inhibited by miR-16 in the DNA damage signaling pathway. *Cancer Research*, *70*(18), 7176–7186.

Zhang, C., Wang, Y., Zhou, Z., Zhang, J., & Tian, Z. (2009). Sodium butyrate upregulates expression of NKG2D ligand MICA/B in HeLa and HepG2 cell lines and increases their susceptibility to NK lysis. *Cancer Immunology, Immunotherapy*, *58*(8), 1275–1285.

Zhang, X., Wu, J., & Luan, Y. (2017). Tip60: Main functions and its inhibitors. *Mini Reviews in Medicinal Chemistry*, *17*(8), 675–682.

Zhang, X., Yashiro, M., Ren, J., & Hirakawa, K. (2006). Histone deacetylase inhibitor, trichostatin A, increases the chemosensitivity of anticancer drugs in gastric cancer cell lines. *Oncology Reports*, *16*(3), 563–568.

Zhu, Y., Armstrong, J. L., Tchkonia, T., & Kirkland, J. L. (2014). Cellular senescence and the senescent secretory phenotype in age-related chronic diseases. *Current Opinion in Clinical Nutrition and Metabolic Care*, *17*(4), 324–328.

Zhu, S., Denman, C. J., Cobanoglu, Z. S., Kiany, S., Lau, C. C., Gottschalk, S. M., et al. (2015). The narrow-spectrum HDAC inhibitor entinostat enhances NKG2D expression without NK cell toxicity, leading to enhanced recognition of cancer cells. *Pharmaceutical Research*, *32*(3), 779–792.

Zhu, Y., Tchkonia, T., Pirtskhalava, T., Gower, A. C., Ding, H., Giorgadze, N., et al. (2015). The Achilles' heel of senescent cells: From transcriptome to senolytic drugs. *Aging Cell*, *14*(4), 644–658.

Zhu, Y., Yu, F., Jiao, Y., Feng, J., Tang, W., Yao, H., et al. (2011). Reduced miR-128 in breast tumor-initiating cells induces chemotherapeutic resistance via Bmi-1 and ABCC5. *Clinical Cancer Research*, *17*(22), 7105–7115.

Zucconi, B. E., Luef, B., Xu, W., Henry, R. A., Nodelman, I. M., Bowman, G. D., et al. (2016). Modulation of p300/CBP acetylation of nucleosomes by bromodomain ligand I-CBP112. *Biochemistry*, *55*(27), 3727–3734.

CHAPTER TWO

VDAC Regulation: A Mitochondrial Target to Stop Cell Proliferation

Diana Fang*, Eduardo N. Maldonado*,†,1

*Medical University of South Carolina, Charleston, SC, United States
†Hollings Cancer Center, Medical University of South Carolina, Charleston, SC, United States
[1]Corresponding author: e-mail address: maldona@musc.edu

Contents

1. Introduction	42
1.1 Mitochondria, Energy Production, and Biosynthesis	42
1.2 Bioenergetics and Biosynthesis in the Warburg Phenotype	44
1.3 Mechanisms to Suppress Mitochondrial ATP Production: A Drive on Glycolysis	47
2. VDAC Channels and Mitochondrial Metabolism	48
2.1 The MOM: A VDAC-Containing Interphase to Modulate Cellular Bioenergetics	48
2.2 VDAC Structure and Regulation of the Conductance	49
3. VDAC–Tubulin Interaction	50
3.1 VDAC Inhibition by Free Tubulin	50
3.2 VDAC–Tubulin Modulation of Cellular Bioenergetics During Cell Cycle	52
4. Tumor Metabolic Flexibility: Advantages of Targeting Metabolism in Chemotherapy	52
5. VDAC–Tubulin Antagonists: A Strategy for Opening VDAC	55
5.1 Erastin and VDAC–Tubulin Antagonists	55
5.2 VDAC Opening, Glycolysis, and Reactive Oxygen Species Formation	56
5.3 VDAC-Dependent Metabolic Hits: Anti-Warburg Effect and Oxidative Stress	59
6. Concluding Remarks	61
Acknowledgment	61
References	61

Abstract

Cancer metabolism is emerging as a chemotherapeutic target. Enhanced glycolysis and suppression of mitochondrial metabolism characterize the Warburg phenotype in cancer cells. The flux of respiratory substrates, ADP, and Pi into mitochondria and the release of mitochondrial ATP to the cytosol occur through voltage-dependent anion channels (VDACs) located in the mitochondrial outer membrane. Catabolism of respiratory substrates in the Krebs cycle generates NADH and FADH$_2$ that enter the electron transport chain (ETC) to generate a proton motive force that maintains mitochondrial membrane potential ($\Delta\Psi$) and is utilized to generate ATP. The ETC is also the major cellular source of mitochondrial reactive oxygen species (ROS). αβ-Tubulin heterodimers decrease VDAC conductance in lipid bilayers. High constitutive levels of cytosolic free tubulin in intact cancer cells close VDAC decreasing mitochondrial $\Delta\Psi$ and mitochondrial metabolism. The VDAC–tubulin interaction regulates VDAC opening and globally controls mitochondrial metabolism, ROS formation, and the intracellular flow of energy. Erastin, a VDAC-binding molecule lethal to some cancer cell types, and erastin-like compounds identified in a high-throughput screening antagonize the inhibitory effect of tubulin on VDAC. Reversal of tubulin inhibition of VDAC increases VDAC conductance and the flux of metabolites into and out of mitochondria. VDAC opening promotes a higher mitochondrial $\Delta\Psi$ and a global increase in mitochondrial metabolism leading to high cytosolic ATP/ADP ratios that inhibit glycolysis. VDAC opening also increases ROS production causing oxidative stress that, in turn, leads to mitochondrial dysfunction, bioenergetic failure, and cell death. In summary, antagonism of the VDAC–tubulin interaction promotes cell death by a "double-hit model" characterized by reversion of the proproliferative Warburg phenotype (anti-Warburg) and promotion of oxidative stress.

1. INTRODUCTION
1.1 Mitochondria, Energy Production, and Biosynthesis

Mitochondria, classically described as the "power house" of cells in textbooks, produce about 95% or more of the total ATP in nonproliferating cells. Respiratory substrates including fatty acids, pyruvate, and certain amino acids are fully oxidized in mitochondria. The highly efficient ATP-generating machine requires the entrance of respiratory substrates, ADP, and Pi through the mitochondrial outer membrane (MOM). Metabolites that reach the intermembrane space are further transported into the matrix by numerous different carriers located in the mitochondrial inner membrane (MIM). Respiratory substrates in the matrix, catabolized in the Krebs or tricarboxylic acid cycle, generate NADH and $FADH_2$ which enter the electron transport chain (ETC) comprised of protein Complexes I–IV. The Krebs cycle is fueled by the intermediate acetyl-coenzyme A (AcCoA) which is generated by the oxidation of glucose-derived pyruvate, β-oxidation of fatty acids and from the catabolism of the amino acids leucine, isoleucine, glycine, serine, and tryptophan. In oxidative phosphorylation (OXPHOS), the transfer of electrons from NADH and $FADH_2$ to reduce the final acceptor molecular O_2 to H_2O, drives proton translocation across the MIM by Complexes I, III, and IV. Proton accumulation in the intermembrane space generates a negative transmembrane $\Delta\Psi$ (mitochondrial $\Delta\Psi$) and positive ΔpH, the components of the proton motive force (Δp).

Δp then drives ATP synthesis from ADP and Pi by Complex V (F_1F_0-ATP synthase) (Nicholls & Ferguson, 2013; Fig. 1).

From a biosynthetic perspective, the Krebs cycle does not only contribute NADH and $FADH_2$ to the ETC to produce ATP but also generates metabolites involved in cell signaling and biosynthesis. AcCoA is essential

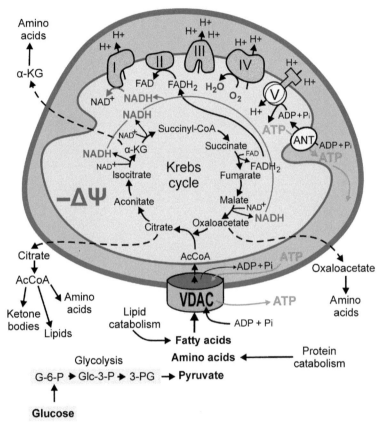

Fig. 1 Mitochondrial metabolism. Flux of metabolites including fatty acids, certain amino acids, pyruvate, ADP, and Pi across the mitochondrial outer membrane occurs through VDAC. Catabolism of respiratory substrates in the Krebs cycle generates NADH and $FADH_2$, which fuel the electron transport chain (Complexes I–IV) and supports oxidative phosphorylation. The Krebs cycle also produces metabolic intermediaries released to the cytosol for the biosynthesis of proteins and lipids. Proton pumping by the respiratory chain across mitochondrial inner membrane (MIM) generates mitochondrial $\Delta\Psi$. Protons moving back across MIM into the matrix drive ATP synthesis from ADP and Pi by the F_1-F_0-ATP synthase (Complex V). Mitochondrial ATP is exported from the matrix by the adenine nucleotide transporter (ANT) and released to the cytosol through VDAC.

for the synthesis of lipids, ketone bodies, amino acids, and cholesterol; citrate is used for lipid biosynthesis and oxaloacetate and α-ketoglutarate for the synthesis of some nonessential amino acids (Frezza, 2017; Fig. 1).

An important feature of mitochondrial metabolism is that full oxidation of respiratory substrates maximizes the yield of ATP per mole of substrate and leaves no residual carbon backbones. The dynamic range of ATP synthesized depends on the cell type, catabolic, and anabolic activities. In general, in most cell types that do not undergo uncontrolled cell division the biosynthesis of macromolecules depends mostly on the turnover of protein, lipids, and nucleic acids. Moreover, differentiated cells with the possible exception of neurons and cardiac myocytes only proliferate to maintain the population within a defined number of replication cycles. By contrast, the frequent cell division in cancer cells poses a metabolic challenge. Replicating cells adjust the bioenergetics status and biosynthetic machinery to support the high demand of biomass generation for daughter cells.

1.2 Bioenergetics and Biosynthesis in the Warburg Phenotype

Energy production, energy utilization, and supply of substrates for metabolic reactions including biosynthesis are finely regulated in all eukaryotic cells. Cells that do not proliferate operate at a different dynamic equilibrium of supply and demand for energy and substrates than cells that undergo continuous divisions.

One of the differences in the bioenergetics of tumor cells was initially observed in the early 20th century by Warburg who made the seminal observation that tumors produce more lactic acid than differentiated cells even in the presence of oxygen. The tumor metabolic phenotype, termed the Warburg effect after his contribution, is characterized by enhanced glycolysis and suppression of mitochondrial metabolism even in the presence of physiological levels of oxygen (Warburg, 1956; Warburg, Wind, & Negelein, 1927). Warburg also proposed that the cause of cancer was an irreversible but incomplete damage to respiration causing low energy production that was compensated by increased conversion of glucose to lactic acid (fermentation). Cells that could successfully increase glucose utilization through successive divisions to overcome the detrimental effects of defective respiration eventually became cancerous (Warburg, 1956). The provocative idea of failing mitochondria as the cause of cancer was quickly challenged by Weinhouse and others who demonstrated concurrent high glycolysis and oxidative metabolism in cancer tissues (Weinhouse, 1956). After the

Warburg–Weinhouse controversy, several investigations confirmed both enhanced glycolysis and active mitochondrial metabolism in cancer cells. Functional mitochondria in tumor cells have been demonstrated by measurements of ATP generation, NADH production, oxygen consumption, and maintenance of mitochondrial $\Delta\Psi$, among other parameters (Lim, Ho, Low, Choolani, & Wong, 2011; Maldonado, Patnaik, Mullins, & Lemasters, 2010; Mathupala, Ko, & Pedersen, 2010; Moreno-Sanchez et al., 2014; Nakashima, Paggi, & Pedersen, 1984; Pedersen, 1978; Singleterry, Sreedhar, & Zhao, 2014).

The relative contribution of mitochondria to total cellular ATP generation by OXPHOS is generally lower in cancer cells compared to differentiated cells. Mitochondria produce about 95% of the total ATP by OXPHOS in differentiated cells, and the remaining 5% is generated by the aerobic catabolism of glucose in the cytosol. By contrast in cancer and other proliferating cells, glycolysis contributes 20%–90% of total ATP production (Griguer, Oliva, & Gillespie, 2005; Nakashima et al., 1984). Enhanced glycolysis in cancer cells has been associated with a high rate of cell proliferation (Griguer et al., 2005; Guppy, Leedman, Zu, & Russell, 2002; Moreno-Sanchez, Rodriguez-Enriquez, Marin-Hernandez, & Saavedra, 2007; Scott et al., 2011). The "glucose avidity" of tumors is currently used to diagnose primary tumors, recurrences, and metastases by positron emission tomography of the glucose analogue ^{18}fluorodeoxyglucose (Zhu, Lee, & Shim, 2011). Although enhanced glycolysis is a feature of almost all cancer cells, the bioenergetic profiles of tumors are remarkably heterogeneous not only among different tumor types but also among cells in the same tumor. Subsets of cells predominantly glycolytic or displaying an oxidative metabolism have been shown in gliomas and large B-cell lymphomas (Beckner et al., 2005; Bouzier, Voisin, Goodwin, Canioni, & Merle, 1998; Caro et al., 2012). The metabolic heterogeneity of tumors raises questions about the influence of metabolic differences on cell proliferation and response to chemotherapy.

Incomplete catabolism of glucose in the cytosol by glycolysis generates only 2 mol of ATP per mole of glucose sparing carbon backbones in the form of lactate, whereas mitochondrial oxidation of 2 mol of pyruvate generated from 1 mol of glucose to CO_2 and H_2O generates about an additional 31 mol of ATP. Although the total amount of mitochondrial ATP calculated considers the currently accepted proton stoichiometries for ATP synthesis, ATP/ADP·Pi exchange, respiration, and the malate/aspartate shuttle, the actual ATP yield could also be less due to proton leak to the mitochondrial matrix (Brand, 2005; Rich, 2003; Rich & Marechal, 2010; Walker, 2013; Wikstrom, Sharma, Kaila, Hosler, & Hummer, 2015). In cancer cells, lower efficiency of ATP generation by aerobic glycolysis appears to be offset by high glycolytic rates (Locasale & Cantley, 2010). Regardless, it is unlikely that ATP generation be a limiting factor for cell proliferation because the ATP demand for cell division is lower than the energy requirements for basal cellular processes (Kilburn, Lilly, & Webb, 1969).

A long-standing question in cancer metabolism since the initial observations of Warburg has been why would cancer cells switch from an oxidative metabolism to a glycolytic phenotype especially considering that changes in ATP production may or may not be crucial for cell division. Would the dynamic range of energy production be strikingly different in highly proliferating cells compared to the nonproliferating cells or does the switch actually serve other purposes?

A dividing cell must double its biomass (proteins, lipids, and nucleic acids) before mitosis. This high biosynthetic demand requires the generation of simple carbon backbones that can be used as building blocks for the synthesis of new macromolecules. Cells expressing the Warburg phenotype partially break down glucose with a high proportion of pyruvate converted into lactate instead of being fully oxidized in mitochondria to CO_2 and H_2O (Fig. 2). In addition, a decreased mitochondrial utilization of glutamine and fatty acids provides additional carbon precursors for biomass formation in proliferating cells (Cairns, 2015; DeBerardinis, Sayed, Ditsworth, & Thompson, 2008; Keibler et al., 2016; Liberti & Locasale, 2016; Lunt & Vander Heiden, 2011). Enhanced glycolysis also provides by-products of glucose catabolism and NADPH. Glucose-6-phosphate, glyceraldehyde-3-phosphate, and 3-phosphoglycerate, as metabolic intermediaries, contribute to the synthesis of nucleotides, lipids, and amino acids, respectively (Fig. 1). Increased NADPH production by the pentose phosphate pathway is used for reductive biosynthesis.

In the Warburg phenotype, mitochondria are not just passive generators of ATP but also essential contributors of metabolites required for cell proliferation. Catabolism of glutamine, pyruvate, and other respiratory substrates generates biosynthetic precursors in the Krebs cycle, including α-ketoglutarate and oxaloacetate for synthesis of nonessential amino acids and citrate exported to the cytosol to be converted into AcCoA and utilized for the synthesis of fatty acid, cholesterol, and amino acids (DeBerardinis & Cheng, 2010; Fig. 1). Recently, one-carbon metabolism, a set of reactions that transfer one-carbon units from serine and glycine, has been shown to be important for de novo synthesis of purines and thymidylate in highly

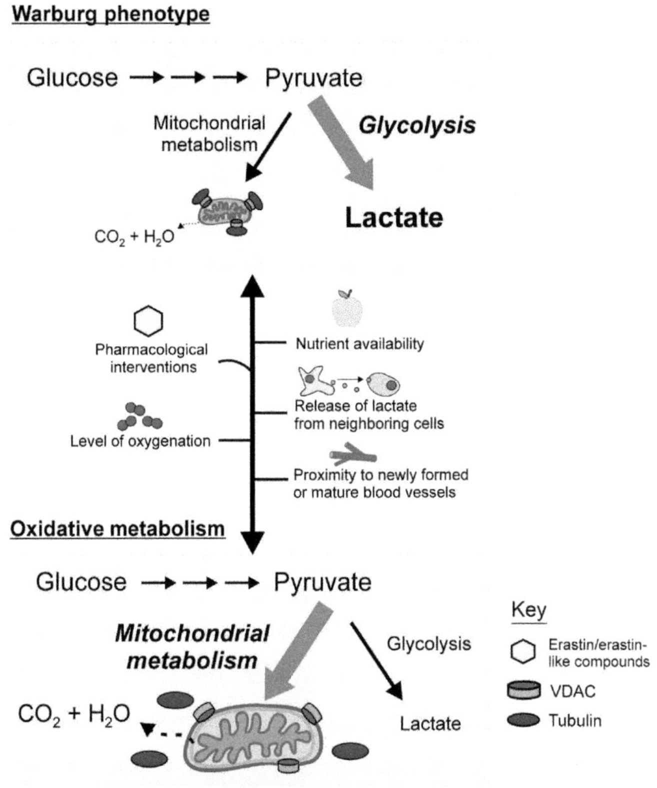

Fig. 2 Metabolic flexibility of tumors and VDAC opening. Cancer cells switch between oxidative and glycolytic bioenergetic profiles depending on nutrient availability, tissue oxygenation, intratumor localization, and pharmacological treatments to inhibit glycolysis or to promote mitochondrial metabolism. In cancer cells, constitutive high free tubulin blocks VDAC conductance, suppresses mitochondrial metabolism, and decreases cytosolic ATP/ADP to favor glycolysis. Reversal of the inhibitory effect of tubulin by VDAC–tubulin antagonists leads to VDAC opening and reversal of the Warburg phenotype.

proliferative tumors (Meise & Vazquez, 2016). In summary, the proproliferative Warburg phenotype is sustained by the differential cytosolic and mitochondrial utilization of glucose and other fuels that determine a bioenergetic profile that favors biosynthesis.

1.3 Mechanisms to Suppress Mitochondrial ATP Production: A Drive on Glycolysis

Mitochondrial oxidation of respiratory substrates by OXPHOS maximizes the yield of ATP per mol of fuels that enter mitochondria. The electrogenic adenine nucleotide translocator (ANT) located in the MIM transport the newly synthesized ATP in the mitochondrial matrix to the cytosol exchanging ATP^{-4} for ADP^{-3} (Fig. 1). In differentiated cells with predominantly oxidative metabolism, cytosolic ATP/ADP ratios can be 50–100 times higher than in the mitochondrial matrix (Schwenke, Soboll, Seitz, & Sies, 1981). Oxidative metabolism favors a high cytosolic ATP/ADP ratio which suppresses glycolysis through inhibition of phosphofructokinase-1 (PFK-1), although other mechanisms may be involved. ATP is a strong allosteric inhibitor, whereas ADP and AMP activate PFK-1 (Mor, Cheung, & Vousden, 2011; Moreno-Sanchez et al., 2007). In cancer cells, suppression of mitochondrial metabolism contributes to a low cytosolic ATP/ADP ratio, which releases this brake on glycolysis.

Recently, we demonstrated that closing of the voltage-dependent anion channels (VDACs) by free tubulin limits the influx of metabolites into mitochondria and limits ATP production, whereas replacement of electrogenic ATP/ADP exchange by ANT with a nonelectrogenic exchange mechanism decreases cytosolic ATP/ADP ratios. These two independent mechanisms contribute to suppress mitochondrial metabolism and to maintain a low cytosolic ATP/ADP ratio favoring aerobic glycolysis in cancer cells (Maldonado et al., 2016, 2013; Maldonado & Lemasters, 2014).

2. VDAC CHANNELS AND MITOCHONDRIAL METABOLISM

2.1 The MOM: A VDAC-Containing Interphase to Modulate Cellular Bioenergetics

The MOM is a functional barrier that physically separates mitochondria from the cytosol (Fig. 1). VDAC, the most abundant protein in the MOM, is the gateway through which most respiratory substrates, ADP, and Pi enter mitochondria and ATP exits. Based on its role in metabolite exchange between mitochondria and the cytosol and its subcellular localization, VDAC opening is proposed to be a master regulator to globally modulate mitochondrial bioenergetics and the intracellular flow of energy (Lemasters & Holmuhamedov, 2006; Maldonado & Lemasters, 2012, 2014; Maldonado et al., 2013).

Interactions between VDAC with tubulin and possibly other proteins, such as hexokinase (Pastorino & Hoek, 2003; Wolf et al., 2011) and posttranslational modifications of VDAC especially phosphorylation by protein kinase A (PKA) and glycogen synthase 3β (GSK3β), modulate the open/closed state of VDAC (Sheldon, Maldonado, Lemasters, Rostovtseva, & Bezrukov, 2011). Single and double knockdown of the three different VDAC isoforms support this concept that VDAC serves as a master regulator of mitochondrial metabolism in cancer cells (Maldonado et al., 2013). Thus, VDAC regulation by free tubulin emerges as a mechanism to block or promote OXPHOS and indirectly regulate glycolysis through the cytosolic ATP/ADP ratio. Ultimately, VDAC–tubulin interactions appear as a new pharmacological target to increase mitochondrial metabolism in cancer cells and to reverse Warburg metabolism (Maldonado, 2017).

2.2 VDAC Structure and Regulation of the Conductance

A protein with pore-forming activity first described in extracts of mitochondria from *Paramecium tetraurelia* (Schein, Colombini, & Finkelstein, 1976) was initially called mitochondrial porin and later renamed VDAC (Colombini, 1979). VDAC present in all eukaryotic cells comprises three isoforms, VDAC1, VDAC2, and VDAC3, encoded by separate genes. VDAC1 and VDAC2 are the main isoforms in most mammalian cells, including cancer cells in which they account for up to 90% of the total. The least abundant isoform, VDAC3, comprising the remaining 10% (De Pinto et al., 2010; Huang, Shah, Bradbury, Li, & White, 2014; Maldonado et al., 2013) is abundant only in testis (Sampson et al., 2001; Sampson, Lovell, & Craigen, 1997).

VDAC in humans and mice is a $\sim$30-kDa protein enclosing an aqueous channel of $\sim$3 nm internal diameter that allows the passage of molecules up to $\sim$5 kDa (Colombini, 1980, 2012; Song & Colombini, 1996). The influx of polar metabolites through VDAC is determined mostly by their charge and size (Colombini, 1980, 2004). Once in the intermembrane space polar metabolites cross the MIM through several specific carriers that utilize electrical, chemical, or electrochemical potential gradients to transport diverse solutes including pyruvate, Pi, ADP, ATP, acylcarnitine, citrate, oxoglutarate, and glutamate. The activity of mitochondrial carriers is finely regulated to allow a sufficient flux of metabolites to adapt to different physiological demands (Palmieri & Pierri, 2010). The availability of solutes to the carriers in the MIM depends on the metabolites produced in the mitochondrial matrix mainly through the Krebs cycle and the metabolites that access the intermembrane space through VDAC in the MOM. Thus, the probability of VDAC in an open or close conformation has a substantial impact on mitochondrial metabolism and cellular bioenergetics.

In the closed state, only small ions like Na^+, K^+, or Cl^- but not most anionic metabolites including respiratory substrates, ATP, ADP, and Pi permeate through VDAC. Structurally, VDAC1 has a barrel configuration with staves formed by 19 β-strands (Hiller, Abramson, Mannella, Wagner, & Zeth, 2010; Ujwal et al., 2008). An additional N-terminal sequence forms the only α-helical segment that appears to move to the center of the channel, blocking the passage of metabolites. Recently, a similar β barrel structure with 19 β-strands has been shown for VDAC2 from zebrafish (Schredelseker et al., 2014). Gating and selectivity of VDAC1 and VDAC2 are highly conserved among mammals (Blachly-Dyson & Forte, 2001).

After the initial discovery, the consensus was that VDAC was constitutively open to the flux of metabolites between the mitochondrial matrix and the cytosol as an "all-time open gate." Extensive research in vitro and in intact cells showed instead that VDAC opening is subjected to modulation. Numerous studies have shown regulation by multiple factors, including hexokinase (Al Jamal, 2005; Azoulay-Zohar, Israelson, Abu-Hamad, & Shoshan-Barmatz, 2004; Nakashima, Paggi, Scott, & Pedersen, 1988), Bcl2 family members (Tsujimoto & Shimizu, 2000), glutamate (Gincel, Silberberg, & Shoshan-Barmatz, 2000), ethanol (Holmuhamedov & Lemasters, 2009; Lemasters & Holmuhamedov, 2006), and NADH (Zizi, Forte, Blachly-Dyson, & Colombini, 1994). VDAC phosphorylation by protein kinases, GSK3β, PKA, and protein kinase C epsilon (PKCε), blocks or inhibits association of VDAC with other proteins, such as Bax and tBid, and also regulates VDAC opening (Azoulay-Zohar et al., 2004; Baines et al., 2003; Das, Wong, Rajapakse, Murphy, & Steenbergen, 2008; Lee, Zizi, & Colombini, 1994; Rostovtseva et al., 2004; Vander Heiden et al., 2000, 2001). PKA-dependent VDAC phosphorylation and GSK3β-mediated VDAC2 phosphorylation increase VDAC conductance (Bera, Ghosh, & Das, 1995; Das et al., 2008; Sheldon et al., 2011). Here, we will focus on the inhibitory effect of free tubulin on VDAC in cancer cells as a regulatory mechanism of VDAC opening and as a pharmacological target (Maldonado et al., 2010, 2013).

3. VDAC–TUBULIN INTERACTION
3.1 VDAC Inhibition by Free Tubulin

Mitochondrial $\Delta\Psi$ in cancer cells is sustained both by the respiratory chain and from hydrolysis of glycolytic ATP by the mitochondrial F_1F_0-ATPase working in reverse. We previously showed that treatment of cancer cells with the microtubule stabilizer paclitaxel or the microtubule destabilizers nocodazole and colchicine decreased and increased, respectively, cytosolic free tubulin. We also showed that low and high cytosolic free tubulin promotes high and low mitochondrial $\Delta\Psi$, respectively (Maldonado et al., 2010). By contrast, in the nonproliferating rat hepatocyte, mitochondrial $\Delta\Psi$ was relatively insensitive to changes in free tubulin levels. The lack of response to pharmacological interventions to stabilize/destabilize microtubules could be explained by the much lower constitutive pool of free tubulin in rat hepatocytes compared to hepatocarcinoma cells. Nonproliferating hepatocytes do not need a reservoir of tubulin for spindle formation at mitosis. Thus, microtubule stabilization with paclitaxel does not increase $\Delta\Psi$ in hepatocytes, because levels of free tubulin are already low, whereas microtubule destabilization still increases tubulin and, in turn, decreases $\Delta\Psi$. These findings imply that VDAC is indeed constitutively open in nonproliferating hepatocytes. By contrast, paclitaxel increases and nocodazole/colchicine decreases $\Delta\Psi$ in tumor cells, leading to the conclusion that endogenous free tubulin partially closes VDAC in tumor cells (Maldonado et al., 2010). We propose that inhibition of VDAC conductance by free tubulin is a mechanism that contributes to the suppression of mitochondrial metabolism in the Warburg phenotype. Our studies performed in intact cancer cells were in agreement with earlier work showing that heterodimeric $\alpha\beta$-tubulin closes VDAC inserted into lipid bilayers and decreases respiration in isolated brain mitochondria and permeabilized synaptosomes (Rostovtseva et al., 2008; Timohhina et al., 2009).

Knockdown studies of VDAC1, VDAC2, and VDAC3 in HepG2 cells further characterized the role of VDAC in mitochondrial metabolism in cancer cells and showed that VDAC sensitivity to tubulin inhibition is isoform dependent. Single knockdown of each of the three VDAC isoforms, especially the minor isoform VDAC3, decreased mitochondrial $\Delta\Psi$, indicating that all VDAC isoforms contribute to $\Delta\Psi$ formation. Knockdown of VDAC3 also decreased cellular ATP and ADP and the $NAD(P)H/NAD(P)^+$ ratio, suggesting that the least abundant isoform VDAC3 contributed most to the maintenance of mitochondrial metabolism (Maldonado et al., 2013). The response of each isoform to tubulin inhibition was characterized by double knockdown of VDAC isoforms in all combinations. All single and double knockdowns partially reversed the suppression of $\Delta\Psi$ induced by free tubulin (Maldonado et al., 2013). Electrophysiology studies of VDAC1 and VDAC2 isolated from double-knockdown HepG2 cells inserted in lipid bilayers showed that both isoforms were almost equally sensitive to tubulin inhibition, whereas VDAC3 was insensitive at tubulin concentrations even five-fold higher than those used to inhibit VDAC1 and

VDAC2 (Maldonado et al., 2013). The voltage gating and the response to dimeric αβ-tubulin of constitutive VDAC isolated from wild-type HepG2 cells compared to VDAC from heart and liver mitochondria were almost identical. These similarities suggest that the differential response to tubulin inhibition in cancer vs nonproliferating cells depends on the amount of cytosolic free tubulin and not on a cell type-specific sensitivity of the channels. The knockdown studies supported the conclusion that VDAC3, at least in HepG2 cells, is constitutively open, whereas VDAC1 and VDAC2 are totally or partially closed by free tubulin.

3.2 VDAC–Tubulin Modulation of Cellular Bioenergetics During Cell Cycle

Biosynthesis of new macromolecules to double the biomass before mitosis occurs during G1, S, and G2 phases of the cell cycle. VDAC–tubulin-dependent suppression of mitochondrial metabolism caused by high constitutive free tubulin would favor the probiosynthetic Warburg phenotype during these growth stages. In fact, in HeLa, NIH3T3, NCI-H292, and other cancer cells, most of the cell cycle lasting 20–30 h is composed of G1, S, and G2 phases, whereas cell division that occurs during the M or mitotic phase lasts only about 30 min (Hahn, Jones, & Meyer, 2009). During mitosis, energy demand increases sharply to support chromosome segregation and cytokinesis. At this point, where null or minimal biosynthesis is expected, a Warburg phenotype would not be beneficial for cell division. Moreover, full oxidation of respiratory substrates with maximum yield of ATP may be required to meet the high energy demands of cell division. An open VDAC would allow maximum OXPHOS activity. A possible sequence of events is that as the spindle forms during prophase, the free tubulin pool decreases abruptly, releasing tubulin inhibition of VDAC. VDAC opening then promotes increased mitochondrial metabolism reverting the Warburg phenotype precisely when the energy demand is maximal. After mitosis, the pool of free tubulin increases again, and cells return to a high glycolytic, proproliferative phenotype during the nonmitotic stages of the cell cycle (Maldonado, 2017; Maldonado & Lemasters, 2012).

4. TUMOR METABOLIC FLEXIBILITY: ADVANTAGES OF TARGETING METABOLISM IN CHEMOTHERAPY

Our work showing constitutive inhibition of VDAC conductance in cancer cells by free tubulin raised questions about the possibility of decreasing cell proliferation by modifying the bioenergetic status of the cell. Cytotoxic chemotherapy is commonly based on drugs that cause cell death or prevent cell growth by inhibiting microtubule function, protein function, or DNA synthesis and replication. Although from a bioenergetics perspective both glycolysis and mitochondria contribute to energy production in cancer cells, most of the attempts to target metabolism as a new approach for cancer therapy have been directed to inhibit glycolysis.

The increasingly recognized heterogeneity of tumor metabolism and the capability of tumor cells to switch bioenergetics profiles from glycolytic to oxidative and vice versa open questions about the feasibility of developing metabolism oriented chemotherapies as sole treatments or as coadjuvants for current chemotherapeutic protocols. The predominance of a glycolytic or oxidative metabolism in cancer cells is determined genetically and by both temporary and long-term epigenetic changes. The relative contribution of glycolysis and OXPHOS can vary substantially over time depending on multiple factors, including availability to different fuels, proximity to newly formed vs mature blood vessels, and the release of soluble factors such as lactate from neighboring cells (Fig. 2).

Inadequate blood perfusion in rapidly growing tumors not only exposes cells to hypoxia but also to a lower supply of nutrients including glucose. Hypoxia can decrease the OXPHOS flux depending on the cell type, time of hypoxic exposure, and environmental conditions. When comparing MCF-7 and HeLa cells that predominantly depend on OXPHOS for ATP supply, prolonged hypoxia increases glycolysis only in MCF-7 cells (Rodriguez-Enriquez et al., 2010). In solid tumors the respiratory chain can still be fully functional at oxygen levels as low as 0.5%, indicating that hypoxic tumor cells exposed to <2% oxygen in rapidly growing and heterogeneously perfused tumors still produce ATP by OXPHOS. Under those conditions, even if pyruvate utilization is compromised mitochondria from tumor cells can utilize glutamine as an energy source so actually both glycolysis and OXPHOS can sustain tumor growth (Mullen et al., 2012).

Nutrient availability not only influences tumor growth but can also promote a switch from aerobic glycolysis to OXPHOS in breast cancer cell lines and lymphoma cells cultured in glucose-free media (Robinson, Dinsdale, MacFarlane, & Cain, 2012; Smolkova et al., 2010). The broad adaptability of tumor cells to oxidize other substrates when glucose or glutamine is limited includes the utilization of lactate, methionine, arginine, cysteine, asparagine, leucine, acetate, and even proteins and lipids from the microenvironment (Chung et al., 2005; Clavell et al., 1986;

Comerford et al., 2014; Commisso et al., 2013; Keenan & Chi, 2015; Kennedy et al., 2013; Kreis, Baker, Ryan, & Bertasso, 1980; Mashimo et al., 2014; Scott, Lamb, Smith, & Wheatley, 2000; Sheen, Zoncu, Kim, & Sabatini, 2011; Sonveaux et al., 2008). Inhibition of Complex III by antimycin and Complex I by piericidin A triggers a compensatory increase in the uptake and consumption of glucose in myoblasts. In these myoblasts total cellular ATP production before and after OXPHOS inhibition was similar, indicating that the loss of ATP generation by OXPHOS was fully compensated by increased glycolytic ATP (Liemburg-Apers, Schirris, Russel, Willems, & Koopman, 2015).

The metabolic heterogeneity and flexibility of tumors and the potential to switch between glycolytic and oxidative metabolisms underscore the relevance of mechanisms that underlie adaptive changes like VDAC regulation. Most research efforts to target tumor metabolism have been directed toward inhibition of glycolysis (Doherty & Cleveland, 2013; Pelicano, Martin, Xu, & Huang, 2006). Only recently mitochondrial metabolism emerged as a chemotherapeutic target with most approaches attempting to inhibit mitochondrial metabolism in cancer cells (Bhat, Kumar, Chaudhary, Yadav, & Chandra, 2015; Weinberg & Chandel, 2015). The observation that the antidiabetic drug metformin decreased the prevalence of certain types of cancer triggered an interest in the role of mitochondrial inhibition as a mechanism to suppress abnormal cell proliferation (Giovannucci et al., 2010; Libby et al., 2009). Although metformin decreases OXPHOS by inhibiting Complex I of the respiratory chain, it also inhibits the mammalian target of rapamycin (mTOR), interferes with folate metabolism, and activates AMP kinase. It is uncertain if the antiproliferative effect of metformin is actually due to OXPHOS inhibition (Jara & Lopez-Munoz, 2015). Other approaches to inhibit mitochondrial metabolism in various cancer cell models include etomoxir to inhibit carnitine O-palmitoyltransferase 1 and subsequently mitochondrial fatty acid oxidation (leukemia); tigecycline to inhibit mitochondrial protein translation (leukemia); glutaminase inhibitors (breast cancer, lymphoma); and the compound VLX600 to inhibit OXPHOS (colon cancer) (Samudio et al., 2010; Skrtic et al., 2011; Wang et al., 2010; Zhang et al., 2014). Instead of inhibiting mitochondrial metabolism, other antiproliferative strategies promote mitochondrial metabolism. For example the pyruvate analog dichloroacetate, which promotes cell killing in several cancer cell lines and in some in vivo models, activates pyruvate dehydrogenase to increase mitochondrial metabolism (Sutendra & Michelakis, 2013).

5. VDAC–TUBULIN ANTAGONISTS: A STRATEGY FOR OPENING VDAC

5.1 Erastin and VDAC–Tubulin Antagonists

Relative closure of VDAC in cancer cells and the broad metabolic consequences of VDAC opening turn the VDAC–tubulin interaction into a novel pharmacological target to increase mitochondrial metabolism. Antagonists of the constitutive inhibition of VDAC by free tubulin would be expected to increase oxidative metabolism and promote an anti-Warburg effect by decreasing glycolysis. In search of VDAC opening drugs we reported the small-molecule erastin as the first described antagonist of the inhibitory effect of free tubulin on VDAC (Maldonado et al., 2013). Erastin selectively induces nonapoptotic cell death in human cells engineered to harbor small T oncoprotein and the oncogenic allele of HRAS, v-Ha-ras Harvey rat sarcoma viral oncogene homologue RAS^{v12} (Dolma, Lessnick, Hahn, & Stockwell, 2003). Erastin-dependent cell death is blocked by antioxidants, such as α-tocopherol, butylated hydroxytoluene, and desferal, but not by pan-caspase inhibitors (Dolma et al., 2003). Other cell lines harboring the v-Ki-ras2 Kirsten rat sarcoma viral oncogene homologue (KRAS) and an activating V600E mutation in v-raf-murine sarcoma viral oncogene homologue B1 (BRAF) are moderately sensitive to erastin. Erastin is proposed to bind to VDAC2 and VDAC3, leading to oxidative stress and cell death in cells with activated RAS–RAF–MEK signaling (Yagoda et al., 2007).

We showed that erastin in HepG2 cells and other cell lines increases mitochondrial $\Delta\Psi$ and prevent depolarization induced by high cytosolic free tubulin, an effect not described previously. In addition, erastin added after microtubule destabilizers also restores mitochondrial $\Delta\Psi$, indicating that erastin prevents and reverses the inhibitory effect of free tubulin on VDAC (Maldonado et al., 2013). Also, erastin completely blocks tubulin inhibition of VDAC without modifying the voltage dependence of channels from HepG2 cells in planar lipid bilayers, confirming that the effect of erastin is specific for tubulin-dependent inhibition of VDAC (Maldonado et al., 2013). Following the identification of erastin as a VDAC–tubulin antagonist, we identified a group of erastin-like anti-Warburg compounds using a cell-based high-throughput drug screening. We tested the 50,080 DIVERSet Chembridge compound library to select small molecules that, similar to erastin, hyperpolarized mitochondria in the presence of microtubule destabilizers and high cytosolic free tubulin. The six lead compounds hyperpolarized mitochondria without causing changes in tubulin polymerization in a dose-dependent fashion. Erastin and the most potent X1 not only increased mitochondrial metabolism but had an anti-Warburg effect evidenced by the decreased lactate release in HepG2 and Huh7 hepatocarcinoma cells and HCC4006 lung carcinoma cell lines (DeHart, Lemasters, & Maldonado, 2017).

5.2 VDAC Opening, Glycolysis, and Reactive Oxygen Species Formation

VDAC opening in cancer cells leads to three main biological effects: increased mitochondrial metabolism caused by augmented entry of substrates into mitochondria, subsequent decreased glycolysis due to a higher cytosolic ATP/ADP ratio, and increased formation of reactive oxygen species (ROS) caused by enhanced activity of the ETC (Fig. 3).

After VDAC opening, flux of respiratory substrates into mitochondria fuels the Krebs cycle to produce NADH that enters the ETC. Electron pairs from NADH flow down the ETC to the final acceptor O_2. Simultaneous with the flow of electrons through the ETC, single electrons leak from Complexes I, II, and III to form the superoxide anion ($O_2^{\bullet-}$) (Chance, Sies, & Boveris, 1979). Although there are other mitochondrial and non-mitochondrial sources of ROS formation, the mitochondrial ETC especially Complex I (site I_Q), Complex II (site II_F), and Complex III (site III_{Qo}) have the highest capacity of ROS production among the seven major mitochondrial sites that produce ROS in mammals (Chen, Vazquez, Moghaddas, Hoppel, & Lesnefsky, 2003; Quinlan et al., 2012; Skulachev, 1996; Tribble, Jones, & Edmondson, 1988). The metabolic fate of $O_2^{\bullet-}$ varies depending on the site of origin. $O_2^{\bullet-}$ formed at Complexes I and II is released to the matrix, whereas $O_2^{\bullet-}$ generated at Complex III is released in large part to the intermembrane space and hence to the cytosol through VDAC (Brand, 2010; Han, Antunes, Canali, Rettori, & Cadenas, 2003; Muller, Liu, & Van, 2004). Superoxide dismutases located in the mitochondrial matrix (manganese-containing enzyme MnSOD or SOD2) and the cytosol (copper- and-zinc-containing enzyme Cu,ZnSOD or SOD1) rapidly convert $O_2^{\bullet-}$ to H_2O_2 (Fridovich, 1997). H_2O_2, a nonradical and the least reactive of ROS species, diffuses across membranes acting as a cell signaling molecule in proproliferative and prosurvival pathways without necessarily disrupting redox homeostasis (Morgan, Sobotta, & Dick, 2011; Veal, Day, & Morgan, 2007). For example, H_2O_2 modulates the prosurvival HIF-1 and MAP/ERK, PI3K/akt/mTOR pathways that favor

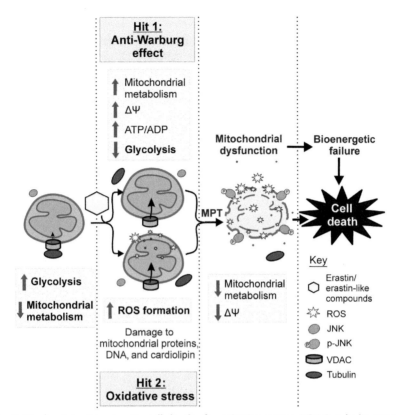

Fig. 3 Mechanisms to promote cell death after VDAC opening. VDAC–tubulin antagonists open VDAC increasing the flux of metabolites into and out of mitochondria leading to increased mitochondrial $\Delta\Psi$, increased mitochondrial metabolism, high cytosolic ATP/ADP ratio, and decreased glycolysis (Hit 1: anti-Warburg effect). VDAC opening also increases ROS formation concurrent with the decrease in glycolysis. Increased ROS damage mitochondrial DNA, cardiolipin, and mitochondrial proteins and activate JNK (Hit 2: oxidative stress). Activated JNK translocates to mitochondria causing mitochondrial dysfunction, bioenergetic failure, and cell death.

tumorigenesis and metastasis (Clerkin, Naughton, Quiney, & Cotter, 2008; Giles, 2006; Ushio-Fukai & Nakamura, 2008). H_2O_2 levels depend on the rate of production and degradation by the enzyme catalase or the interconversion that occurs when H_2O_2 accept an electron from free and loosely bound Fe^{2+} to form the highly reactive hydroxyl radical (OH$^\bullet$) by the Fenton reaction.

The effects of mitochondrial ROS on cellular structures depend on the specific ROS. The lifetimes of H_2O_2 and $O_2^{\bullet-}$ allow them to react both with mitochondria and with extramitochondrial structures. By contrast,

OH$^\bullet$ are so reactive that the effects are almost completely restricted to mitochondria. Both $O_2^{\bullet-}$ and OH$^\bullet$ inactivate mitochondrial proteins, including ATP synthase, NADH oxidase, and NADH dehydrogenase (Zhang, Marcillat, Giulivi, Ernster, & Davies, 1990). Beyond proteins, ROS damage mitochondrial DNA and lipids in the MIM. ROS-dependent peroxidation of cardiolipin, a MIM phospholipid rich in polyunsaturated fatty acids, is considered an early event in apoptosis (Schenkel & Bakovic, 2014).

Cytosolic ROS activate members of the MAPK family of serine/threonine kinases, especially c-Jun N-terminal kinase (JNK), the extracellular signal-regulated kinase (ERK 1/2), and p38 whose signaling can cause mitochondrial dysfunction (Kamata et al., 2005; Son et al., 2011). In fact, treatment with erastin and X1 in HepG2 and Huh7 hepatocarcinoma cells promoted JNK activation and increased ROS formation that caused mitochondrial dysfunction prevented by antioxidants. Moreover, blocking of the translocation of activated JNK to mitochondria prevented mitochondrial dysfunction induced by VDAC–tubulin antagonists (unpublished). Oxidative stress caused by accumulation of ROS also induces mitochondrial permeability transition (MPT) pores in the inner membrane. Opening of MPT pores results in the MIM becoming highly permeable to ions and solutes up to $\sim$1.5 kDa collapsing $\Delta\Psi$ (Zoratti & Szabo, 1995).

ROS levels are higher in tumor cells from cell lines, in animal models of cancer, and in human tumor tissues compared to nonproliferating cells. Increased H_2O_2 formation in cell lines and oxidative-induced DNA modifications and 4-hydroxy-2-nonenal-modified proteins in animal models and human tissues support this concept (Kawanishi, Hiraku, Pinlaor, & Ma, 2006; Szatrowski & Nathan, 1991; Toyokuni, Okamoto, Yodoi, & Hiai, 1995; Uchida, 2003). In cancer cells higher levels of ROS are neutralized by a high content of scavenging enzymes and antioxidants including SODs, catalase, and the glutathione system that reduces protein disulfide bonds (Liou & Storz, 2010; Panieri & Santoro, 2016; Sullivan & Chandel, 2014; Venditti, Di, & Di, 2013). Although ROS in cancer cells are proposed to be cytostatic, to favor tumor growth, or to be cytotoxic, the different levels of ROS that would cause each outcome have not been determined experimentally (Marengo et al., 2016; Panieri & Santoro, 2016; Sullivan & Chandel, 2014). Oxidative stress has been reported to induce mitochondrial dysfunction, cell cycle arrest, senescence, apoptosis, or necrosis (Liou & Storz, 2010).

Chemotherapeutic agents including cisplatin, adriamycin, the anthracyclines doxorubicin, epirubicin, and daunorubicin among others promote oxidative stress and depletion of the antioxidant capacity of tumor cells leading to a tumoricidal effect (Conklin, 2004; Faber, Coudray, Hida, Mousseau, & Favier, 1995; Ladner, Ehninger, Gey, & Clemens, 1989; Weijl et al., 1998). VDAC–tubulin antagonists by opening VDAC promote mitochondrial metabolism which increases the activity of the ETC leading to increased ROS formation. Continued enhanced ROS production eventually overcomes the antioxidant capacity of cancer cells leading to cytotoxicity.

5.3 VDAC-Dependent Metabolic Hits: Anti-Warburg Effect and Oxidative Stress

Metabolic heterogeneity is a complicating factor for the success of cancer chemotherapy (Dang, 2012; Eason & Sadanandam, 2016; Gerlinger et al., 2012; Yun, Johnson, Hanigan, & Locasale, 2012). All cancer cells, even with distinct metabolic signatures, display some level of enhanced glycolysis, suggesting different degrees of contribution by VDAC closure to the suppression of mitochondrial metabolism (Griguer et al., 2005; Guppy et al., 2002; Moreno-Sanchez et al., 2007; Scott et al., 2011). Reversal of the inhibitory effect of tubulin on VDAC triggers two distinct and nearly simultaneous effects: (1) increase of mitochondrial metabolism and activation of OXPHOS with consequent decrease of glycolysis (anti-Warburg effect) and (2) an increase in ROS formation leading to oxidative stress (Fig. 3). Oxidative stress is potentially more deleterious in highly glycolytic cells, which presumably could have lower antioxidant capacity since OXPHOS is not very active in these cells. By contrast, the reversal of the Warburg effect could damage more the highly glycolytic cells that survive oxidative stress and continue proliferating or the low glycolytic cells with a presumably constitutively higher basal level of ROS.

The VDAC–tubulin antagonist erastin and erastin-like compounds cause mitochondrial hyperpolarization followed by mitochondrial depolarization indicative of mitochondrial dysfunction in human hepatocarcinoma cells (unpublished). The initial increase in $\Delta\Psi$ precedes the increase in ROS generation and JNK activation resulting in mitochondrial dysfunction and possibly the onset of MPT. MPT causes nonselective permeabilization of the MIM leading to mitochondrial swelling, loss of $\Delta\Psi$ and ATP synthesis, rupture of the MOM, and cytochrome c release resulting in cell death (Bonora & Pinton, 2014; Green & Kroemer, 2004). MPT is mediated by the irreversible opening of the permeability transition pore complex (PTPC), a multiprotein pore assembled with proteins from both the

MOM and the MIM. VDAC, ANT, cyclophilin D, and the subunit c of the F_1F_0-ATP synthase among other mitochondrial proteins have been included as PTPC-forming proteins, although the molecular identity of the pore remains debatable (Izzo, Bravo-San Pedro, Sica, Kroemer, & Galluzzi, 2016). VDAC, initially considered a main component of the pore, is dispensable for the onset of MPT. Oxidative stress, a well-known inducer of MPT (Bonora & Pinton, 2014; Kowaltowski, Castilho, & Vercesi, 2001; Takeyama, Matsuo, & Tanaka, 1993) promotes MPT even in knockout cells for all VDAC isoforms (Baines, Kaiser, Sheiko, Craigen, & Molkentin, 2007).

The erastin-like anti-Warburg compound X1 also decreases glycolysis as evidenced by a decrease in lactate release (DeHart et al., 2017). The combination of reversal of Warburg metabolism and oxidative stress by the lead compound caused cell death to human hepatocarcinoma cell lines in culture and to xenografted Huh7 hepatocarcinoma cells (Fig. 4). Thus, erastin and

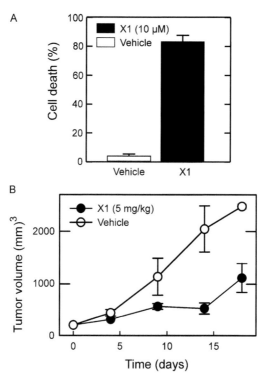

Fig. 4 Erastin-like compound X1 causes cell death in situ and in vivo. Lead compound X1 caused cell death in Huh7 hepatocarcinoma cells in culture and slowed tumor growth in a xenograft model of Huh7 cells in nude mice.

lead erastin-like compounds by causing "two-hits": an anti-Warburg effect and promotion of oxidative stress, represent a potential new class of cancer chemotherapeutic agents (Fig. 3).

6. CONCLUDING REMARKS

The VDAC–tubulin interaction in cancer cells operates as a metabolic switch to control cellular bioenergetics and regulate the Warburg phenotype. VDAC opening exerts a global influence on mitochondrial metabolism increasing OXPHOS and ROS production and indirectly modulating glycolysis. Pharmacological inhibition of the VDAC switch triggers two concurrent and complementary "hits": an anti-Warburg effect that promotes a nonproliferative metabolic phenotype and oxidative stress leading to mitochondrial dysfunction and cell death. VDAC-dependent oxidative stress is expected to promote cell killing in highly glycolytic cells with presumably lower antioxidant defenses and to cause nonlethal cell damage in less glycolytic tumor types. The anti-Warburg effect will decrease or stop cell proliferation in those cells in which increase in ROS formation was sublethal. In summary, the VDAC–tubulin interaction represents a new pharmacological target to turn a proproliferative phenotype into a cytotoxic, mitochondrial-dependent, and prooxidant metabolic profile. VDAC–tubulin antagonists could become a new generation of metabolism-oriented cancer chemotherapy.

ACKNOWLEDGMENT

This work was funded by R01CA184456, GM103542, and ACS 13-041-01-IRG to E.N.M.

REFERENCES

Al Jamal, J. A. (2005). Involvement of porin N,N-dicyclohexylcarbodiimide-reactive domain in hexokinase binding to the outer mitochondrial membrane. *The Protein Journal*, *24*, 1–8.
Azoulay-Zohar, H., Israelson, A., Abu-Hamad, S., & Shoshan-Barmatz, V. (2004). In self-defence: Hexokinase promotes voltage-dependent anion channel closure and prevents mitochondria-mediated apoptotic cell death. *Biochemistry Journal*, *377*, 347–355.
Baines, C. P., Kaiser, R. A., Sheiko, T., Craigen, W. J., & Molkentin, J. D. (2007). Voltage-dependent anion channels are dispensable for mitochondrial-dependent cell death. *Nature Cell Biology*, *9*, 550–555.
Baines, C. P., Song, C. X., Zheng, Y. T., Wang, G. W., Zhang, J., Wang, O. L., et al. (2003). Protein kinase Cepsilon interacts with and inhibits the permeability transition pore in cardiac mitochondria. *Circulatory Research*, *92*, 873–880.

Beckner, M. E., Gobbel, G. T., Abounader, R., Burovic, F., Agostino, N. R., Laterra, J., et al. (2005). Glycolytic glioma cells with active glycogen synthase are sensitive to PTEN and inhibitors of PI3K and gluconeogenesis. *Laboratory Investigation, 85*, 1457–1470.

Bera, A. K., Ghosh, S., & Das, S. (1995). Mitochondrial VDAC can be phosphorylated by cyclic AMP-dependent protein kinase. *Biochemical and Biophysical Research Communications, 209*, 213–217.

Bhat, T. A., Kumar, S., Chaudhary, A. K., Yadav, N., & Chandra, D. (2015). Restoration of mitochondria function as a target for cancer therapy. *Drug Discovery Today, 20*, 635–643.

Blachly-Dyson, E., & Forte, M. (2001). VDAC channels. *IUBMB Life, 52*, 113–118.

Bonora, M., & Pinton, P. (2014). The mitochondrial permeability transition pore and cancer: Molecular mechanisms involved in cell death. *Frontiers in Oncology, 4*, 302.

Bouzier, A. K., Voisin, P., Goodwin, R., Canioni, P., & Merle, M. (1998). Glucose and lactate metabolism in C6 glioma cells: Evidence for the preferential utilization of lactate for cell oxidative metabolism. *Developmental Neuroscience, 20*, 331–338.

Brand, M. D. (2005). The efficiency and plasticity of mitochondrial energy transduction. *Biochemical Society Transactions, 33*, 897–904.

Brand, M. D. (2010). The sites and topology of mitochondrial superoxide production. *Experimental Gerontology, 45*, 466–472.

Cairns, R. A. (2015). Drivers of the warburg phenotype. *Cancer Journal, 21*, 56–61.

Caro, P., Kishan, A. U., Norberg, E., Stanley, I. A., Chapuy, B., Ficarro, S. B., et al. (2012). Metabolic signatures uncover distinct targets in molecular subsets of diffuse large B cell lymphoma. *Cancer Cell, 22*, 547–560.

Chance, B., Sies, H., & Boveris, A. (1979). Hydroperoxide metabolism in mammalian organs. *Physiological Reviews, 59*, 527–605.

Chen, Q., Vazquez, E. J., Moghaddas, S., Hoppel, C. L., & Lesnefsky, E. J. (2003). Production of reactive oxygen species by mitochondria: Central role of complex III. *Journal of Biological Chemistry, 278*, 36027–36031.

Chung, W. J., Lyons, S. A., Nelson, G. M., Hamza, H., Gladson, C. L., Gillespie, G. Y., et al. (2005). Inhibition of cystine uptake disrupts the growth of primary brain tumors. *The Journal of Neuroscience, 25*, 7101–7110.

Clavell, L. A., Gelber, R. D., Cohen, H. J., Hitchcock-Bryan, S., Cassady, J. R., Tarbell, N. J., et al. (1986). Four-agent induction and intensive asparaginase therapy for treatment of childhood acute lymphoblastic leukemia. *The New England Journal of Medicine, 315*, 657–663.

Clerkin, J. S., Naughton, R., Quiney, C., & Cotter, T. G. (2008). Mechanisms of ROS modulated cell survival during carcinogenesis. *Cancer Letters, 266*, 30–36.

Colombini, M. (1979). A candidate for the permeability pathway of the outer mitochondrial membrane. *Nature, 279*, 643–645.

Colombini, M. (1980). Structure and mode of action of a voltage dependent anion-selective channel (VDAC) located in the outer mitochondrial membrane. *The Annals of the New York Academy of Sciences, 341*, 552–563.

Colombini, M. (2004). VDAC: The channel at the interface between mitochondria and the cytosol. *Molecular and Cellular Biochemistry, 256–257*, 107–115.

Colombini, M. (2012). VDAC structure, selectivity, and dynamics. *Biochimica et Biophysica Acta, 1818*, 1457–1465.

Comerford, S. A., Huang, Z., Du, X., Wang, Y., Cai, L., Witkiewicz, A. K., et al. (2014). Acetate dependence of tumors. *Cell, 159*, 1591–1602.

Commisso, C., Davidson, S. M., Soydaner-Azeloglu, R. G., Parker, S. J., Kamphorst, J. J., Hackett, S., et al. (2013). Macropinocytosis of protein is an amino acid supply route in Ras-transformed cells. *Nature, 497*, 633–637.

Conklin, K. A. (2004). Chemotherapy-associated oxidative stress: Impact on chemotherapeutic effectiveness. *Integrative Cancer Therapies, 3*, 294–300.

Dang, C. V. (2012). Links between metabolism and cancer. *Genes & Development, 26*, 877–890.
Das, S., Wong, R., Rajapakse, N., Murphy, E., & Steenbergen, C. (2008). Glycogen synthase kinase 3 inhibition slows mitochondrial adenine nucleotide transport and regulates voltage-dependent anion channel phosphorylation. *Circulation Research, 103*, 983–991.
DeBerardinis, R. J., & Cheng, T. (2010). Q's next: The diverse functions of glutamine in metabolism, cell biology, and cancer. *Oncogene, 29*, 313–324.
DeBerardinis, R. J., Sayed, N., Ditsworth, D., & Thompson, C. B. (2008). Brick by brick: Metabolism and tumor cell growth. *Current Opinion in Genetics & Development, 18*, 54–61.
DeHart, D. N., Lemasters, J. J., & Maldonado, E. N. (2017). Erastin-like anti-warburg agents prevent mitochondrial depolarization induced by free tubulin and decrease lactate formation in cancer cells. *SLAS Discovery, 23*, 23–33.
De Pinto, V., Guarino, F., Guarnera, A., Messina, A., Reina, S., Tomasello, F. M., et al. (2010). Characterization of human VDAC isoforms: A peculiar function for VDAC3? *Biochimica et Biophysica Acta, 1797*, 1268–1275.
Doherty, J. R., & Cleveland, J. L. (2013). Targeting lactate metabolism for cancer therapeutics. *The Journal of Clinical Investigation, 123*, 3685–3692.
Dolma, S., Lessnick, S. L., Hahn, W. C., & Stockwell, B. R. (2003). Identification of genotype-selective antitumor agents using synthetic lethal chemical screening in engineered human tumor cells. *Cancer Cell, 3*, 285–296.
Eason, K., & Sadanandam, A. (2016). Molecular or metabolic reprograming: What triggers tumor subtypes? *Cancer Research, 76*, 5195–5200.
Faber, M., Coudray, C., Hida, H., Mousseau, M., & Favier, A. (1995). Lipid peroxidation products, and vitamin and trace element status in patients with cancer before and after chemotherapy, including adriamycin. A preliminary study. *Biological Trace Element Research, 47*, 117–123.
Frezza, C. (2017). Mitochondrial metabolites: Undercover signalling molecules. *Interface Focus, 7*, 20160100.
Fridovich, I. (1997). Superoxide anion radical (O2-), superoxide dismutases, and related matters. *Journal of Biological Chemistry, 272*, 18515–18517.
Gerlinger, M., Rowan, A. J., Horswell, S., Larkin, J., Endesfelder, D., Gronroos, E., et al. (2012). Intratumor heterogeneity and branched evolution revealed by multiregion sequencing. *The New England Journal of Medicine, 366*, 883–892.
Giles, G. I. (2006). The redox regulation of thiol dependent signaling pathways in cancer. *Current Pharmaceutical Design, 12*, 4427–4443.
Gincel, D., Silberberg, S. D., & Shoshan-Barmatz, V. (2000). Modulation of the voltage-dependent anion channel (VDAC) by glutamate. *Journal of Bioenergetics and Biomembranes, 32*, 571–583.
Giovannucci, E., Harlan, D. M., Archer, M. C., Bergenstal, R. M., Gapstur, S. M., Habel, L. A., et al. (2010). Diabetes and cancer: A consensus report. *CA: A Cancer Journal for Clinicians, 60*, 207–221.
Green, D. R., & Kroemer, G. (2004). The pathophysiology of mitochondrial cell death. *Science, 305*, 626–629.
Griguer, C. E., Oliva, C. R., & Gillespie, G. Y. (2005). Glucose metabolism heterogeneity in human and mouse malignant glioma cell lines. *The Journal of Neuro-Oncology, 74*, 123–133.
Guppy, M., Leedman, P., Zu, X., & Russell, V. (2002). Contribution by different fuels and metabolic pathways to the total ATP turnover of proliferating MCF-7 breast cancer cells. *Biochemistry Journal, 364*, 309–315.
Hahn, A. T., Jones, J. T., & Meyer, T. (2009). Quantitative analysis of cell cycle phase durations and PC12 differentiation using fluorescent biosensors. *Cell Cycle, 8*, 1044–1052.

Han, D., Antunes, F., Canali, R., Rettori, D., & Cadenas, E. (2003). Voltage-dependent anion channels control the release of the superoxide anion from mitochondria to cytosol. *Journal of Biological Chemistry, 278*, 5557–5563.

Hiller, S., Abramson, J., Mannella, C., Wagner, G., & Zeth, K. (2010). The 3D structures of VDAC represent a native conformation. *Trends in Biochemical Sciences, 35*, 514–521.

Holmuhamedov, E., & Lemasters, J. J. (2009). Ethanol exposure decreases mitochondrial outer membrane permeability in cultured rat hepatocytes. *Archives of Biochemistry and Biophysics, 481*, 226–233.

Huang, H., Shah, K., Bradbury, N. A., Li, C., & White, C. (2014). Mcl-1 promotes lung cancer cell migration by directly interacting with VDAC to increase mitochondrial Ca2+ uptake and reactive oxygen species generation. *Cell Death & Disease, 5*, e1482.

Izzo, V., Bravo-San Pedro, J. M., Sica, V., Kroemer, G., & Galluzzi, L. (2016). Mitochondrial permeability transition: New findings and persisting uncertainties. *Trends in Cell Biology, 26*, 655–667.

Jara, J. A., & Lopez-Munoz, R. (2015). Metformin and cancer: Between the bioenergetic disturbances and the antifolate activity. *Pharmacological Research, 101*, 102–108.

Kamata, H., Honda, S., Maeda, S., Chang, L., Hirata, H., & Karin, M. (2005). Reactive oxygen species promote TNFalpha-induced death and sustained JNK activation by inhibiting MAP kinase phosphatases. *Cell, 120*, 649–661.

Kawanishi, S., Hiraku, Y., Pinlaor, S., & Ma, N. (2006). Oxidative and nitrative DNA damage in animals and patients with inflammatory diseases in relation to inflammation-related carcinogenesis. *Biological Chemistry, 387*, 365–372.

Keenan, M. M., & Chi, J. T. (2015). Alternative fuels for cancer cells. *Cancer Journal, 21*, 49–55.

Keibler, M. A., Wasylenko, T. M., Kelleher, J. K., Iliopoulos, O., Vander Heiden, M. G., & Stephanopoulos, G. (2016). Metabolic requirements for cancer cell proliferation. *Cancer Metabolism, 4*, 16.

Kennedy, K. M., Scarbrough, P. M., Ribeiro, A., Richardson, R., Yuan, H., Sonveaux, P., et al. (2013). Catabolism of exogenous lactate reveals it as a legitimate metabolic substrate in breast cancer. *PLoS One, 8*, e75154.

Kilburn, D. G., Lilly, M. D., & Webb, F. C. (1969). The energetics of mammalian cell growth. *Journal of Cell Science, 4*, 645–654.

Kowaltowski, A. J., Castilho, R. F., & Vercesi, A. E. (2001). Mitochondrial permeability transition and oxidative stress. *FEBS Letters, 495*, 12–15, %20.

Kreis, W., Baker, A., Ryan, V., & Bertasso, A. (1980). Effect of nutritional and enzymatic methionine deprivation upon human normal and malignant cells in tissue culture. *Cancer Research, 40*, 634–641.

Ladner, C., Ehninger, G., Gey, K. F., & Clemens, M. R. (1989). Effect of etoposide (VP16-213) on lipid peroxidation and antioxidant status in a high-dose radiochemotherapy regimen. *Cancer Chemotherapy and Pharmacology, 25*, 210–212.

Lee, A. C., Zizi, M., & Colombini, M. (1994). Beta-NADH decreases the permeability of the mitochondrial outer membrane to ADP by a factor of 6. *Journal of Biological Chemistry, 269*, 30974–30980.

Lemasters, J. J., & Holmuhamedov, E. (2006). Voltage-dependent anion channel (VDAC) as mitochondrial governator-thinking outside the box. *Biochimica et Biophysica Acta, 1762*, 181–190.

Libby, G., Donnelly, L. A., Donnan, P. T., Alessi, D. R., Morris, A. D., & Evans, J. M. (2009). New users of metformin are at low risk of incident cancer: A cohort study among people with type 2 diabetes. *Diabetes Care, 32*, 1620–1625.

Liberti, M. V., & Locasale, J. W. (2016). The warburg effect: How does it benefit cancer cells? *Trends in Biochemical Sciences, 41*, 211–218.

Liemburg-Apers, D. C., Schirris, T. J., Russel, F. G., Willems, P. H., & Koopman, W. J. (2015). Mitoenergetic dysfunction triggers a rapid compensatory increase in steady-state glucose flux. *Biophysical Journal, 109*, 1372–1386.

Lim, H. Y., Ho, Q. S., Low, J., Choolani, M., & Wong, K. P. (2011). Respiratory competent mitochondria in human ovarian and peritoneal cancer. *Mitochondrion, 11*, 437–443.

Liou, G. Y., & Storz, P. (2010). Reactive oxygen species in cancer. *Free Radical Research, 44*, 479–496.

Locasale, J. W., & Cantley, L. C. (2010). Altered metabolism in cancer. *BMC Biology, 8*, 88.

Lunt, S. Y., & Vander Heiden, M. G. (2011). Aerobic glycolysis: Meeting the metabolic requirements of cell proliferation. *Annual Review of Cell and Developmental Biology, 27*, 441–464.

Maldonado, E. N. (2017). VDAC-tubulin, an anti-warburg pro-oxidant switch. *Frontiers in Oncology, 7*, 4.

Maldonado, E. N., DeHart, D. N., Patnaik, J., Klatt, S. C., Beck, G. M., & Lemasters, J. J. (2016). ATP/ADP turnover and import of glycolytic ATP into mitochondria in cancer cells is independent of the adenine nucleotide translocator. *Journal of Biological Chemistry, 291*, 19642–19650.

Maldonado, E. N., & Lemasters, J. J. (2012). Warburg revisited: Regulation of mitochondrial metabolism by voltage-dependent anion channels in cancer cells. *The Journal of Pharmacology and Experimental Therapeutics, 342*, 637–641.

Maldonado, E. N., & Lemasters, J. J. (2014). ATP/ADP ratio, the missed connection between mitochondria and the warburg effect. *Mitochondrion, 19*, 78–84, Pt. A.

Maldonado, E. N., Patnaik, J., Mullins, M. R., & Lemasters, J. J. (2010). Free tubulin modulates mitochondrial membrane potential in cancer cells. *Cancer Research, 70*, 10192–10201.

Maldonado, E. N., Sheldon, K. L., DeHart, D. N., Patnaik, J., Manevich, Y., Townsend, D. M., et al. (2013). Voltage-dependent anion channels modulate mitochondrial metabolism in cancer cells: Regulation by free tubulin and erastin. *Journal of Biological Chemistry, 288*, 11920–11929.

Marengo, B., Nitti, M., Furfaro, A. L., Colla, R., Ciucis, C. D., Marinari, U. M., et al. (2016). Redox homeostasis and cellular antioxidant systems: Crucial players in cancer growth and therapy. *Oxidative Medicine and Cellular Longevity, 2016*, 6235641.

Mashimo, T., Pichumani, K., Vemireddy, V., Hatanpaa, K. J., Singh, D. K., Sirasanagandla, S., et al. (2014). Acetate is a bioenergetic substrate for human glioblastoma and brain metastases. *Cell, 159*, 1603–1614.

Mathupala, S. P., Ko, Y. H., & Pedersen, P. L. (2010). The pivotal roles of mitochondria in cancer: Warburg and beyond and encouraging prospects for effective therapies. *Biochimica et Biophysica Acta, 1797*, 1225–1230.

Meiser, J., & Vazquez, A. (2016). Give it or take it: The flux of one-carbon in cancer cells. *The FEBS Journal, 283*, 3695–3704.

Mor, I., Cheung, E. C., & Vousden, K. H. (2011). Control of glycolysis through regulation of PFK1: Old friends and recent additions. *Cold Spring Harbor Symposia on Quantitative Biology, 76*, 211–216.

Moreno-Sanchez, R., Marin-Hernandez, A., Saavedra, E., Pardo, J. P., Ralph, S. J., & Rodriguez-Enriquez, S. (2014). Who controls the ATP supply in cancer cells? Biochemistry lessons to understand cancer energy metabolism. *The International Journal of Biochemistry & Cell Biology, 50*, 10–23.

Moreno-Sanchez, R., Rodriguez-Enriquez, S., Marin-Hernandez, A., & Saavedra, E. (2007). Energy metabolism in tumor cells. *The FEBS Journal, 274*, 1393–1418.

Morgan, B., Sobotta, M. C., & Dick, T. P. (2011). Measuring E(GSH) and H2O2 with roGFP2-based redox probes. *Free Radical Biology & Medicine, 51*, 1943–1951.

Mullen, A. R., Wheaton, W. W., Jin, E. S., Chen, P. H., Sullivan, L. B., Cheng, T., et al. (2012). Reductive carboxylation supports growth in tumour cells with defective mitochondria. *Nature, 481*, 385–388.
Muller, F. L., Liu, Y., & Van, R. H. (2004). Complex III releases superoxide to both sides of the inner mitochondrial membrane. *Journal of Biological Chemistry, 279*, 49064–49073.
Nakashima, R. A., Paggi, M. G., & Pedersen, P. L. (1984). Contributions of glycolysis and oxidative phosphorylation to adenosine 5′-triphosphate production in AS-30D hepatoma cells. *Cancer Research, 44*, 5702–5706.
Nakashima, R. A., Paggi, M. G., Scott, L. J., & Pedersen, P. L. (1988). Purification and characterization of a bindable form of mitochondrial bound hexokinase from the highly glycolytic AS-30D rat hepatoma cell line. *Cancer Research, 48*, 913–919.
Nicholls, D. G., & Ferguson, S. J. (2013). *Bioenergetics 4*. London: Elsevier.
Palmieri, F., & Pierri, C. L. (2010). Mitochondrial metabolite transport. *Essays in Biochemistry, 47*, 37–52.
Panieri, E., & Santoro, M. M. (2016). ROS homeostasis and metabolism: A dangerous liason in cancer cells. *Cell Death & Disease, 7*, e2253.
Pastorino, J. G., & Hoek, J. B. (2003). Hexokinase II: The integration of energy metabolism and control of apoptosis. *Current Medicinal Chemistry, 10*, 1535–1551.
Pedersen, P. L. (1978). Tumor mitochondria and the bioenergetics of cancer cells. *Progress in Experimental Tumor Research, 22*, 190–274.
Pelicano, H., Martin, D. S., Xu, R. H., & Huang, P. (2006). Glycolysis inhibition for anticancer treatment. *Oncogene, 25*, 4633–4646.
Quinlan, C. L., Orr, A. L., Perevoshchikova, I. V., Treberg, J. R., Ackrell, B. A., & Brand, M. D. (2012). Mitochondrial complex II can generate reactive oxygen species at high rates in both the forward and reverse reactions. *Journal of Biological Chemistry, 287*, 27255–27264.
Rich, P. R. (2003). The molecular machinery of Keilin's respiratory chain. *Biochemical Society Transactions, 31*, 1095–1105.
Rich, P. R., & Marechal, A. (2010). The mitochondrial respiratory chain. *Essays in Biochemistry, 47*, 1–23.
Robinson, G. L., Dinsdale, D., MacFarlane, M., & Cain, K. (2012). Switching from aerobic glycolysis to oxidative phosphorylation modulates the sensitivity of mantle cell lymphoma cells to TRAIL. *Oncogene, 31*, 4996–5006.
Rodriguez-Enriquez, S., Carreno-Fuentes, L., Gallardo-Perez, J. C., Saavedra, E., Quezada, H., Vega, A., et al. (2010). Oxidative phosphorylation is impaired by prolonged hypoxia in breast and possibly in cervix carcinoma. *The International Journal of Biochemistry & Cell Biology, 42*, 1744–1751.
Rostovtseva, T. K., Antonsson, B., Suzuki, M., Youle, R. J., Colombini, M., & Bezrukov, S. M. (2004). Bid, but not bax, regulates VDAC channels. *The Journal of Biological Chemistry, 279*, 13575–13583.
Rostovtseva, T. K., Sheldon, K. L., Hassanzadeh, E., Monge, C., Saks, V., Bezrukov, S. M., et al. (2008). Tubulin binding blocks mitochondrial voltage-dependent anion channel and regulates respiration. *Proceedings of the National Academy of Sciences of the United States of America, 105*, 18746–18751.
Sampson, M. J., Decker, W. K., Beaudet, A. L., Ruitenbeek, W., Armstrong, D., Hicks, M. J., et al. (2001). Immotile sperm and infertility in mice lacking mitochondrial voltage-dependent anion channel type 3. *Journal of Biological Chemistry, 276*, 39206–39212.
Sampson, M. J., Lovell, R. S., & Craigen, W. J. (1997). The murine voltage-dependent anion channel gene family. Conserved structure and function. *Journal of Biological Chemistry, 272*, 18966–18973.

Samudio, I., Harmancey, R., Fiegl, M., Kantarjian, H., Konopleva, M., Korchin, B., et al. (2010). Pharmacologic inhibition of fatty acid oxidation sensitizes human leukemia cells to apoptosis induction. *The Journal of Clinical Investigation, 120*, 142–156.

Schein, S. J., Colombini, M., & Finkelstein, A. (1976). Reconstitution in planar lipid bilayers of a voltage-dependent anion-selective channel obtained from paramecium mitochondria. *The Journal of Membrane Biology, 30*, 99–120.

Schenkel, L. C., & Bakovic, M. (2014). Formation and regulation of mitochondrial membranes. *International Journal of Cell Biology, 2014*, 709828.

Schredelseker, J., Paz, A., Lopez, C. J., Altenbach, C., Leung, C. S., Drexler, M. K., et al. (2014). High resolution structure and double electron-electron resonance of the zebrafish voltage-dependent anion channel 2 reveal an oligomeric population. *Journal of Biological Chemistry, 289*, 12566–12577.

Schwenke, W. D., Soboll, S., Seitz, H. J., & Sies, H. (1981). Mitochondrial and cytosolic ATP/ADP ratios in rat liver in vivo. *Biochemistry Journal, 200*, 405–408.

Scott, L., Lamb, J., Smith, S., & Wheatley, D. N. (2000). Single amino acid (arginine) deprivation: Rapid and selective death of cultured transformed and malignant cells. *British Journal of Cancer, 83*, 800–810.

Scott, D. A., Richardson, A. D., Filipp, F. V., Knutzen, C. A., Chiang, G. G., Ronai, Z. A., et al. (2011). Comparative metabolic flux profiling of melanoma cell lines: Beyond the Warburg effect. *Journal of Biological Chemistry, 286*, 42626–42634.

Sheen, J. H., Zoncu, R., Kim, D., & Sabatini, D. M. (2011). Defective regulation of autophagy upon leucine deprivation reveals a targetable liability of human melanoma cells in vitro and in vivo. *Cancer Cell, 19*, 613–628.

Sheldon, K. L., Maldonado, E. N., Lemasters, J. J., Rostovtseva, T. K., & Bezrukov, S. M. (2011). Phosphorylation of voltage-dependent anion channel by serine/threonine kinases governs its interaction with tubulin. *PLoS One, 6*, e25539.

Singleterry, J., Sreedhar, A., & Zhao, Y. (2014). Components of cancer metabolism and therapeutic interventions. *Mitochondrion, 17C*, 50–55.

Skrtic, M., Sriskanthadevan, S., Jhas, B., Gebbia, M., Wang, X., Wang, Z., et al. (2011). Inhibition of mitochondrial translation as a therapeutic strategy for human acute myeloid leukemia. *Cancer Cell, 20*, 674–688.

Skulachev, V. P. (1996). Role of uncoupled and non-coupled oxidations in maintenance of safely low levels of oxygen and its one-electron reductants. *Quarterly Reviews of Biophysics, 29*, 169–202.

Smolkova, K., Bellance, N., Scandurra, F., Genot, E., Gnaiger, E., Plecita-Hlavata, L., et al. (2010). Mitochondrial bioenergetic adaptations of breast cancer cells to aglycemia and hypoxia. *Journal of Bioenergetics and Biomembranes, 42*, 55–67.

Son, Y., Cheong, Y. K., Kim, N. H., Chung, H. T., Kang, D. G., & Pae, H. O. (2011). Mitogen-activated protein kinases and reactive oxygen species: How can ROS activate MAPK pathways? *Journal of Signal Transduction, 2011*, 792639.

Song, J., & Colombini, M. (1996). Indications of a common folding pattern for VDAC channels from all sources. *Journal of Bioenergetics and Biomembranes, 28*, 153–161.

Sonveaux, P., Vegran, F., Schroeder, T., Wergin, M. C., Verrax, J., Rabbani, Z. N., et al. (2008). Targeting lactate-fueled respiration selectively kills hypoxic tumor cells in mice. *The Journal of Clinical Investigation, 118*, 3930–3942.

Sullivan, L. B., & Chandel, N. S. (2014). Mitochondrial reactive oxygen species and cancer. *Cancer Metabolism, 2*, 17.

Sutendra, G., & Michelakis, E. D. (2013). Pyruvate dehydrogenase kinase as a novel therapeutic target in oncology. *Frontiers in Oncology, 3*, 38.

Szatrowski, T. P., & Nathan, C. F. (1991). Production of large amounts of hydrogen peroxide by human tumor cells. *Cancer Research, 51*, 794–798.

Takeyama, N., Matsuo, N., & Tanaka, T. (1993). Oxidative damage to mitochondria is mediated by the Ca(2+)-dependent inner-membrane permeability transition. *Biochemistry Journal*, *294*(Pt. 3), 719–725.

Timohhina, N., Guzun, R., Tepp, K., Monge, C., Varikmaa, M., Vija, H., et al. (2009). Direct measurement of energy fluxes from mitochondria into cytoplasm in permeabilized cardiac cells in situ: Some evidence for mitochondrial interactosome. *Journal of Bioenergetics and Biomembranes*, *41*, 259–275.

Toyokuni, S., Okamoto, K., Yodoi, J., & Hiai, H. (1995). Persistent oxidative stress in cancer. *FEBS Letters*, *358*, 1–3.

Tribble, D. L., Jones, D. P., & Edmondson, D. E. (1988). Effect of hypoxia on tert-butylhydroperoxide-induced oxidative injury in hepatocytes. *Molecular Pharmacology*, *34*, 413–420.

Tsujimoto, Y., & Shimizu, S. (2000). VDAC regulation by the bcl-2 family of proteins. *Cell Death and Differentiation*, *7*, 1174–1181.

Uchida, K. (2003). 4-Hydroxy-2-nonenal: A product and mediator of oxidative stress. *Progress in Lipid Research*, *42*, 318–343.

Ujwal, R., Cascio, D., Colletier, J. P., Faham, S., Zhang, J., Toro, L., et al. (2008). The crystal structure of mouse VDAC1 at 2.3 Å resolution reveals mechanistic insights into metabolite gating. *Proceedings of the National Academy of Sciences of the United States of America*, *105*, 17742–17747.

Ushio-Fukai, M., & Nakamura, Y. (2008). Reactive oxygen species and angiogenesis: NADPH oxidase as target for cancer therapy. *Cancer Letters*, *266*, 37–52.

Vander Heiden, M. G., Chandel, N. S., Li, X. X., Schumacker, P. T., Colombini, M., & Thompson, C. B. (2000). Outer mitochondrial membrane permeability can regulate coupled respiration and cell survival. *Proceedings of the National Academy of Sciences of the United States of America*, *97*, 4666–4671.

Vander Heiden, M. G., Li, X. X., Gottleib, E., Hill, R. B., Thompson, C. B., & Colombini, M. (2001). Bcl-xL promotes the open configuration of the voltage-dependent anion channel and metabolite passage through the outer mitochondrial membrane. *The Journal of Biological Chemistry*, *276*, 19414–19419.

Veal, E. A., Day, A. M., & Morgan, B. A. (2007). Hydrogen peroxide sensing and signaling. *Molecular Cell*, *26*, 1–14.

Venditti, P., Di, S. L., & Di, M. S. (2013). Mitochondrial metabolism of reactive oxygen species. *Mitochondrion*, *13*, 71–82.

Walker, J. E. (2013). The ATP synthase: The understood, the uncertain and the unknown. *Biochemical Society Transactions*, *41*, 1–16.

Wang, J. B., Erickson, J. W., Fuji, R., Ramachandran, S., Gao, P., Dinavahi, R., et al. (2010). Targeting mitochondrial glutaminase activity inhibits oncogenic transformation. *Cancer Cell*, *18*, 207–219.

Warburg, O. (1956). On the origin of cancer cells. *Science*, *123*, 309–314.

Warburg, O., Wind, F., & Negelein, E. (1927). The metabolism of tumors in the body. *The Journal of General Physiology*, *8*, 519–530.

Weijl, N. I., Hopman, G. D., Wipkink-Bakker, A., Lentjes, E. G., Berger, H. M., Cleton, F. J., et al. (1998). Cisplatin combination chemotherapy induces a fall in plasma antioxidants of cancer patients. *Annals of Oncology*, *9*, 1331–1337.

Weinberg, S. E., & Chandel, N. S. (2015). Targeting mitochondria metabolism for cancer therapy. *Nature Chemical Biology*, *11*, 9–15.

Weinhouse, S. (1956). On respiratory impairment in cancer cells. *Science*, *124*, 267–269.

Wikstrom, M., Sharma, V., Kaila, V. R., Hosler, J. P., & Hummer, G. (2015). New perspectives on proton pumping in cellular respiration. *Chemical Reviews*, *115*, 2196–2221.

Wolf, A., Agnihotri, S., Micallef, J., Mukherjee, J., Sabha, N., Cairns, R., et al. (2011). Hexokinase 2 is a key mediator of aerobic glycolysis and promotes tumor growth in human glioblastoma multiforme. *The Journal of Experimental Medicine*, *208*, 313–326.

Yagoda, N., Von, R. M., Zaganjor, E., Bauer, A. J., Yang, W. S., Fridman, D. J., et al. (2007). RAS-RAF-MEK-dependent oxidative cell death involving voltage-dependent anion channels. *Nature, 447,* 864–868.

Yun, J., Johnson, J. L., Hanigan, C. L., & Locasale, J. W. (2012). Interactions between epigenetics and metabolism in cancers. *Frontiers in Oncology, 2,* 163.

Zhang, X., Fryknas, M., Hernlund, E., Fayad, W., De, M. A., Olofsson, M. H., et al. (2014). Induction of mitochondrial dysfunction as a strategy for targeting tumour cells in metabolically compromised microenvironments. *Nature Communications, 5,* 3295.

Zhang, Y., Marcillat, O., Giulivi, C., Ernster, L., & Davies, K. J. (1990). The oxidative inactivation of mitochondrial electron transport chain components and ATPase. *Journal of Biological Chemistry, 265,* 16330–16336.

Zhu, A., Lee, D., & Shim, H. (2011). Metabolic positron emission tomography imaging in cancer detection and therapy response. *Seminars in Oncology, 38,* 55–69.

Zizi, M., Forte, M., Blachly-Dyson, E., & Colombini, M. (1994). NADH regulates the gating of VDAC, the mitochondrial outer membrane channel. *The Journal of Biological Chemistry, 269,* 1614–1616.

Zoratti, M., & Szabo, I. (1995). The mitochondrial permeability transition. *Biochimica et Biophysica Acta, 1241,* 139–176.

CHAPTER THREE

Acquired Resistance to Drugs Targeting Tyrosine Kinases

Steven A. Rosenzweig[1]

Hollings Cancer Center, Medical University of South Carolina, Charleston, SC, United States
[1]Corresponding author: e-mail address: rosenzsa@musc.edu

Contents

1. Introduction	73
2. Inhibition of Bcr-Abl and Nonreceptor Tyrosine Kinases	74
2.1 Mechanisms of Acquired vs Intrinsic Resistance to TKIs	75
2.2 Acquired Resistance to Abl Kinase TKIs	75
3. Receptor and Nonreceptor Tyrosine Kinases Activate Common Pathways	77
4. Receptor TKIs and the EGFR Family	80
4.1 Lapatinib, a Dual Kinase Inhibitor of EGFR and HER2, and Afatinib, a Covalent ErbB1 RTKI	82
4.2 Lapatinib-Induced Kinome Reprogramming and Its Role in Resistance	83
5. Epigenetic Mechanisms of Resistance	84
5.1 Resistance to Receptor TKIs vs Receptor-Targeted Antibodies: IGF-1R	86
5.2 Other mAbs and Acquired Resistance: Trastuzumab	87
6. IGF-1R and Dependence Receptors in Drug Resistance	89
7. Conclusions and Future Perspective	91
Acknowledgments	92
References	92

Abstract

Resistance to chemotherapeutic drugs exemplifies the greatest hindrance to effective treatment of cancer patients. The molecular mechanisms responsible have been investigated for over 50 years and have revealed the lack of a single cause, but instead, multiple mechanisms including induced expression of membrane transporters that pump drugs out of cells (multidrug resistance (MDR) phenotype), changes in the glutathione system, and altered metabolism. Treatment of cancer patients/cancer cells with chemotherapeutic agents and/or molecularly targeted drugs is accompanied by acquisition of resistance to the treatment administered. Chemotherapeutic agent resistance was initially assumed to be due to induction of mutations leading to a resistant phenotype. While this has occurred for molecularly targeted drugs, it is clear that drugs selectively targeting tyrosine kinases (TKs) cause the acquisition of mutational changes and resistance to inhibition. The first TK to be targeted, Bcr-Abl, led to the generation of several drugs including imatinib, dasatinib, and sunitinib that provided a rich understanding of this phenomenon. It became clear that mutations alone were not the only cause of resistance. Additional mechanisms were involved, including alternative splicing, alternative/compensatory signaling pathways, and epigenetic changes. This review will focus on resistance to tyrosine kinase inhibitors (TKIs), receptor TK (RTK)-directed antibodies, and antibodies that inactivate specific RTK ligands. New approaches and concepts aimed at avoiding the generation of drug resistance will be examined. Many RTKs, including the IGF-1R, are dependence receptors that induce ligand-independent apoptosis. How this signaling paradigm has implications on therapeutic strategies will also be considered.

ABBREVIATIONS

Akt	Ak (mouse strain)—thymoma
AXL	a receptor tyrosine kinase
Bcr-Abl	breakpoint cluster-Abelson tyrosine kinase
CML	chronic myelogenous leukemia
c-Src	cellular sarcoma
DACH1	dachshund homolog 1 (Drosophila)
DFG	Asp-Phe-Gly
DTP	drug-tolerant persisters
DTEP	drug-tolerant expanded persisters
ECD	extracellular domain
EGF	epidermal growth factor
EGFR	epidermal growth factor receptor
ERK	extracellular-regulated kinase
FGFR	fibroblast growth factor receptor
FLT3	FMS-like tyrosine kinase 3
GIST	gastrointestinal stromal tumor
HB-EGF	heparin-binding epidermal growth factor
HDAC	histone deacetylase
HER2	human epidermal growth factor receptor 2
HGF	hepatocyte growth factor
IGF	insulin-like growth factor
IGF-F1-1	cyclic hexadecapeptide, IGF antagonist
IGF-1R	insulin-like growth factor-1 receptor
IGFBP	insulin-like growth factor binding protein
IQGAP1	Ras GTPase-activating-like protein
JAK	Janus kinase
KD	kinase domain
mAb	monoclonal antibody
MAPK	mitogen-activated protein kinase
MDR	multidrug resistance
MET	MNNG HOS transforming gene
mTOR	mammalian target of rapamycin

NMR	nuclear magnetic resonance
NSCLC	nonsmall cell lung cancer
PDGFR	platelet-derived growth factor
PFKFB2	6-phosphofructo-2-kinase/fructose-2,6-biphosphatase 2
PI3K	phosphoinositide 3-kinase
PTB	phosphotyrosine binding domain
PTEN	phosphatase and tensin homolog
Raf	Ras family member
Ras	rat sarcoma
RTK	receptor tyrosine kinase
SH2	src homology 2 domain
Sos	son of sevenless
STAT	signal transducer and activator of transcription
TGFα	transforming growth factor α
TKI	tyrosine kinase inhibitor
VEGFR	vascular endothelial growth factor receptor

1. INTRODUCTION

Cancer cells frequently exhibit resistance to the growth inhibitory and cytotoxic actions of chemotherapeutic drugs reflecting the potential to undergo a rapid form of molecular evolution as a means of developing a survival strategy. In this context, multiple mechanism(s) have been demonstrated as being responsible for the observed cancer cell chemoresistance/drug tolerance. These include acquiring mutations that enable survival, the "switching" between different receptor-driven signaling pathways, and the induction of transporter protein expression enabling efflux of drugs out of the cell. As more therapeutic strategies and molecularly targeted drugs that inhibit specific TKs are developed, we have been provided the opportunity to probe deeper into the processes involved in drug resistance; it is clear that cancer cells have developed additional mechanisms of chemoresistance. In this review, the molecular basis for resistance to tyrosine kinase inhibitors (TKIs) will be discussed and how this compares to the resistance to receptor TKIs (RTKIs). Also included is a discussion of acquired resistance to monoclonal antibodies (mAbs) targeting RTKs and how the resistance mechanisms compare to RTKIs. These analyses will be considered in the context of tumor heterogeneity whereby cells populating any given tumor are heterogeneous and that natural selection by drug dosing is a key mechanism in this process.

2. INHIBITION OF Bcr-Abl AND NONRECEPTOR TYROSINE KINASES

From a historic perspective, Gleevec (STI-571; imatinib) was the first therapeutically successful Abl TKI introduced for the treatment of chronic myeloid leukemia (CML). Accordingly, it has served as an instructional model for rational drug design of nonreceptor, as well as receptor TKIs since its FDA approval in 2001. Given the fact that the structures of the tyrosine kinase family are highly homologous, particularly within the ATP-binding sites of their kinase domains (KDs), and these compounds competed with ATP for binding to the ATP-binding site, there was early concern that they would lack specificity for a single tyrosine kinase and fail to be effective drugs. The long succession of selective receptor and nonreceptor TKIs developed since imatinib proved this to be incorrect. Based on elegant crystallographic studies of Abl kinase in the presence of imatinib (then referred to as STI-571 or CGP 57148) a mechanism of inhibition was determined, in which imatinib binds to the ATP-binding site of the KD, stabilizing the inactive non-ATP-binding conformation of the Abl activation loop, thereby "locking" the kinase in the off position and preventing transphosphorylation (Marcucci, Perrotti, & Caligiuri, 2003; Schindler et al., 2000).

In order to determine the rationale for imatinib's specificity and potency, Lin and coworkers carried out a detailed analysis of the binding energies of imatinib for Abl and Src kinases (Lin, Meng, Jiang, & Roux, 2013). The premise for this comparison stemmed from the fact that these tyrosine kinases have 47% sequence identity, yet Src is not effectively inhibited by imatinib. Crystal studies of the Abl–imatinib complex revealed that the Asp-Phe-Gly (DFG) present near the N-terminal end of the KD activation loop adopts a "DFG-out" conformation corresponding to inactivation of the catalytic subunit (Schindler et al., 2000). It was initially assumed that Src and other TKs did not adopt the DFG-out conformation, precluding their abilities to bind imatinib. This view was eventually proven to be incorrect with the observation of c-Src bound to imatinib in the DFG-out state (Seeliger et al., 2007). The model then changed to one of induced-fit being responsible for imatinib preferentially interacting with Abl over Src kinase based on NMR analysis of enzyme–inhibitor complexes (Agafonov, Wilson, Otten, Buosi, & Kern, 2014). This was further confirmed based on analyses of common ancestral forms of these kinases and X-ray structural analysis. It was postulated that affinity for imatinib increases as these kinases evolve toward Abl and lost as they evolve toward Src (Wilson et al., 2015).

More recent next-generation TKIs targeting Abl include bosutinib (Bosulif®) and ponatinib (Iclusig®). Bosutinib was FDA approved in 2012 for use in treating adults with chronic, accelerated, and blast-phase CML or Philadelphia chromosome-positive ALL. Bosutinib was developed to overcome resistance; that end result did not occur. However, ponatinib was FDA approved in November 2016 for the treatment of the above forms of AML as well as T315I-positive CML (all phases). Unfortunately, ponatinib has significant cardiovascular side effects and toxicities leading to life-threatening blood clots and severe vascular occlusive effects. These actions are consistent with the fact that in addition to Abl, ponatinib has inhibitory effects on upward of 40 tyrosine kinases (O'Hare et al., 2009), including FGFR1–4 (Gozgit et al., 2012), thereby contributing to ponatinib's polypharmacology profile and toxicities.

2.1 Mechanisms of Acquired vs Intrinsic Resistance to TKIs

Over the years, a number of mechanisms have been identified as contributing to or causing resistance to receptor and nonreceptor TKIs. Preexisting, primary, or intrinsic resistance refers to those mechanisms present in cells before they were ever exposed to an inhibitor (Jänne, Gray, & Settleman, 2009). Extrinsic or acquired mechanisms are represented by pharmacokinetic parameters that influence the efficacy of drugs as well as molecular changes in drug targets that alter drug efficacy. The tumor microenvironment (TME) has more recently been shown to contribute to both intrinsic and acquired drug resistance (Klemm & Joyce, 2015). The TME has been well accepted in contributing to tumorigenesis via angiogenesis/vascular remodeling and the release of factors from various stromal cells, recruitment of cancer stem cells, and the effects of cancer-activated fibroblasts (Hanahan & Coussens, 2012). It is becoming increasingly evident that the TME also impacts therapeutic response including a role for the ECM in supporting primary and metastatic niches which can affect drug response (Lu, Weaver, & Werb, 2012).

2.2 Acquired Resistance to Abl Kinase TKIs

For patients treated with imatinib, the primary cause for relapse/resistance to imatinib is reactivation of the Bcr-Abl kinase as a result of the appearance of point mutation(s) within the KD (O'Hare et al., 2005). These mutations alter imatinib action without significantly reducing ATP binding or kinase function (Deininger, Buchdunger, & Druker, 2005). Identification of the sites of point mutations in Bcr-Abl resulting from imatinib therapy, or the second-line Abl kinase inhibitors dasatinib and nilotinib and their impact on kinase function have been well characterized by a number of investigative teams (O'Hare, Eide, & Deininger, 2007).

A number of kinase domain point mutations have been identified and characterized for their effects on Bcr-Abl function in vitro and sensitivity to dasatinib and nilotinib; these analyses have been reviewed elsewhere (O'Hare et al., 2007). The natural evolution of KD mutations in TKIs is typified by the T315I mutation in Abl, a key contact site for imatinib. T315I represents mutation of the "gatekeeper" residue in Abl and results in conferring resistance to the Abl inhibitors, imatinib, dasatinib, and nilotinib (Barouch-Bentov & Sauer, 2011). A key feature of gatekeeper mutations such as T315I in Abl is that they typically have no effect on kinase activity. Rather, they block TKI access to the hydrophobic pocket within the activation loop via steric hindrance which in turn blocks inhibitor binding via loss of the necessary hydrogen bonding required to form a stable enzyme–inhibitor complex (Zhang, Yang, & Gray, 2009). Additional point mutations located within the ATP-binding loop prevent Abl from assuming a high-affinity conformation capable of binding imatinib. Activation loop mutations are thought to stabilize the active conformation, which imatinib is unable to bind. Of note, a number of activation loop mutations were inhibitable with the second-generation Bcr-Abl kinase inhibitors such as nilotinib (Weisberg et al., 2005) and dasatinib, a dual Src/Abl inhibitor (Shah et al., 2004), as a result of their increased affinity for Abl kinase compared to imatinib. Dasatinib has a 300-fold greater potency than imatinib and it binds to the catalytically active conformation of Abl, further enabling its ability to inhibit imatinib-resistant mutants (Shah et al., 2004). In differentiating between intrinsic and acquired resistance, Zhang et al. raise the issue that gatekeeper mutations may be preexisting rather than acquired (Zhang et al., 2009).

The point mutations identified in the Bcr-Abl KD result in resistance to imatinib as a result of reduced KD flexibility, limiting its ability to form an inactive conformation necessary for imatinib binding and inhibition (Burgess, Skaggs, Shah, Lee, & Sawyers, 2005). On this basis, second-generation inhibitors were developed with the goal of increased potency above that of imatinib. Indeed, mutations found to be resistant to dasatinib are present within contact sites (Burgess et al., 2005), while nilotinib-induced point mutations were also resistant to imatinib (Ray, Cowan-Jacob, Manley, Mestan, & Griffin, 2007).

In contrast, in vitro induction of imatinib resistance is often associated with Bcr-Abl mRNA and protein overexpression, which is not always associated with gene amplification. Elevated P-glycoprotein expression and multidrug resistance (MDR)-based drug efflux, as seen with many chemotherapeutics, have also been observed for imatinib (Mahon et al., 2000), and the activation of integrin and/or growth factor receptor signaling pathways has been described as mechanisms responsible for imatinib refractoriness (Deininger et al., 2005).

3. RECEPTOR AND NONRECEPTOR TYROSINE KINASES ACTIVATE COMMON PATHWAYS

Receptor and nonreceptor tyrosine kinases utilize a variety of common effector proteins and pathways to mediate their downstream effects in normal cells and cancer cells. A key family of RTKs in tumorigenesis and therapeutic strategies in multiple cancer sites is the epidermal growth factor receptor (EGFR) also referred to as HER1 (human epidermal growth factor receptor 1) or ErbB1 family (based on their relatedness to the avian viral erythroblastosis oncogene) and is comprised of four members HER1–4 or ErbB1–4. Ligand binding leads to a conformational change in the 3D structure of the EGFR, its increased lateral mobility in the plasma membrane, homo- or heterodimerization, and transphosphorylation of its partnering receptor's intracellular domain. The phosphorylated receptor dimer, through interactions of its phosphotyrosines, binds to effectors containing Src homology 2 (SH2) and phosphotyrosine-binding (PTB) domains activating downstream pathways (Roskoski, 2014) including Ras-MAPK (ERK), phosphoinositide 3-kinase (PI3K)/Akt, and STAT activation downstream of the JAK nonreceptor tyrosine kinase. Of note, activation of the IGF-1R can result in "receptor cross talk" as a result to protease activation and the shedding of membrane-tethered EGFR ligands. Alternatively, activation of the HIF-1 transcription factor resulting in VEGF expression and secretion can, in turn, activate the EGFR and/or VEGFR, respectively (Fig. 1; Rosenzweig, 2009; Rosenzweig & Atreya, 2010; Slomiany et al., 2007; Slomiany & Rosenzweig, 2006). Fig. 2 illustrates signaling pathways regulated by Bcr-Abl are common to those regulated by RTKs and other nonreceptor tyrosine kinase leading to enhanced cell proliferation, tumorigenesis, invasion, and metastasis (Steelman et al., 2004).

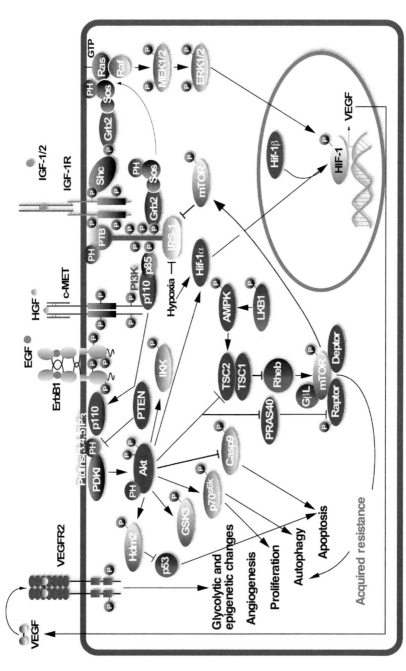

Fig. 1 Receptor tyrosine kinase signaling pathway cross talk. Following ligand-induced receptor transphosphorylation, growth factor receptor tyrosine kinases such as the ErbB1 and IGF-1R recruit effector molecules containing SH2 or PTB domains to initiate a downstream cascade activating the Ras-ERK or PI3-K/Akt pathways, which impinge upon a number of additional pathways and activities including mTOR regulation.

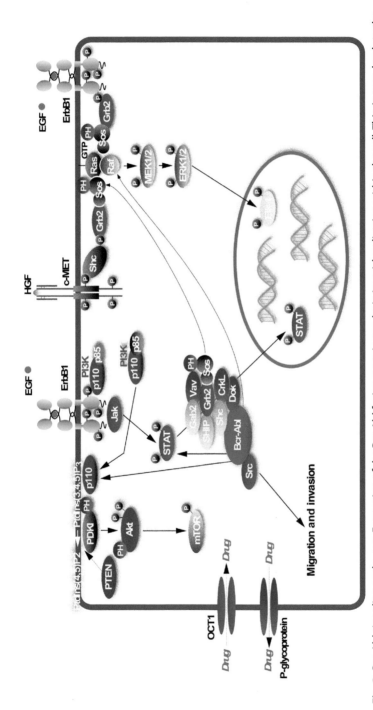

Fig. 2 Bcr-Abl signaling pathways. Formation of the Bcr-Abl fusion protein results in its mislocalization within the cell. This, in turn, leads to the phosphorylation and activation of a number of pathways common to receptor tyrosine kinases. Also shown is signaling by ErbB1 (EGFR).

The existence of overlapping or "redundant" pathways across receptor and nonreceptor kinases provides insight as to how compensatory signaling pathways may take the place of those RTK pathways inhibited by a given molecularly targeted RTKI. These mechanisms, in addition to kinase mutations, represent important ways in which cancer cells become resistant to targeted therapeutics and will be reviewed below starting with Bcr-Abl TKIs and extending to a discussion of EGF and IGF-1 receptors. While this review is focused on receptor and nonreceptor TKIs and mechanisms of acquired resistance, it should be realized that there are currently inhibitors being evaluated or in clinical trials that target one or more kinase depicted in Figs. 1 and 2 (Liu, Cheng, Roberts, & Zhao, 2009; Rosenzweig & Atreya, 2010). This further underscores the polypharmacology approach taken by some companies as an anticancer or antiaging therapeutic strategy.

4. RECEPTOR TKIs AND THE EGFR FAMILY

As observed with chemotherapeutic agents that lack targeting specificity, rationally designed drugs (molecularly targeted drugs; TKIs and mAbs) that selectively target receptor and nonreceptor tyrosine kinases can also result in acquired resistance. As with Abl kinase TKIs, considerable experience has been gained in the study of drugs that target the EGFR family both in terms of acquired resistance and in defining sensitivity to drug. It was determined early on in the experience with sensitivity to the EGFR RTKIs gefitinib and erlotinib that drug-sensitive patient populations could be selected for therapy based on the presence of an activating mutation in the EGFR (Lynch et al., 2004; Paez et al., 2004; Pao et al., 2004). For example, ~10% of all nonsmall cell lung cancer (NSCLC) patients in the United States—with a higher percentage in East Asia—exhibit gain-of-function mutations within the EGFR KD. These are attributable to a single amino acid substitution of arginine (R) for leucine (L) at position 858 (nucleotide 2573 TmG in exon 21) or an exon 19 in-frame deletion, removing the tetrapeptide Leu-Arg-Glu-Ala (Pao et al., 2005). Despite early positive responses to therapeutic intervention, most of these patients eventually developed resistance to erlotinib and gefitinib, as seen with acquired resistance to imatinib treatment in CML. The underlying cause for resistance was eventually determined to be caused by secondary mutations as observed in the Abl KD (Shah et al., 2002). These "loss-of-inhibition" mutations were found in over half of the patients exhibiting acquired resistance to imatinib, and were clustered within the ATP-binding and activation loops of the Abl

KD, resulting in blocking imatinib binding to Abl (Shah et al., 2002). A single nucleotide change in the EGFR resulting in replacing a threonine with methionine at residue 790 (T790M) is typically observed. It is notable that this represents a gatekeeper mutation analogous to other gatekeeper mutations such as T670I in c-Kit, T674I in PDGFRα, and T315I in Abl described earlier, which, in addition to causing resistance in CML, is responsible for acquired resistance to imatinib in gastrointestinal stromal tumors (GISTs) (Heinrich et al., 2003; Pao et al., 2005).

Based on the previous experience with acquired resistance to imatinib, a number of investigators (Lynch et al., 2004; Paez et al., 2004; Pao et al., 2004) examined the EGFR kinase domain, spanning exons 18–24 in patients who were initially responsive to RTKI treatment, but whose tumors progressed over time. Pao et al. (2005) examined the EGFR KD in five patients with acquired resistance to EGFR TKIs and found the presence of a second mutation in exon 20 at residue 790 (T790M). The net effect of replacing threonine with the bulkier and more hydrophobic methionine residue is loss of the TKI-binding cleft created by the threonine residue, thereby eliminating this druggable site. This mechanism is common to multiple kinases including Abl, Src, FLT3 (FMS-like tyrosine kinase 3), platelet-derived growth factor β (PDGFRβ), and the fibroblast growth factor receptor (FGFR) (reviewed in Jänne et al., 2009). Moreover, this substitution, located within the ATP-binding pocket, results in a greater affinity of the EGFR for ATP, reducing the potency of ATP-competitive drugs (Yun et al., 2008). Significantly, this mutation was not detected in tumor tissue from untreated patients, underscoring the selection for this somatic mutation by TKI treatment (Jänne et al., 2009). These findings underscore both the desire and need to carry out genomic studies on patients, which provides an advantage in screening patients for their drug sensitivities as well as their potential and/or eventual drug resistance (Jänne et al., 2009).

In addition to the acquired resistance in TKI-sensitive tumors stemming from the generation of secondary mutation(s) in the EGFR, additional mechanisms of acquired resistance have been described. Two such examples are overexpression of the MET receptor or of its ligand, hepatocyte growth factor (HGF), accounting for acquired resistance in a small percentage of tumors (Bean et al., 2007; Engelman et al., 2007). Additional studies using cell culture models of EGFR-acquired resistance have confirmed that MET overexpression and phosphorylation can compensate for loss of EGFR (Mueller, Yang, Haddad, Ethier, & Boerner, 2010). In this case, it was shown that MET served as a coreceptor for the EGFR and that the physical link between these two proteins resulted in MET activation in the absence of HGF, but in the presence of c-Src kinase activity (Mueller et al., 2010). A study of gefitinib-resistant cell lines and human lung adenocarcinoma specimens showed that HGF overexpression (coupled with MET activation) leads to PI-3 kinase pathway restoration in the absence of MET amplification or T790M mutation of the EGFR (Yano et al., 2008). An important observation was that HGF expressed by tumor stromal cells affects gefitinib resistance in mutant EGFR-expressing tumor cells (Yano et al., 2008). This underscores the role the TME (see above) plays in contributing to acquired resistance in what is referred to as noncell autonomous drug resistance mechanisms vs cell autonomous mechanisms, the latter occurring independently of cells in the TME, alterations in drug metabolism, angiogenesis, epigenetic changes, or other considerations (Jänne et al., 2009; Ji, 2010).

4.1 Lapatinib, a Dual Kinase Inhibitor of EGFR and HER2, and Afatinib, a Covalent ErbB1 RTKI

The RTKI lapatinib (Tykerb®) is a dual-action TKI that inhibits both the EGFR (ErbB1) and HER2 (ErbB2). The ErbB2/Neu receptor is overexpressed in 25% of all breast cancers where it typically servers as the primary driver of tumor cell growth in most of these cancers which exhibit addiction to ErbB2. Another aspect of ErbB2 is it lacks a known ligand. Owing to its stabilized conformation in the ligand-activated state, overexpression of the ErbB2 gene leads to ligand-independent heterodimerization, tonic stimulation of signaling pathways, and cell proliferation (De Keulenaer, Doggen, & Lemmens, 2010). In ErbB2-amplified cells, the principal signaling unit is comprised of ErbB2/ErbB3/PI3K complexes leading to Akt activation (Junttila et al., 2009). Lapatinib inhibits both ligand-dependent and -independent signaling and, as seen for the EGFR RTKIs erlotinib and gefitinib, can induce acquired resistance following its initial clinical benefit. As with the EGFR the gatekeeper residue T798M analogous to T790M in the EGFR gatekeeper residue is responsible (Roskoski, 2014). Rexer and coworkers demonstrated that stable expression of breast cancer cell lines with HER2 containing the gatekeeper mutation T798M resulted in resistance to lapatinib and trastuzumab (described below) and overexpression of the EGFR ligands EGF, TGFα, amphiregulin, and HB-EGF. These cells were sensitized to trastuzumab treatment by coaddition of the anti-EGFR antibody cetuximab or lapatinib consistent with increased EGFR ligands as the cause for drug resistance (Rexer et al., 2013). Afatinib (Gilotrif®) is an irreversible EGFR TKI that covalently inserts into the active site at residue Cys797. Afatinib is capable of irreversibly inhibiting the gatekeeper mutation T790M in the EGFR and irreversibly inhibits ErbB2 (Cys805) and ErbB4 (Cys803) (Roskoski, 2014).

4.2 Lapatinib-Induced Kinome Reprogramming and Its Role in Resistance

As indicated, TKI or RTKI inhibition is beneficial following its initiation but that resistance to the drug usually develops. Cell adaptation responses whereby gatekeeper and other residues in the ATP binding site/activation loop are mutated, limit drug efficacy. In addition to directly impacting the targeted kinase one often observes a compensation for this loss via the induction of alternative kinases or amplification of ligands for alternative growth factor receptors enabling a bypass of the inhibited pathway. The application of combination therapies has been proposed as a means around this phenomenon. As detailed above, ErbB2 is overexpressed in 25% of breast cancers where it heterodimerizes with ErbB3 leading to ErbB3 upregulation as a contributor to lapatinib resistance (Garrett et al., 2011). This is often accompanied by increases in multiple receptor and nonreceptor tyrosine kinases contributing to resistance ranging from IGF-1R, FGFR2, MET, FAK, and Src family kinases (Rexer & Arteaga, 2012). Stuhlmiller and coworkers observed that tumors are capable of evading the long-term effects of kinase-targeting drugs by launching an adaptive kinase upregulating alternative kinases or by overcoming inhibition by reactivating the targeted pathway (Stuhlmiller et al., 2015). They refer to this as "adaptive kinome reprogramming" and in the case of ErbB2 signaling showed lapatinib treatment of breast cancer cell line. Given the heterogeneity of the adaptive kinome response in different cell lines, multiple alternative inhibitors would have to be used during the course of intermittent therapy. To circumvent this issue, these investigators tested inhibitors of factors that modify or associate with chromatin (epigenetic enzymes) and determined that the BET family of bromodomains (Delmore et al., 2011) were inhibited by JQ1 and caused the suppression of lapatinib-induced kinome reprogramming. This suppression of lapatinib-induced kinase expression blocked cell growth and caused lapatinib inhibition to be durable response (Stuhlmiller et al., 2015).

Related to kinome reprogramming, a recent study examined the proteome, kinome, and phosphoproteome of ErbB2 overexpressing lapatinib-resistant breast cancer cells employing mass spectrometric analysis. The results obtained both confirmed the occurrence of kinome alterations including overexpression of AXL kinase and reactivation of PI3K and further extended these findings with the demonstration that lapatinib resistance is associated with phosphorylation-based reprogramming of glycolysis (Ruprecht et al., 2017). Lapatinib-sensitive and -resistant cells have been investigated by numerous laboratories and shown to undergo metabolic reprogramming of glycolysis during resistance (Komurov et al., 2012). However, the report by Ruprecht and coworkers revealed a considerable amount of glycolytic enzyme posttranslational modification occurred following the inhibition of phosphorylation of the key regulator of the rate-limiting step in glycolysis, 6-phosphofructo-2-kinase/fructose-2,6-biphosphatase 2 (PFKFB2) on S-466 (Ruprecht et al., 2017). Of particular note is that the metabolic reprogramming observed was also supportive of the invasive/metastatic phenotype including the release of large amounts of glutamate, a known stimulator of growth and invasion (Li & Hanahan, 2013). This may have therapeutic ramifications in the future.

5. EPIGENETIC MECHANISMS OF RESISTANCE

Epigenetic alterations have also been shown to affect resistance mechanisms in addition to their well-known effects on tumor induction and development. Histone acetyltransferases acetylate histone N-terminal lysine residues promoting chromatin expansion and transcription factor access to promoter regions. Histone deacetylases (HDACs) catalyze the removal of acetyl groups from histone lysines resulting in DNA/histone complex compaction that blocks transcription factor access to binding sites decreasing gene transcription. Blockade of this modification with HDAC inhibitors favors growth arrest, differentiation, and apoptosis (Bolden, Peart, & Johnstone, 2006). Consequently, HDAC inhibitors such as vorinostat have antitumor activity and are effective as cancer therapeutic drugs (Lane & Chabner, 2009).

Epigenetic mechanisms may further participate in RTKI resistance mechanisms. One example is the EGFR which along with many other RTKs requires the chaperone protein heat shock protein 90 (Hsp90) for its proper folding and function. The HDAC inhibitor LBH589 (panobinostat) increases Hsp90 acetylation thereby decreasing its association with EGFRs, causing downregulation of survival signaling proteins and inducing cell death (Edwards, Li, Atadja, Bhalla, & Haura, 2007). Accordingly, the EGFR is sensitive to the actions of HDAC inhibitors. Consistent with this, in cells lacking EGFR dependence, LBH589 has a negligible effect on apoptosis causing cell cycle arrest instead. A 10-fold increase in LBH589 dose was required to deplete EGFR and Akt in cells lacking EGFR mutations. Cotreatment of cells with the EGFR TKI erlotinib and LBH589 resulted in synergistic actions on lung cancer cells dependent on EGFRs for growth and/or survival. This suggests that EGFR mutation status may be predictive of a positive response to LBH589 and other HDAC inhibitors (Edwards et al., 2007).

Taken together, these observations underscore the notion that drug-resistant cell populations may be selected via multiple mechanisms ranging from drug efflux, modulation of drug metabolism, secondary mutation of the target protein, induction of alternate signaling pathways, and the induction of epigenetic mechanisms (Trumpp & Wiestler, 2008). An additional mechanism to consider is the selection of drug refractory cancer stem cell populations or cancer-initiating cells; their existence also underscores the well-known cellular heterogeneity present within a tumor that enhances a tumor's ability to adapt to a changing environment (Dannenberg & Berns, 2010). Consistent with the idea that cancer cell populations within a tumor are heterogeneous, Sharma et al. (2010) recently described a subpopulation of PC9 cells (an EGFR mutant NSCLC cell line) that were reversibly drug tolerant and labeled as "drug-tolerant persisters" (DTPs). These cells remained viable under conditions that killed-off the majority of cell populations. DTPs were detected following expansion of single drug-sensitive cells and their phenotype remained reversible. Because DTPs occur at frequencies higher than expected as a result of mutation, it was reasoned that epigenetic regulatory mechanisms may be responsible (Sharma et al., 2010). While DTPs are quiescent cells, owing to their heterogeneity, a small percentage (~20%) of these cells exhibit normal proliferation responses when grown in the presence of drug and were thereby termed "drug-tolerant expanded persisters" (DTEPs).

In an effort to define the underlying mechanisms of the drug-tolerant state, Sharma et al. determined that the resistant cells retained the sensitizing EGFR mutation and did not acquire the T790M "gatekeeper" mutation or MET gene amplification, suggesting an alternative modification may be occurring (Sharma et al., 2010). Using genome-wide gene expression analysis of parental PC9, DTP, and DTEP cells, significant expression differences were identified among the three cell lines. The DTPs and DTEPs exhibited a single gene elevation, KDM5A/RBP2/Jarid1A (KDM5A), a histone H3KA demethylase (Fattaey et al., 1993; Klose et al., 2007). Importantly, silencing KDM5A in PC9 cells reduced the number of DTEPs generated in response to cisplatin challenge without affecting PC9 cell proliferation. It was thus concluded that KDM5A expression was a necessary requirement for induction of reversible drug tolerance (Sharma et al., 2010). Because KDM5A is known to interact with HDACs (Klose et al., 2007), HDAC inhibition was tested for its ability to phenocopy KDM5A knockdown in PC9 cells. Addition of the HDACI/II inhibitor, trichostatin A, caused the rapid death of DTPs and DTEPs without having an effect on parental PC9 cells and this was verified by demonstrating HDAC inhibitor cotreatment of PC-9 cells in the presence of an EGFR TKI eliminated the emergence of DTEPs, suggesting that drug-tolerant cell populations are susceptible to HDAC inhibition. More recently, several groups have reported the successful identification of selective inhibitor of the KDM5 family of histone demethylases (Gale et al., 2016; Vinogradova et al., 2016). One of these compounds, CPI-455, behaved as predicted; it inhibited KDM5, elevated levels of H3K4 trimethylation, and decreased the number of DTPs in multiple cancer cell lines (Vinogradova et al., 2016). The development of this inhibitor has confirmed that removal of the DTP subpopulation of cancer cells may further reduce relapse/acquired resistance in these cancer sites.

It is noteworthy that as an alternative to the use of HDAC inhibitors, treatment of cells with the IGF-1R TKI, NVP-AEW541 (Rosenzweig & Atreya, 2010), was capable of inhibiting the emergence of DTEPs, suggesting that IGF-1R signaling can lead to chromatin modifications resulting from altered KDM5A activity or expression. A small percentage of DTEPs harboring the T790M EGFR mutation arose during treatment of PC9 cells with NVP-AEW541 and erlotinib, consistent with mutational mechanisms being responsible for mediating the pathway to drug resistance.

5.1 Resistance to Receptor TKIs vs Receptor-Targeted Antibodies: IGF-1R

As detailed above, Abl kinase and the EGFR provide clear examples of how KD and gatekeeper mutations affect the sensitivity to drugs and the resistance to small-molecule TKIs; additional mechanisms may also be in place for these and other RTKs and nonreceptor TKs. An example highlighting this is the insulin-like growth factor-1 receptor (IGF-1R), which was at one point a major focus of targeted therapeutic strategies, and a large number of TKIs and antibodies were developed to target this receptor in various cancer sites (reviewed in Rosenzweig & Atreya, 2010). The IGF-1R is a prosurvival, antiapoptotic signaling growth factor receptor tyrosine kinase that is frequently overexpressed in cancer, with little evidence available demonstrating that it has a propensity for undergoing mutational change,

SNPs, or true gene amplification. The small-molecule, dual-kinase IGF-1R/insulin receptor (IR) TKI, BMS-754807, was reported to inhibit IGF-1R signaling in vitro and in in vivo animal models (Huang et al., 2010). Huang et al. (2010) generated two drug-resistant rhabdomyosarcoma cell lines from parental Rh41 cells: Rh41-807R with acquired resistance to BMS-754807 and Rh41-MAB391R cells with acquired resistance to an IGF-1R blocking antibody, MAB391, in an effort to determine the mechanisms responsible for the acquired resistance to TKIs and mAbs targeting the IGF-1R. By applying gene expression profiling and DNA copy number analyses both unique and common mechanisms of resistance were identified. In common, both resistant cell lines upregulated alternate signaling pathways, but the pathways induced differed in each case. PDGFRα was amplified, overexpressed, and constitutively activated in Rh41-807R cells; knockdown of PDGFRα resulted in resensitization of the cells to BMS-754807. Interestingly, AXL expression levels were upregulated in Rh41-MAB391R cells; in contrast, this pathway was downregulated in Rh41-807R cells. Although both inhibitors target the IGF-1R, their mechanisms of action significantly differ, presumably contributing to the observed distinctions in the mechanisms of acquired resistance described. Whether the mechanisms involved depend upon mutational or epigenetic pathways has not been determined. A main difference in the actions of these agents is that small-molecule TKIs are able to access all intracellular compartments, unlike mAbs, enabling them to bind to and potentially influence multiple proteins (protein kinases in particular) besides the RTK to which they are targeted. Specific to the IGF-1R, which, as mentioned, typically lacks mutations or amplification in cancer, induction of alternate compensatory pathways over mutational changes may be the more expected outcome. Acquired resistance to trastuzumab (see below) occurs whether it is administered as monotherapy or as the more common combination therapy with a standard of care chemotherapeutic (Slamon et al., 2001).

5.2 Other mAbs and Acquired Resistance: Trastuzumab

The Human EGF Receptor-2 (HER-2, erbB2/neu) is overexpressed in 20%–25% of metastatic breast cancers (Slamon et al., 1987). Trastuzumab (herceptin) is a humanized mAb directed against subdomain IV of the ErbB2 extracellular domain (ECD), that is, in current use as a targeted therapy in cases where HER2 is shown to be overexpressed (Nahta & Esteva, 2006). The mechanism by which trastuzumab action leads to tumor regression is not completely known; it is known that treatment of tumor cells with trastuzumab results in reduced ErbB2 signaling, cell cycle arrest, reduced proliferation, ErbB2 endocytosis, and downregulation (Nahta & Esteva, 2006). Mechanistically, trastuzumab binds to subdomain IV of ErbB2 to block the ligand-independent signaling mediated by ErbB2–ErbB3 heterodimers—the principal signaling unit in ErbB2-overexpressing cells (Junttila et al., 2009). Pertuzumab (Perjeta®) is a humanized monoclonal antibody that binds to subdomain II of ErbB2 and sterically hindering the binding pocket required for ligand-dependent dimerization and downstream signaling (Badache & Hynes, 2004; Franklin et al., 2004) (Fig. 3).

Whether used as monotherapy or in combination therapy, patients who initially exhibited a positive response to trastuzumab eventually exhibit acquired resistance (Slamon et al., 2001). A number of underlying mechanisms may be responsible for acquired trastuzumab resistance. One clear possibility is mutation of the ErbB2 ECD, precluding trastuzumab binding to the HER2 ECD, similar to mutational events seen in response to EGFR TKIs (see above). Alternatively, elevated EGFR:ErbB3 heterodimers, EGFR homodimers, or loss of ErbB2 could be responsible for a loss of trastuzumab sensitivity. Akt and/or PI3K activation (Yakes et al., 2002)

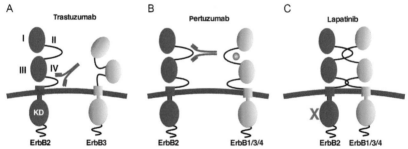

Fig. 3 Mechanism of action of ErbB2 inhibitors. (A) Trastuzumab is a humanized mAb binds to subdomain IV of ErbB2 blocking ErbB2 dimerization and ErbB2–ErbB3 complex formation which represents the major signaling unit in ErbB2 overexpressing cancer cells. This represents inhibition of ligand-independent heterodimers and signaling through PI3K and Akt. (B) Pertuzumab is a humanized mAb selective for subdomain II of ErbB2 through which dimerization occurs with other ErbB family members. Pertuzumab treatment blocks ligand-induced heterodimerization and signaling. (C) Lapatinib is a RTKI selective for ErbB1 and ErbB2. Lapatinib is an ATP-competitive inhibitor that binds to the kinase domain of Erb1 and ErbB2 to inhibit kinase activity resulting in blockade of ligand-dependent and -independent signaling. *After De Keulenaer, G. W., Doggen, K., & Lemmens, K. (2010). The vulnerability of the heart as a pluricellular paracrine organ: Lessons from unexpected triggers of heart failure in targeted ErbB2 anticancer therapy. Circulation Research, 106(1), 35–46.*

or loss of PTEN activity (Nagata et al., 2004) can also lead to trastuzumab resistance. Induction of alternate signaling pathways has been observed in trastuzumab resistance, in particular, elevation of IGF-1R signaling (Lu, Zi, Zhao, Mascarenhas, & Pollak, 2001). This is similar to the induction of the redundant MET pathway (Bean et al., 2007; Engelman et al., 2007; Jänne et al., 2009) in EGFR TKI resistance. Indeed, IGF-1R levels were found to be increased in herceptin-resistant breast cancer cell lines; treatment with the IGF-1R TKI, NVP-AEW541, restored sensitivity to trastuzumab (Browne et al., 2011). It has also been reported that trastuzumab treatment of trastuzumab-sensitive SKBR3 breast cancer cells induces insulin-like growth factor-binding protein-3 (IGFBP-3) secretion which blocks autocrine- and paracrine-expressed IGF-1/2 access to the IGF-1R causing growth inhibition (Dokmanovic, Shen, Bonacci, Hirsch, & Wu, 2011).

Induction of IGF-1R signaling has also been implicated in acquired resistance to EGFR TKIs. Generation of gefitinib-resistant A431 squamous cancer cells was associated with the loss of IGFBP-3 and IGFBP-4 expression leading to increased IGF access to the IGF-1R (Guix et al., 2008). Treatment of cells with recombinant IGFBP-3 restored gefitinib sensitivity and cotreatment of mice bearing A431 xenografts with gefitinib and an IGF-1R targeting mAb blocked tumor growth, whereas either treatment alone had no effect on tumor growth (Guix et al., 2008).

The scaffold protein IQGAP1 has been reported to interact with ErbB2 to mediate trastuzumab resistance (White, Li, Dillon, & Sacks, 2011). Herceptin-resistant human breast epithelial cells were shown to overexpress IQGAP1, with reduction of IQGAP1 levels resulting in restoration of trastuzumab sensitivity (White et al., 2011). The tumor suppressor DACH1 can downregulate EGFRs, and cyclin D1 exhibited loss of its suppressor activity in response to IGF-1 stimulation which suggests that IGF-dependent cancer cells are capable of escaping the tumor-suppressive effects of DACH1 (DeAngelis, Wu, Pestell, & Baserga, 2011).

6. IGF-1R AND DEPENDENCE RECEPTORS IN DRUG RESISTANCE

For a while the IGF-1R was the focus of a number of therapeutic strategies aimed at targeting a number of solid tumors (Rosenzweig & Atreya, 2010). The IGF-1R is an important regulator of prosurvival, antiapoptotic signaling that has surfaced as a significant target in multiple cancers. A key reason for this relates to the fact that the IGF-1R is a potent activator of Akt which is consistent with the findings that inhibition of mTOR signaling by rapamycin frequently results in the loss of feedback inhibition of IGF-1R signaling, in turn, leading to Akt activation (Wan, Harkavy, Shen, Grohar, & Helman, 2007). Similar findings to these findings have been reported by a number of laboratories and support the strategy of combination therapy with rapamycin analogs plus an IGF-1R targeting TKI or mAb (Kolb et al., 2012). In addition to its involvement in the acquired resistance to EGFR TKIs and herceptin (Jameson et al., 2011 and described above), IGF-1R signaling was reported to regulate RON receptor activation by direct physical interaction in pancreatic cancer cells, suggesting that RON activation may be involved in acquired resistance to IGF-1R therapies (Jaquish et al., 2011). IGF-1Rs have been found to be downstream of RTKs (Ahmad, Farnie, Bundred, & Anderson, 2004) and G-protein-coupled receptors with cross talk occurring at the receptor level, as well as via downstream effectors (Rozengurt, Sinnett-Smith, & Kisfalvi, 2010). For example, IGF-1R cross talk between with neurotensin receptors was shown to be Src dependent, providing evidence for IGF-1R-dependent regulation of inflammatory signaling in human colonic epithelial cells (Zhao et al., 2011).

As indicated, the IGF-1R is well known for its prosurvival antiapoptotic signaling paradigm mediated by PI3K/Akt signaling. The IGF-1R and IR were both shown to be "dependence receptors" (Boucher et al., 2010). Dependence receptors are so named based on the fact that when they are unliganded, they promote apoptosis; therefore, cells expressing them are dependent upon ligands for survival (Goldschneider & Mehlen, 2010). There are over 12 members of this family, which includes a wide variety of membrane receptor proteins including the p75 neurotrophin receptor, MET, RET, ALK, EphA4, integrin α5β1, and the androgen receptor (Goldschneider & Mehlen, 2010). They lack specific homology domains; however, many dependence receptors possess caspase cleavage sites enabling them to recruit and bind to caspases, which may reflect their mechanism of action (Mehlen, 2010).

Double knockout (DKO) cells lacking both IGF-1Rs and IRs were resistant to apoptosis via intrinsic or extrinsic pathway stimulation (Boucher et al., 2010). The dependence pathway is receptor dependent, ligand independent, and required for cells to undergo apoptosis. Therefore, reexpression of one of these receptors cells enables cells to undergo apoptosis in the absence of ligand; reexpression of a kinase-dead mutant also results in the ability of cells to undergo apoptosis (Boucher et al., 2010). The impact of this pathway with respect to cancer therapeutics is likely of significance.

If the IGF-1R is required/permissive to cell death, then mechanisms that downregulate IGF-1Rs or remove them from the cell surface may have the unwanted effect of promoting cancer cell survival. IGF-1R TKIs, on the other hand, have no effect on receptor expression, but may promote a TK-independent cell death signaling paradigm. There have been a number of reports describing the TK-independent activation of IGF-1R signaling pathways. IGF-1 treatment of smooth muscle cells was reported to result in extracellular-regulated kinase 1/2 phosphorylation/activity in the presence of IGF-1R TKIs (Perrault, Wright, Storie, Hatherell, & Zahradka, 2011). This observation further supports the concept that this activity was independent of IGF-1R tyrosine kinase signaling.

In a study of the IGF-1R, we tested the effects of IGF-F1-1, a cyclic hexadecapeptide identified by phage display screening technology and developed to be an IGFBP-mimetic (Robinson & Rosenzweig, 2006) based on its ability to block IGF-1 action. We found that IGF-F1-1, which interacts with the IGFBP-binding domain on IGF-1, inhibited IGF-1 binding to MCF-7 cells but also increased Akt activation, S-phase transition, and thymidine incorporation into DNA without stimulating IGF-1R tyrosine phosphorylation/tyrosine kinase activation. Paradoxically, these activities could be blocked by the IGF-1R/IR TKI NVP-AEW541 (Robinson & Rosenzweig, 2006). The signaling mechanisms responsible for these actions along with the TK-independent apoptotic signaling of the IGF-1R, although not well understood, could provide insight into future cancer therapeutic strategies. Based on the IGF-1R being a dependence receptor, novel approaches to cancer therapeutics that promote apoptosis via the unliganded IGF-1R may be developed. Using this paradigm, combination therapy comprised of inactivating RTKs in conjunction with either an antagonist that blocks endogenous ligand binding, the use of a decoy receptor (Pavet, Portal, Moulin, Herbrecht, & Gronemeyer, 2011) or an alternative method for ligand inactivation/removal may have merit as a future therapeutic strategy. These observations provide a rationale for targeting RTKs in a way that does not induce their endocytosis and downregulation in future therapeutic strategies.

7. CONCLUSIONS AND FUTURE PERSPECTIVE

The experience gained through administering receptor and nonreceptor TKI therapeutics has led to the realization that selecting patient populations sensitive to a particular inhibitor—based on the presence of a specific mutation or the existence of oncogene addiction—provides a key therapeutic advantage toward treatment success. There have also been attempts to predict patient populations that may become resistant to targeted therapeutics such as erlotinib (Goodin, 2006; Van Schaeybroeck et al., 2006), with women, Asian patients having adenocarcinoma and never-smokers, all being more likely to positively respond to erlotinib and gefitinib treatment due to select mutations within the EGFR TK domain or exhibiting EGFR amplification (Tsao et al., 2005). While erlotinib and gefitinib sensitivity may predict responsiveness, this does not, in turn, translate to survival. The same unpredictability has been seen with IGF-1R TKIs. Here, acquired resistance to NVP-AEW541 in a mouse model of metastatic alveolar rhabdomyosarcoma was caused by ERK reactivation and HER2 overexpression instead of the predicted induction of PDGFRα (Abraham et al., 2011). There is a possibility that this is caused by HER2:IGF-1R heterodimerization and receptor cross-phosphorylation by alternate ligands. In this particular instance, combination therapy with lapatinib and an IGF-1R TKI was more effective than either drug alone. The physical association of heterologous receptors adds a new dimension to current and future therapeutic strategies. In addition to the identification of RTK heterodimerization the future clearly holds promise for the development of new RTKIs and mAbs, as well as the identification of new cancer-related receptors belonging to the dependence receptor family. While autocrine/paracrine signaling by these receptors maintains normal cell and tissue growth and physiology, ligand overexpression will lead to tumor survival. Thus, future therapies may focus on targeting RTK ligands in order to enhance apoptotic signaling. It is clear that each tumor, owing to its heterogeneity and the contributions of the tumor microenvironment to tumor progression, will require personalized therapeutic strategies for each patient.

ACKNOWLEDGMENTS
This work was supported by NIH grant CA134845 (S.A.R.) and NIH P30 CA138313 awarded to Hollings Cancer Center.

REFERENCES
Abraham, J., Prajapati, S. I., Nishijo, K., Schaffer, B. S., Taniguchi, E., Kilcoyne, A., et al. (2011). Evasion mechanisms to Igf1r inhibition in rhabdomyosarcoma. *Molecular Cancer Therapeutics, 10*(4), 697–707.
Agafonov, R. V., Wilson, C., Otten, R., Buosi, V., & Kern, D. (2014). Energetic dissection of gleevec's selectivity toward human tyrosine kinases. *Nature Structural & Molecular Biology, 21*(10), 848–853.

Ahmad, T., Farnie, G., Bundred, N. J., & Anderson, N. G. (2004). The mitogenic action of insulin-like growth factor I in normal human mammary epithelial cells requires the epidermal growth factor receptor tyrosine kinase. *The Journal of Biological Chemistry, 279*(3), 1713–1719.

Badache, A., & Hynes, N. E. (2004). A new therapeutic antibody masks ErbB2 to its partners. *Cancer Cell, 5*(4), 299–301.

Barouch-Bentov, R., & Sauer, K. (2011). Mechanisms of drug resistance in kinases. *Expert Opinion on Investigational Drugs, 20*(2), 153–208.

Bean, J., Brennan, C., Shih, J.-Y., Riely, G., Viale, A., Wang, L., et al. (2007). MET amplification occurs with or without T790M mutations in EGFR mutant lung tumors with acquired resistance to gefitinib or erlotinib. *Proceedings of the National Academy of Sciences of the United States of America, 104*(52), 20932–20937.

Bolden, J. E., Peart, M. J., & Johnstone, R. W. (2006). Anticancer activities of histone deacetylase inhibitors. *Nature Reviews. Drug Discovery, 5*(9), 769–784.

Boucher, J., Macotela, Y., Bezy, O., Mori, M. A., Kriauciunas, K., & Kahn, C. R. (2010). A kinase-independent role for unoccupied insulin and IGF-1 receptors in the control of apoptosis. *Science Signaling, 3*(151), ra87.

Browne, B. C., Crown, J., Venkatesan, N., Duffy, M. J., Clynes, M., Slamon, D., et al. (2011). Inhibition of IGF1R activity enhances response to trastuzumab in HER-2-positive breast cancer cells. *Annals of Oncology, 22*(1), 68–73.

Burgess, M. R., Skaggs, B. J., Shah, N. P., Lee, F. Y., & Sawyers, C. L. (2005). Comparative analysis of two clinically active BCR-ABL kinase inhibitors reveals the role of conformation-specific binding in resistance. *Proceedings of the National Academy of Sciences of the United States of America, 102*(9), 3395–3400.

Dannenberg, J. H., & Berns, A. (2010). Drugging drug resistance. *Cell, 141*(1), 18–20.

DeAngelis, T., Wu, K., Pestell, R., & Baserga, R. (2011). The type 1 insulin-like growth factor receptor and resistance to DACH1. *Cell Cycle, 10*(12), 1956–1959.

Deininger, M., Buchdunger, E., & Druker, B. J. (2005). The development of imatinib as a therapeutic agent for chronic myeloid leukemia. *Blood, 105*(7), 2640–2653.

De Keulenaer, G. W., Doggen, K., & Lemmens, K. (2010). The vulnerability of the heart as a pluricellular paracrine organ: Lessons from unexpected triggers of heart failure in targeted ErbB2 anticancer therapy. *Circulation Research, 106*(1), 35–46.

Delmore, J. E., Issa, G. C., Lemieux, M. E., Rahl, P. B., Shi, J., Jacobs, H. M., et al. (2011). BET bromodomain inhibition as a therapeutic strategy to target c-Myc. *Cell, 146*(6), 904–917.

Dokmanovic, M., Shen, Y., Bonacci, T. M., Hirsch, D. S., & Wu, W. J. (2011). Trastuzumab regulates IGFBP-2 and IGFBP-3 to mediate growth inhibition: Implications for the development of predictive biomarkers for trastuzumab resistance. *Molecular Cancer Therapeutics, 10*(6), 917–928.

Edwards, A., Li, J., Atadja, P., Bhalla, K., & Haura, E. B. (2007). Effect of the histone deacetylase inhibitor LBH589 against epidermal growth factor receptor-dependent human lung cancer cells. *Molecular Cancer Therapeutics, 6*(9), 2515–2524.

Engelman, J. A., Zejnullahu, K., Mitsudomi, T., Song, Y., Hyland, C., Park, J. O., et al. (2007). MET amplification leads to gefitinib resistance in lung cancer by activating ERBB3 signaling. *Science, 316*(5827), 1039–1043.

Fattaey, A. R., Helin, K., Dembski, M. S., Dyson, N., Harlow, E., Vuocolo, G. A., et al. (1993). Characterization of the retinoblastoma binding proteins RBP1 and RBP2. *Oncogene, 8*(11), 3149–3156.

Franklin, M. C., Carey, K. D., Vajdos, F. F., Leahy, D. J., de Vos, A. M., & Sliwkowski, M. X. (2004). Insights into ErbB signaling from the structure of the ErbB2-pertuzumab complex. *Cancer Cell, 5*(4), 317–328.

Gale, M., Sayegh, J., Cao, J., Norcia, M., Gareiss, P., Hoyer, D., et al. (2016). Screen-identified selective inhibitor of lysine demethylase 5A blocks cancer cell growth and drug resistance. *Oncotarget, 7*(26), 39931–39944.

Garrett, J. T., Olivares, M. G., Rinehart, C., Granja-Ingram, N. D., Sanchez, V., Chakrabarty, A., et al. (2011). Transcriptional and posttranslational up-regulation of HER3 (ErbB3) compensates for inhibition of the HER2 tyrosine kinase. *Proceedings of the National Academy of Sciences of the United States of America, 108*(12), 5021–5026.

Goldschneider, D., & Mehlen, P. (2010). Dependence receptors: A new paradigm in cell signaling and cancer therapy. *Oncogene, 29*(13), 1865–1882.

Goodin, S. (2006). Erlotinib: Optimizing therapy with predictors of response? *Clinical Cancer Research, 12*(10), 2961–2963.

Gozgit, J. M., Wong, M. J., Moran, L., Wardwell, S., Mohemmad, Q. K., Narasimhan, N. I., et al. (2012). Ponatinib (AP24534), a multitargeted pan-FGFR inhibitor with activity in multiple FGFR-amplified or mutated cancer models. *Molecular Cancer Therapeutics, 11*(3), 690–699.

Guix, M., Faber, A. C., Wang, S. E., Olivares, M. G., Song, Y., Qu, S., et al. (2008). Acquired resistance to EGFR tyrosine kinase inhibitors in cancer cells is mediated by loss of IGF-binding proteins. *The Journal of Clinical Investigation, 118*(7), 2609–2619.

Hanahan, D., & Coussens, L. M. (2012). Accessories to the crime: Functions of cells recruited to the tumor microenvironment. *Cancer Cell, 21*(3), 309–322.

Heinrich, M. C., Corless, C. L., Demetri, G. D., Blanke, C. D., von Mehren, M., Joensuu, H., et al. (2003). Kinase mutations and imatinib response in patients with metastatic gastrointestinal stromal tumor. *Journal of Clinical Oncology, 21*(23), 4342–4349.

Huang, F., Hurlburt, W., Greer, A., Reeves, K. A., Hillerman, S., Chang, H., et al. (2010). Differential mechanisms of acquired resistance to insulin-like growth factor-I receptor antibody therapy or to a small-molecule inhibitor, BMS-754807, in a human rhabdomyosarcoma model. *Cancer Research, 70*(18), 7221–7231.

Jameson, M. J., Beckler, A. D., Taniguchi, L. E., Allak, A., Vanwagner, L. B., Lee, N. G., et al. (2011). Activation of the insulin-like growth factor-1 receptor induces resistance to epidermal growth factor receptor antagonism in head and neck squamous carcinoma cells. *Molecular Cancer Therapeutics, 10*(11), 2124–2134.

Jänne, P. A., Gray, N., & Settleman, J. (2009). Factors underlying sensitivity of cancers to small-molecule kinase inhibitors. *Nature Reviews Drug Discovery, 8*(9), 709–723.

Jaquish, D. V., Yu, P. T., Shields, D. J., French, R. P., Maruyama, K. P., Niessen, S., et al. (2011). IGF1-R signals through the RON receptor to mediate pancreatic cancer cell migration. *Carcinogenesis, 32*(8), 1151–1156.

Ji, H. (2010). Mechanistic insights into acquired drug resistance in epidermal growth factor receptor mutation-targeted lung cancer therapy. *Cancer Science, 101*(9), 1933–1938.

Junttila, T. T., Akita, R. W., Parsons, K., Fields, C., Lewis Phillips, G. D., Friedman, L. S., et al. (2009). Ligand-independent HER2/HER3/PI3K complex is disrupted by trastuzumab and is effectively inhibited by the PI3K inhibitor GDC-0941. *Cancer Cell, 15*(5), 429–440.

Klemm, F., & Joyce, J. A. (2015). Microenvironmental regulation of therapeutic response in cancer. *Trends in Cell Biology, 25*(4), 198–213.

Klose, R. J., Yan, Q., Tothova, Z., Yamane, K., Erdjument-Bromage, H., Tempst, P., et al. (2007). The retinoblastoma binding protein RBP2 is an H3K4 demethylase. *Cell, 128*(5), 889–900.

Kolb, E. A., Gorlick, R., Maris, J. M., Keir, S. T., Morton, C. L., Wu, J., et al. (2012). Combination testing (stage 2) of the anti-IGF-1 receptor antibody IMC-A12 with rapamycin by the pediatric preclinical testing program. *Pediatric Blood & Cancer, 58*(5), 729–735.

Komurov, K., Tseng, J. T., Muller, M., Seviour, E. G., Moss, T. J., Yang, L., et al. (2012). The glucose-deprivation network counteracts lapatinib-induced toxicity in resistant ErbB2-positive breast cancer cells. *Molecular Systems Biology, 8*, 596.

Lane, A. A., & Chabner, B. A. (2009). Histone deacetylase inhibitors in cancer therapy. *Journal of Clinical Oncology, 27*(32), 5459–5468.

Li, L., & Hanahan, D. (2013). Hijacking the neuronal NMDAR signaling circuit to promote tumor growth and invasion. *Cell, 153*(1), 86–100.

Lin, Y. L., Meng, Y., Jiang, W., & Roux, B. (2013). Explaining why gleevec is a specific and potent inhibitor of Abl kinase. *Proceedings of the National Academy of Sciences of the United States of America, 110*(5), 1664–1669.

Liu, P., Cheng, H., Roberts, T. M., & Zhao, J. J. (2009). Targeting the phosphoinositide 3-kinase pathway in cancer. *Nature Reviews. Drug Discovery, 8*(8), 627–644.

Lu, P., Weaver, V. M., & Werb, Z. (2012). The extracellular matrix: A dynamic niche in cancer progression. *The Journal of Cell Biology, 196*(4), 395–406.

Lu, Y., Zi, X., Zhao, Y., Mascarenhas, D., & Pollak, M. (2001). Insulin-like growth factor-I receptor signaling and resistance to trastuzumab (Herceptin). *Journal of the National Cancer Institute, 93*(24), 1852–1857.

Lynch, T. J., Bell, D. W., Sordella, R., Gurubhagavatula, S., Okimoto, R. A., Brannigan, B. W., et al. (2004). Activating mutations in the epidermal growth factor receptor underlying responsiveness of nonsmall-cell lung cancer to gefitinib. *The New England Journal of Medicine, 350*(21), 2129–2139.

Mahon, F. X., Deininger, M. W., Schultheis, B., Chabrol, J., Reiffers, J., Goldman, J. M., et al. (2000). Selection and characterization of BCR-ABL positive cell lines with differential sensitivity to the tyrosine kinase inhibitor STI571: Diverse mechanisms of resistance. *Blood, 96*(3), 1070–1079.

Marcucci, G., Perrotti, D., & Caligiuri, M. A. (2003). Understanding the molecular basis of imatinib mesylate therapy in chronic myelogenous leukemia and the related mechanisms of resistance. Commentary re: A. N. Mohamed et al., The effect of imatinib mesylate on patients with Philadelphia chromosome-positive chronic myeloid leukemia with secondary chromosomal aberrations. Clin. Cancer Res., 9: 1333-1337, 2003. *Clinical Cancer Research, 9*(4), 1248–1252.

Mehlen, P. (2010). Dependence receptors: The trophic theory revisited. *Science Signaling, 3*(151), pe47.

Mueller, K. L., Yang, Z. Q., Haddad, R., Ethier, S. P., & Boerner, J. L. (2010). EGFR/Met association regulates EGFR TKI resistance in breast cancer. *Journal of Molecular Signaling, 5*, 8.

Nagata, Y., Lan, K. H., Zhou, X., Tan, M., Esteva, F. J., Sahin, A. A., et al. (2004). PTEN activation contributes to tumor inhibition by trastuzumab, and loss of PTEN predicts trastuzumab resistance in patients. *Cancer Cell, 6*(2), 117–127.

Nahta, R., & Esteva, F. J. (2006). Herceptin: Mechanisms of action and resistance. *Cancer Letters, 232*(2), 123–138.

O'Hare, T., Eide, C. A., & Deininger, M. W. N. (2007). Bcr-Abl kinase domain mutations, drug resistance, and the road to a cure for chronic myeloid leukemia. *Blood, 110*(7), 2242–2249.

O'Hare, T., Shakespeare, W. C., Zhu, X., Eide, C. A., Rivera, V. M., Wang, F., et al. (2009). AP24534, a pan-BCR-ABL inhibitor for chronic myeloid leukemia, potently inhibits the T315I mutant and overcomes mutation-based resistance. *Cancer Cell, 16*(5), 401–412.

O'Hare, T., Walters, D. K., Stoffregen, E. P., Jia, T., Manley, P. W., Mestan, J., et al. (2005). In vitro activity of Bcr-Abl inhibitors AMN107 and BMS-354825 against clinically relevant imatinib-resistant Abl kinase domain mutants. *Cancer Research, 65*(11), 4500–4505.

Paez, J. G., Janne, P. A., Lee, J. C., Tracy, S., Greulich, H., Gabriel, S., et al. (2004). EGFR mutations in lung cancer: Correlation with clinical response to gefitinib therapy. *Science, 304*(5676), 1497–1500.

Pao, W., Miller, V. A., Politi, K. A., Riely, G. J., Somwar, R., Zakowski, M. F., et al. (2005). Acquired resistance of lung adenocarcinomas to gefitinib or erlotinib is associated with a second mutation in the EGFR kinase domain. *PLoS Medicine, 2*(3), e73.

Pao, W., Miller, V., Zakowski, M., Doherty, J., Politi, K., Sarkaria, I., et al. (2004). EGF receptor gene mutations are common in lung cancers from "never smokers" and are associated with sensitivity of tumors to gefitinib and erlotinib. *Proceedings of the National Academy of Sciences of the United States of America, 101*(36), 13306–13311.

Pavet, V., Portal, M. M., Moulin, J. C., Herbrecht, R., & Gronemeyer, H. (2011). Towards novel paradigms for cancer therapy. *Oncogene, 30*(1), 1–20.

Perrault, R., Wright, B., Storie, B., Hatherell, A., & Zahradka, P. (2011). Tyrosine kinase-independent activation of extracellular-regulated kinase (ERK) 1/2 by the insulin-like growth factor-1 receptor. *Cellular Signalling, 23*(4), 739–746.

Ray, A., Cowan-Jacob, S. W., Manley, P. W., Mestan, J., & Griffin, J. D. (2007). Identification of BCR-ABL point mutations conferring resistance to the Abl kinase inhibitor AMN107 (nilotinib) by a random mutagenesis study. *Blood, 109*(11), 5011–5015.

Rexer, B. N., & Arteaga, C. L. (2012). Intrinsic and acquired resistance to HER2-targeted therapies in HER2 gene-amplified breast cancer: Mechanisms and clinical implications. *Critical Reviews in Oncogenesis, 17*(1), 1–16.

Rexer, B. N., Ghosh, R., Narasanna, A., Estrada, M. V., Chakrabarty, A., Song, Y., et al. (2013). Human breast cancer cells harboring a gatekeeper T798M mutation in HER2 overexpress EGFR ligands and are sensitive to dual inhibition of EGFR and HER2. *Clinical Cancer Research, 19*(19), 5390–5401.

Robinson, S. A., & Rosenzweig, S. A. (2006). Paradoxical effects of the phage display-derived peptide antagonist IGF-F1-1 on insulin-like growth factor-1 receptor signaling. *Biochemical Pharmacology, 72*(1), 53.

Rosenzweig, S. A. (2009). Receptor cross-talk. In M. Schwab (Ed.), *Encyclopedia of cancer* (2nd ed., p. 3235). Heidelberg: Springer Verlag GmbH.

Rosenzweig, S. A., & Atreya, H. S. (2010). Defining the pathway to insulin-like growth factor system targeting in cancer. *Biochemical Pharmacology, 80*(8), 1115–1124.

Roskoski, R., Jr. (2014). The ErbB/HER family of protein-tyrosine kinases and cancer. *Pharmacological Research, 79*, 34–74.

Rozengurt, E., Sinnett-Smith, J., & Kisfalvi, K. (2010). Crosstalk between insulin/insulin-like growth factor-1 receptors and G protein-coupled receptor signaling systems: A novel target for the antidiabetic drug metformin in pancreatic cancer. *Clinical Cancer Research, 16*(9), 2505–2511.

Ruprecht, B., Zaal, E. A., Zecha, J., Wu, W., Berkers, C. R., Kuster, B., et al. (2017). Lapatinib resistance in breast cancer cells is accompanied by phosphorylation-mediated reprogramming of glycolysis. *Cancer Research, 77*(8), 1842–1853.

Schindler, T., Bornmann, W., Pellicena, P., Miller, W. T., Clarkson, B., & Kuriyan, J. (2000). Structural mechanism for STI-571 inhibition of Abelson tyrosine kinase. *Science, 289*(5486), 1938–1942.

Seeliger, M. A., Nagar, B., Frank, F., Cao, X., Henderson, M. N., & Kuriyan, J. (2007). c-Src binds to the cancer drug imatinib with an inactive Abl/c-kit conformation and a distributed thermodynamic penalty. *Structure, 15*(3), 299–311.

Shah, N. P., Nicoll, J. M., Nagar, B., Gorre, M. E., Paquette, R. L., Kuriyan, J., et al. (2002). Multiple BCR-ABL kinase domain mutations confer polyclonal resistance to the tyrosine kinase inhibitor imatinib (STI571) in chronic phase and blast crisis chronic myeloid leukemia. *Cancer Cell, 2*(2), 117–125.

Shah, N. P., Tran, C., Lee, F. Y., Chen, P., Norris, D., & Sawyers, C. L. (2004). Overriding imatinib resistance with a novel ABL kinase inhibitor. *Science, 305*(5682), 399–401.

Sharma, S. V., Lee, D. Y., Li, B., Quinlan, M. P., Takahashi, F., Maheswaran, S., et al. (2010). A chromatin-mediated reversible drug-tolerant state in cancer cell subpopulations. *Cell, 141*(1), 69–80.

Slamon, D. J., Clark, G. M., Wong, S. G., Levin, W. J., Ullrich, A., & McGuire, W. L. (1987). Human breast cancer: Correlation of relapse and survival with amplification of the HER-2/neu oncogene. *Science, 235*(4785), 177–182.

Slamon, D. J., Leyland-Jones, B., Shak, S., Fuchs, H., Paton, V., Bajamonde, A., et al. (2001). Use of chemotherapy plus a monoclonal antibody against HER2 for metastatic breast cancer that overexpresses HER2. *The New England Journal of Medicine, 344*(11), 783–792.

Slomiany, M. G., Black, L. A., Kibbey, M. M., Tingler, M. A., Day, T. A., & Rosenzweig, S. A. (2007). Insulin-like growth factor-1 receptor and ligand targeting in head and neck squamous cell carcinoma. *Cancer Letters, 248*(2), 269–279.

Slomiany, M. G., & Rosenzweig, S. A. (2006). Hypoxia-inducible factor-1-dependent and -independent regulation of insulin-like growth factor-1-stimulated vascular endothelial growth factor secretion. *The Journal of Pharmacology and Experimental Therapeutics, 318*(2), 666–675.

Steelman, L. S., Pohnert, S. C., Shelton, J. G., Franklin, R. A., Bertrand, F. E., & McCubrey, J. A. (2004). JAK/STAT, Raf/MEK/ERK, PI3K/Akt and BCR-ABL in cell cycle progression and leukemogenesis. *Leukemia, 18*(2), 189–218.

Stuhlmiller, T. J., Miller, S. M., Zawistowski, J. S., Nakamura, K., Beltran, A. S., Duncan, J. S., et al. (2015). Inhibition of lapatinib-induced kinome reprogramming in ERBB2-positive breast cancer by targeting BET family bromodomains. *Cell Reports, 11*(3), 390–404.

Trumpp, A., & Wiestler, O. D. (2008). Mechanisms of disease: Cancer stem cells—Targeting the evil twin. *Nature Clinical Practice. Oncology, 5*(6), 337–347.

Tsao, M. S., Sakurada, A., Cutz, J. C., Zhu, C. Q., Kamel-Reid, S., Squire, J., et al. (2005). Erlotinib in lung cancer—Molecular and clinical predictors of outcome. *The New England Journal of Medicine, 353*(2), 133–144.

Van Schaeybroeck, S., Kyula, J., Kelly, D. M., Karaiskou-McCaul, A., Stokesberry, S. A., Van Cutsem, E., et al. (2006). Chemotherapy-induced epidermal growth factor receptor activation determines response to combined gefitinib/chemotherapy treatment in non-small cell lung cancer cells. *Molecular Cancer Therapeutics, 5*(5), 1154–1165.

Vinogradova, M., Gehling, V. S., Gustafson, A., Arora, S., Tindell, C. A., Wilson, C., et al. (2016). An inhibitor of KDM5 demethylases reduces survival of drug-tolerant cancer cells. *Nature Chemical Biology, 12*(7), 531–538.

Wan, X., Harkavy, B., Shen, N., Grohar, P., & Helman, L. J. (2007). Rapamycin induces feedback activation of Akt signaling through an IGF-1R-dependent mechanism. *Oncogene, 26*(13), 1932–1940.

Weisberg, E., Manley, P. W., Breitenstein, W., Brüggen, J., Cowan-Jacob, S. W., Ray, A., et al. (2005). Characterization of AMN107, a selective inhibitor of native and mutant Bcr-Abl. *Cancer Cell, 7*(2), 129–141.

White, C. D., Li, Z., Dillon, D. A., & Sacks, D. B. (2011). IQGAP1 protein binds human epidermal growth factor receptor 2 (HER2) and modulates trastuzumab resistance. *The Journal of Biological Chemistry, 286*(34), 29734–29747.

Wilson, C., Agafonov, R. V., Hoemberger, M., Kutter, S., Zorba, A., Halpin, J., et al. (2015). Kinase dynamics. Using ancient protein kinases to unravel a modern cancer drug's mechanism. *Science, 347*(6224), 882–886.

Yakes, F. M., Chinratanalab, W., Ritter, C. A., King, W., Seelig, S., & Arteaga, C. L. (2002). Herceptin-induced inhibition of phosphatidylinositol-3 kinase and Akt is required for antibody-mediated effects on p27, cyclin D1, and antitumor action. *Cancer Research, 62*(14), 4132–4141.

Yano, S., Wang, W., Li, Q., Matsumoto, K., Sakurama, H., Nakamura, T., et al. (2008). Hepatocyte growth factor induces gefitinib resistance of lung adenocarcinoma with epidermal growth factor receptor-activating mutations. *Cancer Research, 68*(22), 9479–9487.

Yun, C.-H., Mengwasser, K. E., Toms, A. V., Woo, M. S., Greulich, H., Wong, K.-K., et al. (2008). The T790M mutation in EGFR kinase causes drug resistance by increasing the affinity for ATP. *Proceedings of the National Academy of Sciences of the United States of America, 105*(6), 2070–2075.

Zhang, J., Yang, P. L., & Gray, N. S. (2009). Targeting cancer with small molecule kinase inhibitors. *Nature Reviews. Cancer, 9*(1), 28–39.

Zhao, D., Bakirtzi, K., Zhan, Y., Zeng, H., Koon, H. W., & Pothoulakis, C. (2011). Insulin-like growth factor-1 receptor transactivation modulates the inflammatory and proliferative responses of neurotensin in human colonic epithelial cells. *The Journal of Biological Chemistry, 286*(8), 6092–6099.

CHAPTER FOUR

Extracellular-Regulated Kinases: Signaling From Ras to ERK Substrates to Control Biological Outcomes

Scott T. Eblen[1]
Medical University of South Carolina, Charleston, SC, United States
[1]Corresponding author: e-mail address: eblen@musc.edu

Contents

1. Introduction — 100
2. Identification of Extracellular-Regulated Kinases — 101
3. Ras to MAP Kinase Kinases — 102
   3.1 Ras Activation at the Plasma Membrane — 102
   3.2 Ras Activation by Sos — 105
   3.3 Ras Mutational Activation — 105
   3.4 Rafs — 106
   3.5 MAP or ERK Kinases (MEKs) — 107
4. ERK1 and ERK2 — 109
   4.1 ERK Activation by Dual Phosphorylation — 109
   4.2 ERK1 vs ERK2: Who Gets Top Billing? — 109
   4.3 Cytoplasmic Anchors and Nuclear Translocation — 110
   4.4 To Dimerization or Not to Dimerize? — 112
   4.5 Regulation of ERK Activation by Cell Adhesion — 113
5. ERK Substrates — 116
   5.1 Overview of ERK Signaling to Cellular Substrates — 116
   5.2 Substrate Interaction Domains — 116
   5.3 Substrate Specificity: Whodunit? — 117
   5.4 Substrate Identification Techniques — 118
   5.5 Feedback Phosphorylation — 119
   5.6 Signaling to Transcription Factors for Cell Proliferation — 122
   5.7 ERK Signaling to Focal Adhesions — 124
   5.8 ERK Regulation of RNA Processing — 126
   5.9 ERK Regulation of Protein Synthesis — 127
6. Concluding Remarks — 129
Acknowledgments — 130
References — 130

Abstract

The extracellular-regulated kinases ERK1 and ERK2 are evolutionarily conserved, ubiquitous serine–threonine kinases that are involved in regulating cellular signaling in both normal and pathological conditions. Their expression is critical for development and their hyperactivation is a major factor in cancer development and progression. Since their discovery as one of the major signaling mediators activated by mitogens and *Ras* mutation, we have learned much about their regulation, including their activation, binding partners and substrates. In this review I will discuss some of what has been discovered about the members of the Ras to ERK pathway, including regulation of their activation by growth factors and cell adhesion pathways. Looking downstream of ERK activation I will also highlight some of the many ERK substrates that have been discovered, including those involved in feedback regulation, cell migration and cell cycle progression through the control of transcription, pre-mRNA splicing and protein synthesis.

1. INTRODUCTION

Intracellular signaling is an important mechanism by which cells can respond to their environment and extracellular cues. Cells can sense their environment and modify gene expression, mRNA splicing, protein expression and protein modifications in order to respond to these extracellular cues. Many of these intracellular signaling pathways are affected by activation of some form of extracellular receptor, including receptor tyrosine kinases, cytokine receptors, G-protein-coupled receptors and adhesion receptors, among others. Changes at the cell surface in receptor expression or mutation, as well as ligand abundance can have profound effects on signals that are transferred throughout the cytoplasm into the nucleus. In disease states such as cancer, signaling pathways that stimulate cell proliferation and migration can be hyperactivated, while pathways that inhibit proliferation or signal apoptosis can be lost. Much of the increase in signaling from cell surface receptors in cancer cells comes from receptor tyrosine kinases, which, upon activation with a ligand or a mutation, dimerize and autophosphorylate tyrosine residues. These phosphorylated tyrosines serve as docking sites for the recruitment of adapter and signaling proteins to the receptor, generating signaling complexes at the plasma membrane and stimulating intracellular signal transduction. Intracellular, membrane-associated tyrosine kinases, such as those of the Janus kinase and Src families, act downstream of some cell surface receptors to increase signaling from multiple receptor types, enhancing tyrosine phosphorylation of signaling proteins and the formation of signaling complexes. Upon receptor endocytosis, many of these cell surface receptors continue to signal from endosomes and can either be recycled to the cell surface or degraded in lysosomes. Intracellular signaling from the plasma membrane often occurs through the activation of serine–threonine protein kinases that are part of signaling cascades that can activate subsequent kinases or directly target substrates, changing protein conformation and stimulating or inhibiting the function of the substrate. These signaling pathways can have diverse effects on the cell, including promoting the cancerous phenotypes of increased proliferation, migration, invasion, drug resistance and metastasis.

2. IDENTIFICATION OF EXTRACELLULAR-REGULATED KINASES

One of the main signaling pathways downstream of RTKs and other cell surface receptors that is involved in stimulating many aspects of the cancer cell phenotype is the extracellular-regulated kinase (ERK) pathway, primarily p44ERK1 and p42ERK2. ERKs are the original members of the larger family of mitogen-activated protein kinases (MAP kinases) that include primarily the Jun N-terminal kinases (JNKs), p38, and ERK5 MAP kinases, whose activity is stimulated in response to a variety of factors, including extracellular stimuli and cell stress (Cargnello & Roux, 2011; Hotamisligil & Davis, 2016; Plotnikov, Zehorai, Procaccia, & Seger, 2011; Schaeffer & Weber, 1999; Weston & Davis, 2007). The ERKs were first cloned by Cobb's lab in 1990 and 1991 (Boulton, Nye, et al., 1991; Boulton, Yancopoulos, et al., 1990). ERKs 1 and 2 had previously been observed by many other labs to be proteins that were quickly phosphorylated by a number of extracellular stimuli and viral infection. Studies by Weber and colleagues had observed tyrosine phosphorylation of proteins of 42 and 40 kilodaltons (kDa) upon epidermal growth factor (EGF) or platelet-derived growth factor (PDGF) stimulation or Rous sarcoma virus infection of chicken embryo fibroblasts (Cooper, Nakamura, Hunter, & Weber, 1983; Nakamura, Martinez, & Weber, 1983), and Hunter's group similarly showed tyrosine phosphorylation of two proteins around 42 kDa in response to a variety of mitogens in chicken embryo cells and mouse fibroblasts (Cooper, Sefton, & Hunter, 1984). These proteins were also observed by many other groups to be threonine and tyrosine phosphorylated in response to a variety of extracellular stimuli, particularly EGF and insulin, and by the protein phosphatase 1 and 2A inhibitor okadaic acid, and to have a microtubule-associated serine/threonine kinase activity

(Ahn & Krebs, 1990; Ahn, Weiel, Chan, & Krebs, 1990; Boulton, Gregory, & Cobb, 1991; Boulton, Gregory, et al., 1990; Boulton, Nye, et al., 1991; Haystead et al., 1989, 1990; Kohno & Pouyssegur, 1986; Ray & Sturgill, 1987, 1988). Two additional splice variants of ERK1 have been identified via inclusion of exon 7, which is in frame in some species, resulting in the larger p46ERK1b, which contains a 26-amino acid insertion within a region in the C-terminus of ERK1 and was shown to be differentially expressed and had differential regulation by MEK1 (Yung, Yao, Hanoch, & Seger, 2000). In other species, exon 7 contains a stop codon and results in the smaller, 40 kDa ERK1c, that may regulate Golgi fragmentation under certain conditions in cultured cells (Aebersold et al., 2004). Both splice variants are expressed at lower levels and with a more limited tissue distribution than p44ERK1 and appear to play only minor roles in ERK1 biological function.

3. Ras TO MAP KINASE KINASES

3.1 Ras Activation at the Plasma Membrane

ERKs are rapidly activated after cell stimulation by a number of extracellular stimuli including those which activate receptor tyrosine kinases (Fig. 1).

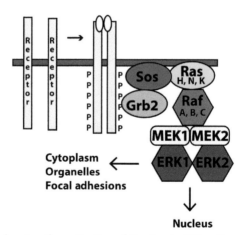

Fig. 1 Schematic showing the activation of the Ras to ERK pathway by growth factor binding to a receptor tyrosine kinase. Ligand binding induces receptor dimerization and autophosphorylation. The Grb2 adapter protein binds to activated receptors and increases association of the guanine nucleotide exchange factor Sos to Ras, resulting in Ras loading of GTP and activation. Ras enhances membrane recruitment and activation of the Raf protein kinases, which activate MEKs, leading to ERK activation. Once activated, ERKs phosphorylate cytoplasmic substrates and translocate to the nucleus to phosphorylate nuclear targets.

Binding of a growth factor to a tyrosine kinase receptor results in receptor dimerization and autophosphorylation on multiple tyrosine residues, generating binding sites for adapter and signaling proteins. One of these adapter proteins is Grb2 (growth factor receptor-bound protein 2), which recruits the Ras guanine nucleotide exchange factor Sos (Son of Sevenless) to activated receptors. The Grb2/Sos complex recruits a member of the Ras family of small GTPases to activated receptors. Ras proteins exist as a superfamily of small GTPases and include the H-Ras (Harvey sarcoma viral oncogene), K-Ras (Kirsten sarcoma viral oncogene) and N-Ras (neuroblastoma oncogene) proteins, which signal to the ERK pathway. The Ras superfamily also contains other small GTPases such as the Rho, Rab, Ran and Arf families of proteins, which each group contain several members (Colicelli, 2004; Wennerberg, Rossman, & Der, 2005). Each set of Ras family proteins regulates an aspect of cell biology, including signaling, nuclear import, control of the actin cytoskeleton and membrane trafficking, with some of these Ras proteins carrying out their function through downstream activation of protein kinases. I will focus on the action of Ras proteins at the plasma membrane that activate the ERK cascade downstream of receptor tyrosine kinases; that being K-Ras, H-Ras and N-Ras. The *K-Ras* gene has two splice variants, K-RasA and K-RasB, with K-RasB having higher expression and enzymatic activity. Knockout studies in mice have shown that both *H-Ras* and *N-Ras* are not required for overall mouse development, viability or fertility, even when both are knocked out at the same time, although fewer mice than normal survive embryogenesis in the double knockouts (Esteban et al., 2001). These results suggest that *K-Ras* is the primary *Ras* gene that is required for normal mouse development, although *N-Ras* may have some role in viability. Initial knockout of *K-Ras* in mice showed that embryos died between day 12.5 and term and that at day 11.5 they showed motor neuron cell death in the medulla and cervical spinal cord (Koera et al., 1997). Additionally, at day 15.5 these knockout mice had thin ventricular walls. These results demonstrated a role for the entire *K-Ras* gene for proper heart and neuronal development. However, this study did not take into account the potential differential effects of deletion of *K-RasA* vs *K-RasB*. A subsequent study knocked out only *K-RasA* and found that these mouse displayed normal viability and fertility (Plowman et al., 2003). These results demonstrated an essential role for the *K-RasB* gene in mouse development.

In unstimulated cells and quiescent cells, Ras proteins primarily exist in an inactive state at the plasma membrane, bound to GDP (guanosine diphosphate), having hydrolyzed the gamma phosphate off of GTP from a previous state of protein activation. In order to achieve membrane localization, which is necessary for Ras proteins to become activated and signal, Ras proteins undergo a complex series of posttranslational modifications that increase their hydrophobicity, allowing them to associate with the lipid bilayer. Ras proteins are synthesized with a CAAX motif at the C-terminus, where C is cysteine, A is an aliphatic amino acid, and X is any amino acid at this C-terminal position. This sequence serves as a recognition motif for Ras modification, first by proteolytic cleavage of the AAX sequence by Ras-converting enzyme (Rce1). This occurs for all four Ras isoforms mentioned earlier. These Ras proteins then undergo addition of a 15 carbon farnesyl group to the now C-terminal cysteine residue, catalyzed by a farnesyltransferase. This modified cysteine then undergoes methylation by a isoprenylcysteine carboxyl methyltransferase. K-RasB undergoes no additional isoprenoid modification and becomes membrane-localized with the help of the farnesyl group and a polylysine sequence just N-terminal to the terminal cysteine in a region called the hypervariable region that enhances the interaction of K-RasB with anionic phospholipids within membranes. This polylysine domain has been shown to ensure K-RasB membrane binding and is important for its transforming ability (Hancock, Paterson, & Marshall, 1990; Jackson, Li, Buss, Der, & Cochrane, 1994). Mutation of the lysines to arginines, preserving the positive charges of these residues, allowed K-RasB to maintain its full transforming potential in NIH3T3 and Rat1 cell lines in vitro. Mutation of just two of the six lysines with glutamines reduced K-RasB membrane localization by 10%, but inhibited K-RasB transforming potential by 90%, demonstrating an essential role for this polybasic sequence in K-RasB function (Jackson et al., 1994). H-Ras, N-Ras and K-RasA do not contain the polybasic region of K-RasB and undergo additional modification of a nearby cysteine, which becomes palmitoylated by a palmitoyltransferase, enhancing membrane association. Efforts in drug discovery have sought to inhibit these isoprenoid modifications to Ras proteins, particularly with the development of farnesyltransferase inhibitors, as this modification is common to all four Ras isoforms. These efforts have been largely unsuccessful due to inhibition of farnesylation stimulating the addition of geranylgeranyl isoprene to Ras in place of farnesylation, allowing for Ras membrane localization and signaling. Some reports have concluded that Ras proteins dimerize upon binding GTP and that this dimerization plays a role in activation of some downstream effectors, including Raf. Structural studies have shown that K-RasB undergoes dimerization upon binding GTP, but is a monomer in its GDP-bound state, and determined the regions involved in dimerization (Muratcioglu et al., 2015). While Ras monomers can bind to Raf, dimerized Ras may help promote the dimerization of Raf (see below), which is important for Raf activation and subsequent activation of ERK (Freeman, Ritt, & Morrison, 2013; Rajakulendran, Sahmi, Lefrancois, Sicheri, & Therrien, 2009). Since farnesyltransferase inhibitors have proven ineffective in inhibiting Ras signaling, these dimerization interfaces of Ras could be viable targets of therapeutic intervention in order to prevent Ras dimer formation and downstream signaling.

3.2 Ras Activation by Sos

Upon activation of a receptor tyrosine kinase at the membrane and subsequent recruitment of the Grb2/Sos complex, Sos associates with inactive Ras and stimulates the release of GDP from Ras, promoting GTP binding. After binding of GTP, Ras undergoes conformational changes in the Switch I and II domains in the effector lobe of Ras to create a protein–protein interaction surface that then associates with signaling proteins, activating signaling pathways from the plasma membrane and distributing them throughout the cytoplasm and into the nucleus. Ras induces activation of many its primary targets, or effector molecules, by stimulating their recruitment to the plasma membrane, where they undergo conformational changes or activation by additional membrane-associated proteins, often by protein phosphorylation. These downstream effectors include the Raf proteins (discussed later), phospholipase C, phosphoinositide 3-kinase, RalGDS (Ral guanine nucleotide dissociation stimulator), and Tiam1, among others (Rajalingam, Schreck, Rapp, & Albert, 2007).

3.3 Ras Mutational Activation

In cancer cells, enhanced growth factor receptor expression or mutation, or increased ligand abundance can lead to enhanced receptor dimerization and phosphorylation, promoting increased signaling from Ras to its effectors. Naturally occurring mutations in *Ras*, particularly at glycine 12, glycine 13, and glutamine 61, inhibit the intrinsic GTPase activity of Ras, resulting in prolonged signaling in the absence of cellular stimulation due to an inability to cleave the gamma phosphate of GTP and return Ras to its basal activation state. *Ras* mutations occur in 30% of all human cancers and up to 95% in cancers such as pancreatic cancer (Bryant, Mancias, Kimmelman, & Der, 2014).

Ras mutations, particularly in *K-rasB*, act to constitutively upregulate signaling pathways in cancer cells that lead to enhanced proliferation, migration, invasion, metastasis and survival.

3.4 Rafs

One of the main effectors of Ras activation and the primary mediators of downstream signaling to the ERKs is the Raf family of protein kinases: C-Raf (also called Raf-1), B-Raf and A-Raf. These are serine/threonine kinases whose main role in cells is phosphorylation and activation of the MAP kinase kinases MEK1 and MEK2. Rafs contain regions important for binding to Ras in their N-terminus and a C-terminal kinase domain. Upon activation of Ras, Raf proteins are recruited to the plasma membrane for activation, in part by conformational changes induced by Ras that result in C-Raf and B-Raf heterodimerization (Weber, Slupsky, Kalmes, & Rapp, 2001) and allow for other proteins to modify the activation of Raf through changes in phosphorylation and interactions with binding proteins. Initial findings suggested that the purpose of Ras in Raf activation was merely for plasma membrane localization, as addition of a CAAX sequence to the C-terminus of C-Raf resulted in full activation of the kinase (Stokoe, Macdonald, Cadwallader, Symons, & Hancock, 1994). In this study, oncogenic K-Ras did not further stimulate Raf-CAAX activation nor did a dominant-negative form of Ras inhibit its activity. Once recruited to the plasma membrane, Raf-CAAX did not stay there, but was found in the cytoplasm and associated with the cytoskeleton (Stokoe et al., 1994). However, a subsequent study found that Raf association with Ras does enhance Raf-CAAX activation (Mineo, Anderson, & White, 1997). Raf proteins undergo multiple phosphorylations that can either stimulate or inhibit kinase activity, as well as affect binding to associated proteins, including Ras, MEK, and 14-3-3 proteins (Dhillon & Kolch, 2002). Two sites whose phosphorylation is required for the kinase activity of C-Raf are Ser338 and Tyr341 (Chaudhary et al., 2000; King et al., 1998; Mason et al., 1999). Ser338 is a reported substrate for PAK1 and PAK3, while Src family kinase overexpression induces phosphorylation of Tyr341. Tyr341 phosphorylation overcomes the autoinhibition of C-Raf activity by its regulatory domain (Chaudhary et al., 2000). Importantly, B-Raf naturally contains a phosphomimetic aspartate residue at position 448, analogous to Tyr341 in C-Raf, contributing to its enhanced basal kinase activity. In addition, B-Raf mutations within the kinase domain, particularly the V600E mutation, are observed in several human cancers, particularly melanoma, and are active targets for therapeutic intervention (Davies et al., 2002; Turski et al., 2016).

3.5 MAP or ERK Kinases (MEKs)

MEK1 and MEK2 are downstream targets of the Rafs and other protein kinases that fall into the mitogen-activated protein kinase kinase category. MEK1 and MEK2 are phosphorylated by MAP kinase kinase kinases on two serine residues in their activation loop which correspond to serines 218 and 222 in MEK1, and phosphorylation of both residues is required for full activation of MEK (Alessi et al., 1994). P21-activated kinase-1 (PAK1) phosphorylation of MEK1 on Ser298 acts as a priming phosphorylation in some instances to facilitate Raf phosphorylation and activation of MEK1, particularly during cellular adhesion (see below) (Frost et al., 1997; Slack-Davis et al., 2003), and can also stimulate autophosphorylation of MEK1 on its activating sites (Park, Eblen, & Catling, 2007). MEK1 and MEK2, as well as other MAP kinase kinases that phosphorylate the JNK, p38, and ERK5 pathways, lie in a small group of kinases that can phosphorylate serine/threonine residues as well as tyrosine residues and are known as dual-specificity kinases. The only substrates of MEK1 and MEK2 are ERKs 1 and 2, providing enhanced fidelity of signaling through the ERK cascade and preventing coactivation of other MAP kinase family members or other substrates. MEK1 and MEK2 are similar in structure and sequence identity, although MEK1 appears to play a stronger role than MEK2 in activation of ERK in cells. Interestingly, both kinases contain an additional proline-rich sequence inserted into the kinase domain between residues 270–307, within kinase subdomains IX and X, that is not found in other kinases within the MAP kinase kinases (Catling, Schaeffer, Reuter, Reddy, & Weber, 1995; Schaeffer et al., 1998). This sequence in MEK1 contains phosphorylation sites not present in MEK2 and has been shown to be necessary for MEK1 to interact with C-Raf and for MEK1 activation in cells. Deletion of the proline-rich sequence from mutationally activated MEK1 makes it unable to transform Rat1 fibroblasts in culture (Catling et al., 1995), suggesting additional roles for this sequence. Seeking a functional explanation for differences between MEK1 and MEK2 mediated by this proline-rich insert within the kinase domain Schaeffer and coworkers (Schaeffer et al., 1998) performed a two-hybrid screen with MEK1 as bait and identified the first mammalian scaffold protein of the ERK pathway, MP1 (MEK Partner 1).

MP1 specifically interacted with the proline-rich sequence of MEK1, but not a proline-rich sequence deletion mutant or with MEK2 in coimmunoprecipitation experiments. In cells, MP1 enhanced the specific binding and activation of ERK1 with MEK1 and enhanced ERK1 activation of the nuclear substrate Elk1 in a luciferase reporter assay. Increased association or activation of ERK2 with MEK1 was not observed (Schaeffer et al., 1998). MP1 overexpression functionally increased ERK1 activation in response to H-RasV12 (Sharma et al., 2005). Independently, a two-hybrid screen of a protein called p14, which localizes to late endosomes and early lysosomes, identified MP1 as a binding partner of p14 (Wunderlich et al., 2001). Further study showed that p14 recruited signaling complexes containing MEK1, ERK1 and MP1 to endosomal compartments to promote signaling. P14 knockdown inhibited efficient ERK signaling to p90RSK (ribosomal protein S6 kinase) in the cytoplasm and inhibited ERK signaling to an Elk1 luciferase reporter in a dose-dependent manner, corresponding to the extent of p14 knockdown (Teis, Wunderlich, & Huber, 2002).

MEK1 contains three lysine residues just three amino acids in from the N-terminal domain and nearby hydrophobic residues that are required for binding to aspartate and hydrophobic residues in the common docking domain of ERK1 and ERK2, allowing for a direct kinase to substrate interaction (Bardwell, Frankson, & Bardwell, 2009; Tanoue, Adachi, Moriguchi, & Nishida, 2000). These residues are important in holding ERK in the cytoplasm in the absence of cellular or mutagenic stimulation of the pathway, which occurs via the presence of a leucine-rich nuclear export signal in the N-terminus of MEK, from residues 32–44 (Fukuda, Gotoh, Gotoh, & Nishida, 1996). The nuclear export signal accounts for the apparent constitutive localization of MEK in the cytoplasm that was observed in earlier studies (Lenormand et al., 1993). This export signal allows for the cytoplasmic localization of MEK via nuclear export by the CRM1 nuclear export protein, which recognizes leucine-rich sequences in proteins and contributes to their removal from the nucleus (Fukuda, Asano, et al., 1997). Mutational disruption of the nuclear export sequence of mutationally active MEK1-induced morphological changes and greatly enhanced the transforming ability of MEK1 in NIH3T3 cells (Fukuda, Gotoh, Adachi, Gotoh, & Nishida, 1997). Deletion of the nuclear export sequence in catalytically inactive MEK1 results in its nuclear translocation upon serum stimulation, while deletion of the nuclear export sequence in mutationally activated MEK1 resulted in its constitutive localization to the nucleus, even under serum-free conditions (Jaaro, Rubinfeld, Hanoch, & Seger, 1997).

These results provide evidence that upon stimulation of the ERK pathway in cells, activated MEK is also translocated to the nucleus, but quickly exported back to the cytoplasm via its nuclear export signal.

4. ERK1 AND ERK2

4.1 ERK Activation by Dual Phosphorylation

ERKs are proline-directed serine–threonine kinases that require a proline at the +1 position relative to the phosphorylation site and substrates often contain a proline at the −2 position relative to the site of phosphorylation, to give a consensus of PXS/TP or simply S/TP, where X is any amino acid (Gonzalez, Raden, & Davis, 1991). Human ERK1 and ERK2 are 87% identical in their primary protein sequences and only 2 kDa different in size, while across mammals ERKs 1 and 2 are 84% identical. ERK1 contains an additional 17-amino acid sequence near the N-terminus and two additional amino acids at the C-terminus. Evolutionarily, one or both forms of ERK, or an ancestral form or ERK, are present in most if not all species, including invertebrates (Busca, Pouyssegur, & Lenormand, 2016), and in mammals the ERKs are ubiquitously expressed. ERKs 1 and 2 are on separate genes encoded on chromosomes 16 and 22, respectively. ERKs are dually phosphorylated by MEK1 and MEK2 on a TEY sequence in the phosphorylation lip corresponding to Thr202 and Tyr204 in ERK1 and Thr183 and Tyr185 in ERK2, stimulating a 1000-fold activation of intrinsic kinase activity (Ahn et al., 1991; Payne et al., 1991). Phosphorylation of ERK by MEK results in conformational changes in both the phosphorylation lip and the P+1 binding pocket, inducing kinase activation (Canagarajah, Khokhlatchev, Cobb, & Goldsmith, 1997). Dual phosphorylation is required for full activity and mutation of either site generates a dominant-negative molecule that can inhibit signaling by binding to MEKs and downstream targets.

4.2 ERK1 vs ERK2: Who Gets Top Billing?

One of the major questions that has been asked of the kinases is whether they have specific functions within the cell, organism, or in human disease, or are they functionally redundant proteins that support each other to carry out common goals. Early studies that performed genetic knockout of *ERK2* in mice resulted in embryonic lethality (Hatano et al., 2003; Saba-El-Leil et al., 2003; Yao et al., 2003), while knockout of *ERK1* resulted in live births of mice that were fertile and initially only exhibited a deficiency in maturation of CD4+CD8+ thymocytes (Pages et al., 1999). These strong phenotypic differences suggested that ERK2 may play a more significant role in mouse development or that temporal or quantitative expression of the genes differed during development. ERK2 protein is generally expressed at higher levels than ERK1 levels in most mammalian tissues (Fremin, Saba-El-Leil, Levesque, Ang, & Meloche, 2015). Numerous studies in cells and mice using knockdown and knockout strategies have suggested specific individual biological functions for ERK1 or ERK2 based on observations when one protein or the other is lost. The question of the importance of ERK1 vs ERK2 in cells was expertly addressed in a recent review article that carefully analyzed the data in a number of papers that showed differences on cellular and mouse effects upon loss of expression of either ERK1 or ERK2 (Busca et al., 2016). After careful analysis of these studies, Busca et al. concluded that ERK1 and ERK2 were functionally redundant and that phenotypic effects that were observed upon loss of one gene/protein or the other were due to overall levels of ERK protein and activity in cells and tissues.

4.3 Cytoplasmic Anchors and Nuclear Translocation

Early studies of ERK localization determined that in serum-starved cells, ERK localized primarily in the cytoplasm, but underwent nuclear translocation upon serum or growth factor stimulation (Chen, Sarnecki, & Blenis, 1992; Gonzalez et al., 1993; Lenormand et al., 1993). Ectopic expression of ERK often leads to erroneous nuclear localization in serum-starved cells, owing to ERK's relatively small size, which allowed for it to enter the nucleus by passive diffusion through the nuclear pore (Fukuda, Gotoh, & Nishida, 1997; Gonzalez et al., 1993). Proteins of around 45 kDa or less can often pass through the nuclear pore by passive diffusion, which is independent of ATP or a nuclear transport protein, such as the importins. This mislocalization of overexpressed ERK in serum-starved cells suggested that when there was overabundant ERK expressed in a cell, mislocalization to the nucleus could be caused by saturation a cytosolic factor that was responsible for cytoplasmic retention of ERK in unstimulated cells. This lead to the discovery of proteins that act as cytoplasmic anchors for ERKs, holding ERKs in the cytoplasm until cellular stimulation and ERK activation by phosphorylation (Brunet et al., 1999; Fukuda, Gotoh, Adachi, et al., 1997; Fukuda, Gotoh, & Nishida, 1997). Coexpression of these proteins with ERK resulted in cytoplasmic retention of ERK in the unstimulated state and nuclear translocation upon cellular stimulation. These proteins include the ERK activator MEK (Fukuda, Gotoh, & Nishida, 1997) and MAP kinase phosphatase 3 (MKP-3), which can dephosphorylate ERKs to inactivate them (Brunet et al., 1999; Karlsson, Mathers, Dickinson, Mandl, & Keyse, 2004). Overexpression of active MKP-3 results in ERK inactivation, while overexpression of a phosphatase inactive mutant of MKP-3 inhibited ERK signaling to nuclear, but not to cytoplasmic substrates (Brunet et al., 1999), demonstrating a clear cytoplasmic anchor function for MKP-3 irrespective of its phosphatase activity. Additionally, inactive ERK has been shown to interact with the actin cytoskeleton via its interaction with the scaffold protein IQGAP1 (IQ motif-containing GTPase-activating protein 1), suggesting that the actin cytoskeleton can play a role in cytoplasmic anchoring of ERK (Vetterkind, Poythress, Lin, & Morgan, 2013). IQGAP1 can also recruit B-Raf, MEK1, and MEK2 to actin filaments in order to facilitate ERK activation in response to a cell stimulus (Ren, Li, & Sacks, 2007).

ERKs do not contain a classic nuclear localization or nuclear export sequence that regulates their localization within the cell; however, activation of the pathway stimulates the translocation of ERK to the nucleus through both energy-dependent and -independent mechanisms (Adachi, Fukuda, & Nishida, 1999; Ranganathan, Yazicioglu, & Cobb, 2006), and localization between the cytoplasmic and nuclear compartments is a dynamic process (Costa et al., 2006). Phosphorylation of ERK and its potential dimerization (discussed later) are not alone required for nuclear translocation of active ERK in response to cellular stimulation. Along with dual phosphorylation of ERK by MEK, ERK is further dually phosphorylated on an SPS sequence starting at amino acid 244 by casein kinase 2 (CK2), independently of the phosphorylation by MEK (Chuderland, Konson, & Seger, 2008; Plotnikov, Chuderland, Karamansha, Livnah, & Seger, 2011). These latter phosphorylations enhance ERK interaction with importin 7, a nuclear import protein, in response to EGF or TPA and enhance ERK nuclear translocation. Alanine mutation of the SPS phosphorylation sequence in ERK prevents its nuclear accumulation upon TPA stimulation (Chuderland et al., 2008), while mutation of the serines to glutamic acids stimulates nuclear accumulation in the absence of cell stimulation. Schevzov et al. (2015) further showed that the tropomyosin Tm5NM1, which associates with actin filaments, is important for both ERK interaction with importin 7 and ERK nuclear translocation after its phosphorylation by CK2.

ERK retention in the nucleus requires new protein synthesis, as stimulation of nuclear import in the presence of protein synthesis inhibitors inhibits nuclear accumulation of ERK (Lenormand, Brondello, Brunet, & Pouyssegur, 1998). The ERK-specific dual-specificity phosphatase DUSP5 has been shown to be a nuclear anchor and inactivator of ERK, terminating the nuclear ERK signal and regulating the duration of ERK in the nucleus (Kucharska, Rushworth, Staples, Morrice, & Keyse, 2009; Mandl, Slack, & Keyse, 2005; Volmat, Camps, Arkinstall, Pouyssegur, & Lenormand, 2001). Although DUSP5 is a substrate for ERK, ERK regulates the stability of DUSP5 protein through a kinase-independent mechanism. Nuclear export of ERK can be inhibited by leptomycin B (Adachi, Fukuda, & Nishida, 2000), an inhibitor of the Crm1 nuclear export protein. ERK export requires MEK, as coinjection of MEK and ERK protein into the nucleus of cells results in rapid export of both proteins, but coinjection of the ERK-binding fragment of MEK with ERK inhibited ERK export (Adachi et al., 2000).

4.4 To Dimerization or Not to Dimerize?

Khokhlatchev et al. (1998) used microinjection of TEY dually phosphorylated ERK2 to show that dual phosphorylation of ERK was sufficient to induce nuclear translocation, which was enhanced when thiophosphorylated ERK2 was used and was independent of ERK2 kinase activity, the latter of which agreed with previous studies (Gonzalez et al., 1993; Lenormand et al., 1993). Moreover, microinjection of thiophosphorylated ERK2 was also shown to enhance nuclear translocation of coinjected, unphosphorylated ERK2. These and other experiments, including gel filtration and structural analysis, lead to the conclusion that active ERK forms dimers with either active or inactive ERK, favoring the formation of ERK1/ERK1 and ERK2/ERK2 homodimers over heterodimers (Khokhlatchev et al., 1998). Dimerization was shown to occur through association of leucine residues in loop 16 (L16) in the C-terminus of associated ERK2 molecules and through association of a glutamic acid (E343) in this loop with histidine 176 in the phosphorylation lip of the partnering ERK2. However, many studies have failed to observe the dimerization of ERKs, using both purified proteins and fluorescence techniques in cells, including FRET (fluorescence resonance energy transfer) (Burack & Shaw, 2005; Kaoud et al., 2011; Lidke et al., 2010). Adachi and colleagues showed that tyrosine phosphorylation of ERK by MEK is sufficient for release of ERK and that this dissociation is required for ERK nuclear translocation (Adachi et al., 1999). Moreover, they showed that ERK molecules that are deficient in their ability to dimerize can still translocate to the nucleus and that translocation occurs through passive diffusion. They concluded that ERK dimers translocate to the nucleus by active transport and could be inhibited by the nuclear transport inhibitor wheat germ agglutinin or dominant-negative Ran, a Ras family GTPase that regulates active transport of proteins (Adachi et al., 1999). In support of the role of ERK dimers in nuclear translocation, ERK phosphorylation or dimerization mutants did not interact with Tpr, a component of the nuclear pore complex (Vomastek et al., 2008), that we previously showed was an ERK2-binding partner and substrate (Eblen et al., 2003). Knockdown or overexpression of Tpr inhibited ERK nuclear translocation, suggesting that dimerization and Tpr binding play a role in ERK nuclear translocation (Vomastek et al., 2008). However, Lidke et al. concluded that nuclear translocation of ERK was controlled by the rate of ERK phosphorylation by MEK and not dimerization, as an ERK1 dimerization mutant showed slower kinetics of phosphorylation by MEK, but accumulated in the nucleus proportionally at the same rate as wild-type ERK (Lidke et al., 2010). While ERK signaling to the nucleus has been shown to be required for gene transcription and cell cycle entry (Brunet et al., 1999), recent work has highlighted a requirement for ERK dimers in cytoplasmic, but not nuclear ERK signaling. Casar et al. found that cytoplasmic scaffold proteins and ERK dimerization were important for phosphorylation of ERK cytoplasmic, but not nuclear, substrates and that signaling to these dimer-dependent substrates in the cytoplasm was required for cell proliferation and transformation and the development of tumors in vivo (Casar, Pinto, & Crespo, 2008). Moreover, a small-molecule inhibitor of ERK dimerization, but not phosphorylation, inhibited tumor cell proliferation and induced apoptosis in tumors that contained Ras or B-Raf oncogenes, demonstrating a novel class of drugs that could combat tumors driven by these oncogenes (Herrero et al., 2015).

4.5 Regulation of ERK Activation by Cell Adhesion

In order for metastasis of solid tumor cancer cells to occur, cells lose their attachment to the underlying basement membrane and migrate and invade into the surrounding tissue. The acquisition of the ability to grow in an anchorage-independent manner allows cancer cells to enter into the vasculature to metastasize to a distant site, which is one of the hallmarks of cancer cells (Hanahan & Weinberg, 2000, 2011). In normal cells that lose their adherence to a substrate, ERK loses its ability to be activated in response to growth factor stimulation (Lin, Chen, Howe, & Juliano, 1997; Renshaw, Ren, & Schwartz, 1997) apparently as a cellular mechanism to prevent anchorage-independent growth. Upon loss of cell anchorage to a substratum, protein kinase A (PKA) becomes activated and phosphorylates PAK, resulting in a decrease in PAK activity. Inhibition of PKA in detached cells results in slower kinetics of focal adhesion kinase (FAK) dephosphorylation and a slower dissociation of FAK and the focal adhesion protein paxillin (Howe & Juliano, 2000). The requirement for adhesion-dependent activation of ERK can be overcome by the overexpression of either a kinase active mutant of FAK (Renshaw, Price, & Schwartz, 1999) or PAK (Howe & Juliano, 2000). MEK and ERK form a direct complex in the cytoplasm of unstimulated, serum-starved cells through an interaction between a positively charged and hydrophobic sequence in the N-terminal domain of MEK that interacts with a double aspartate and hydrophobic sequence in the common docking domain of ERK (Fukuda, Gotoh, Adachi, et al., 1997; Fukuda, Gotoh, & Nishida, 1997). In experiments that were designed to look at the role of the interaction between MEK and ERK in adherent vs detached cells, detached Cos-1 cells showed a loss of the interaction between MEK1 and ERK2, while MEK2/ERK2 interactions remained intact (Eblen, Slack, Weber, & Catling, 2002). Upon replating these cells onto fibronectin, there was a reengagement of MEK1/ERK2 complexes that allowed for activation of ERK. Formation of these complexes required phosphorylation of MEK1 on Ser298 by PAK1, acting downstream of activation of the small GTPase Rac (Eblen et al., 2002), a member of the Rho family of GTPases that regulates lamellipodia formation in cells and stimulates PAK1 activity. Rac1 L61, a mutationally activated form of Rac1, could stimulate the formation of ERK2 with MEK1, but not MEK2, in adherent cells, and this increase in binding could be inhibited with a dominant-negative form of PAK1 (Eblen et al., 2002). As stated earlier, this phosphorylation of MEK1 on Ser298 is also required for MEK1 activation upon cellular adhesion, facilitating the phosphorylation of MEK1 on Ser218 and 222 by Raf (Slack-Davis et al., 2003). Thus, the Rac/PAK signaling pathway acts as an adhesion-dependent sensor for MEK1 activation, acting through PAK1 phosphorylation of MEK1 to stimulate MEK1 phosphorylation by Raf and complex formation with and activation of ERK2.

In cancer cells, the mechanisms that regulate adhesion dependence can be lost, even in the absence of Ras and Raf mutations. In ovarian cancer, Ras- and Raf-activating mutations are rare in high-grade tumors, but occur in approximately two-thirds of low-grade tumors and are mutually exclusive events (Ardighieri et al., 2014; Heublein et al., 2013; Singer et al., 2003; Tsang et al., 2013; Wong et al., 2010; Zeppernick et al., 2014). Ovarian cancer metastasis occurs from the sloughing of cells from the ovary or fallopian tubes into the peritoneal cavity, which is often accompanied by the formation of intraperitoneal ascites fluid, where detached cancer cells form small spheroids in order to survive until they can attach to distant sites (Auersperg, Edelson, Mok, Johnson, & Hamilton, 1998; Labidi-Galy et al., 2017). Ascites fluid contains a complex mixture of growth factors and cytokines that can stimulate ERK activation in adherent cells. Despite the lack of Ras and Raf mutations in high-grade ovarian cancer, ERK has been shown to be activated in 91% of pleural and peritoneal effusions from ovarian cancer patients (Davidson et al., 2003), suggesting that ovarian cancers, among other cancers, have found mechanisms to overcome the requirement of anchorage dependence for activation of the ERK pathway. In support of this, we showed that in ovarian cancer cell lines there was a high incidence of strong ERK activation when adherent cells lost cell matrix attachment and were put into suspension culture (Al-Ayoubi, Tarcsafalvi, Zheng, Sakati, & Eblen, 2008). Activation of ERK in suspended cells followed a slow but sustained kinetics, with full ERK activation not occurring for 3 h, but was sustained for up to 48 h in suspended culture. Activation in suspended cells was not due to loss of phosphatase activity toward ERK, as addition of U0126 to suspended cells quickly inhibited the levels of phosphorylated ERK. Instead, enhanced ERK activation in suspended cells was caused by stimulation of an autocrine loop that was independent of both FAK activity and PAK phosphorylation of MEK1 on Ser298. Replating of cells onto fibronectin resulted in a gradual loss of ERK activation (Al-Ayoubi et al., 2008), similar to the normal kinetics of ERK activity in noncancer cells upon replating onto fibronectin after suspension (Eblen et al., 2002, 2004; Slack-Davis et al., 2003). Even in ovarian cell lines were there was a low requirement for ERK activity for adherent cell proliferation, as judged by proliferation in the presence of the MEK inhibitor U0126, ERK became active with loss of adhesion and addition of U0126 to cells in soft agar strongly inhibited anchorage-independent colony formation (Al-Ayoubi et al., 2008). In this case, ERK activation in the loss of cell adhesion was stimulated through a cell autonomous mechanism, as conditioned media from suspended cells was able to activate ERK in adherent cells. In ovarian and other tumors, factors secreted into the tumor microenvironment likely play a role in anchorage-independent ERK activation, promoting cell survival and anchorage-independent growth.

5. ERK SUBSTRATES

5.1 Overview of ERK Signaling to Cellular Substrates

Activation of ERKs occurs within minutes of mitogen activation of the cell and can be constitutively activated by mutational activation of growth factor receptors, Ras or Raf. Since their initial discovery, ERK1 and ERK2 have been shown to phosphorylate a plethora of cellular substrates, with over 250 having been identified through various mechanisms, discussed later, and phosphorylation of these substrates regulates the numerous biological functions that ERKs regulate (Fig. 2). Yoon and Seger (2006) published a comprehensive list of substrates in 2006, but in the past 12 years that list of substrates has grown, due in part through the progress of mass spectrometry techniques that can identify global changes in the phosphoproteome upon activation or inhibition of the ERK pathway. Unal, Uhlitz, and Bluthgen (2017) have established an online database of substrates to aid investigators in keeping up with the progress of ERK substrate identification. For the remainder of this chapter, I will discuss how ERKs have been shown to interact with their substrates, techniques that have been used for substrate identification and discuss how ERK phosphorylation of some specific substrates have been shown to regulate cell signaling and cell function.

5.2 Substrate Interaction Domains

The association of ERK with its substrates through binding interactions has greatly facilitated the identification and understanding of how ERKs regulate their substrates. ERKs interact with many of their substrates through a DRS (D-site recruitment site) which contains two essential aspartates in the common docking domain (CD domain) and a hydrophobic grove, which

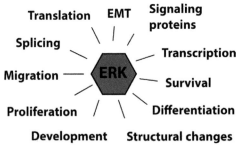

Fig. 2 ERK signaling generates multiple biological responses through direct phosphorylation of its protein substrates.

interact with positively charged and hydrophobic residues in the kinase-interacting motif domain of the substrate, activator (i.e., MEK), and phosphatases, as well as the FRS (F-site recruitment site), which interacts with the FXF/FXFP motif in the F-site or DEF (docking site for ERK, FXF) domain of the substrate or binding partner (Bardwell et al., 2009; Burkhard, Chen, & Shapiro, 2011; Fantz, Jacobs, Glossip, & Kornfeld, 2001; Jacobs, Glossip, Xing, Muslin, & Kornfeld, 1999; Lee et al., 2004; Roskoski, 2012; Sheridan, Kong, Parker, Dalby, & Turk, 2008; Tanoue et al., 2000). These two interaction domains in substrates associate with ERK in different ways: the DEF domain binds ERK near the catalytic cleft and the D domain binds on the opposite side of the kinase. Fernandes, Bailey, Vanvranken, and Allbritton (2007) used substrate peptides to show that addition of a docking site to a peptide enhances the affinity of the peptide for ERK by 200-fold. Careful analysis of protein sequences in putative substrates has been useful in the understanding and characterization of the binding properties of ERK with its substrates, activators such as MEK, and phosphatases that serve in inactivate ERK. Some ERK substrates contain one or both of these binding domains, while others contain neither. However, the presence of binding domains in ERK substrates enhances the ability of ERK to phosphorylate the substrate, as mutation of these sites often lowers their phosphorylation by ERK due to loss of enhanced kinase/substrate interactions (Murphy, MacKeigan, & Blenis, 2004; Murphy, Smith, Chen, Fingar, & Blenis, 2002). Individual mutation of the two substrate-binding domains in ERK can directly compromise the signaling to a specific set of cellular targets that interact with each domain on ERK (Dimitri, Dowdle, MacKeigan, Blenis, & Murphy, 2005).

5.3 Substrate Specificity: Whodunit?

ERKs share their consensus phosphorylation motif, the PXS/TP or S/TP motif, with other members of the MAP kinase family, including the JNKs, p38, and ERK5 MAP kinases. Additionally, other kinases, such as the cyclin-dependent kinases, are proline-directed kinases as well. The presence of the full consensus sequence can often give insight into the potential presence of an ERK phosphorylation site, but does not mean that a given site is phosphorylated by ERK, or any other MAP kinase. Indeed, some S/TP motifs are in the consensus motifs for other kinases and are phosphorylated based on amino acids that surround the serine or threonine that becomes phosphorylated. Chemical inhibitors for individual MAP kinase pathways and phospho-specific antibodies to an individual site are extremely useful in determining the particular kinase pathway that is inducing phosphorylation of a site after stimulation with a given agonist or activation of a signaling pathway. For example, we identified the alternative splicing factor Rbm17 (RNA-binding motif protein 17), also known as SPF45 (splicing factor 45 kDa), as a novel substrate of ERK with phosphorylation occurring on Ser222, a residue that fell within a PRSP sequence, and Thr71, which is in a VDTP sequence, without the proline at the -2 position relative to the threonine phosphorylation site (Al-Ayoubi, Zheng, Liu, Bai, & Eblen, 2012). Phosphorylation of Ser222 of SPF45 was relatively constitutive in cells, but could be induced further by a variety of cellular stimuli in an ERK-dependent manner, whereas Thr71 phosphorylation was absent in resting cells and strongly stimulated by ERK activation. The preponderance of Ser222 phosphorylation could be due to a phosphorylation contribution by another kinase or slow turnover of phosphorylation by phosphatases. In experiments examining the specificity of the phosphorylations by ERK, and not other MAP kinases, we determined through coimmunoprecipitation and in vitro kinase assays that SPF45 was also bound and phosphorylated by JNK and p38. So which one is the true kinase that phosphorylated SPF45? Interestingly, use of chemical inhibitors specific to each kinase pathway and phospho-specific antibodies to the sites revealed that depending on the cellular stimulus, such as H_2O_2, PMA (phorbol 12-myristate 13-acetate), ultraviolet light and anisomycin, one or more MAP kinase pathways were activated and phosphorylated SPF45 on one or both phosphorylation sites. This study highlights the ability of individual extracellular stimuli to induce phosphorylation of a particular MAP kinase target and how pathway-specific inhibitors can allow one to delineate which MAP kinase pathway is targeting an individual substrate under different cellular conditions. This was also the first alternative splicing factor that was shown to be phosphorylated by multiple MAP kinase pathways.

5.4 Substrate Identification Techniques

Since the discovery of the ERK signaling pathway and the recognition of its importance in so many different biological processes, one of the major problems still being addressed is the identification of its cellular substrates and the impact of these phosphorylation events on the protein targets and biological outcomes. Several approaches have been taken to identify ERK substrates. Many substrates have been identified through their association with the ERKs after mitogen stimulation and their changes in migration on one- and two-dimensional gels. Other methods of substrate detection have included the use of pharmacological MEK inhibitors, coimmunoprecipitation of ERK-associated proteins, and two-hybrid screens to identify binding partners (Maekawa, Nishida, & Tanoue, 2002; Waskiewicz, Flynn, Proud, & Cooper, 1997). Additional substrates have been identified by phage display libraries and phosphorylation with recombinant ERK (Fukunaga & Hunter, 1997). We (Al-Ayoubi et al., 2012; Eblen et al., 2003; Eblen, Kumar, & Weber, 2007; Kumar, Eblen, & Weber, 2004; Zheng, Al-Ayoubi, & Eblen, 2010) and others (Carlson et al., 2011) have used a chemical genetics approach pioneered by Shokat to utilize an ERK2 engineered with a point mutation in the "gatekeeper" residue (Shah, Liu, Deirmengian, & Shokat, 1997) that can accept unnatural ATP analogues to specifically label and identify novel ERK targets (Al-Ayoubi et al., 2012; Eblen et al., 2003). This method was expanded by Whites group to utilize a SILAC-based (stable isotype labeling with amino acids in cell culture) proteomic approach for a large-scale identification of ERK substrates, identifying 80 potential substrates, only 13 of which were previously known (Carlson et al., 2011). Several groups have also used various proteomic approaches, including shotgun proteomics and immobilized metal affinity chromatography and 2D-DIGE (two-dimensional difference gel electrophoresis) followed by mass spectrometry, often in combination with pharmacological ERK pathway inhibition (Kosako & Motani, 2017; Kosako et al., 2009; Lewis et al., 2000; Pan, Olsen, Daub, & Mann, 2009). These global approaches primarily look for changes in overall protein phosphorylation and for sites that have a MAP kinase consensus phosphorylation motif, but verification of these proteins and phosphorylation sites requires further follow-up with experiments that directly address the ability of ERK to phosphorylate the proposed target. However, they also give a good overall view of what is happening globally to the proteome and how changes in ERK signaling affect other proteins, even though some are not direct ERK substrates. Indeed, ERKs phosphorylate and activate other protein kinases, collectively called the MAP kinase-activated kinases, including the p90RSKs, MSKs (mitogen- and stress-activated kinases), MNKs (MAP kinase-interacting kinases), and MAPKAPKs (MAP kinase-activated protein kinases) (Cargnello & Roux, 2011).

5.5 Feedback Phosphorylation

ERK activation can be transient or sustained, depending upon the cell stimulus. Part of the mechanism for controlling the duration of ERK activation is through inactivation of upstream components of the signaling pathway through direct phosphorylation by ERKs once they become activated (Fig. 3). ERK phosphorylates its upstream activator MEK1 on Ser292 and Thr386 to inhibit MEK1 activity (Brunet, Pages, & Pouyssegur, 1994; Mansour et al., 1994) and to decrease the duration of MEK1 activation in newly adherent cells (Eblen et al., 2004). The latter inactivation occurs through a mechanism involving MEK1 Ser292 phosphorylation in the proline-rich sequence, with the added negative charge of phosphorylation at this site neutralizing a neighboring arginine at position 293. Arg293 of MEK1 lies in a consensus phosphorylation sequence for PAK1, which phosphorylates MEK1 on Ser298 (Eblen et al., 2004; Frost et al., 1997; Slack-Davis et al., 2003), a priming phosphorylation required for activating Raf phosphorylation of MEK1 on serines 218 and 222 during cell adhesion (Slack-Davis et al., 2003).

Additionally, in NIH3T3 cells stimulated with PDGF, C-Raf was phosphorylated on serines 29, 43, 289, 296, 301, and 642 in a MEK-dependent manner, with all but Ser43, which is not in a serine–proline sequence, being phosphorylated by ERK (Dougherty et al., 2005). ERK phosphorylation of these sites leads to downregulation of C-Raf activity by inhibiting the Ras/Raf interaction. Alanine mutation of these sites increased C-Raf basal activity and prolonged activation in response to PDGF treatment, demonstrating their role in negative feedback regulation of C-Raf. B-Raf has also been shown to be a target of dual ERK phosphorylation on an SPKTP sequence in the C-terminus in response to B-cell antigen receptor activation in B lymphocytes (Brummer, Naegele, Reth, & Misawa, 2003), where B-Raf has been shown to be the major Raf isoform involved in ERK activation (Brummer, Shaw, Reth, & Misawa, 2002). Mutation of the

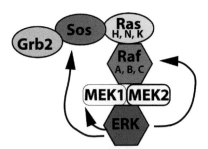

Fig. 3 Activation of ERK results in negative feedback regulation of pathway activation by direct phosphorylation of upstream components of the pathway, namely, MEK1, B-Raf and C-Raf, and Sos.

phosphorylation sites to glutamic acid or aspartic acid inhibited differentiation in PC12 cells, suggesting that these sites are important feedback regulation sites that inhibit Raf activity (Brummer et al., 2003). Morrison's group further showed that B-Raf is phosphorylated by ERK on four residues and that this phosphorylation disrupts the heterodimerization of B-Raf with C-Raf, inhibiting signaling (Ritt, Monson, Specht, & Morrison, 2010). However, feedback phosphorylation on B-Raf by ERK had a greatly reduced ability to inhibit C-Raf activity from cells expressing high-activity mutants of B-Raf, namely, V600E and G469A. These results highlight how some B-Raf mutants can stimulate increased signaling not only through intrinsic increases in kinase activity but also through evasion of negative feedback loops.

Further upstream on the pathway, ERK phosphorylates the Ras GEF Sos1. Initial studies had shown that Sos1 became phosphorylated in response to growth factor stimulation of cells, and ERK was confirmed to phosphorylate some of these sites (Cherniack, Klarlund, & Czech, 1994; Porfiri & McCormick, 1996; Rozakis-Adcock, van der Geer, Mbamalu, & Pawson, 1995), suggesting that Sos1 phosphorylation was a mechanism by which ERK could downregulate signaling from Ras. Further analysis showed that phosphorylation of Sos1 by ERK occurs among the proline-rich SH3 domains in the C-terminal domain of Sos1 on Ser1132, Ser1167, Ser1178, and Ser1193 (Corbalan-Garcia, Yang, Degenhardt, & Bar-Sagi, 1996), although additional kinases may also play a role in phosphorylation of Sos1 on these sites, as at least one site does not contain a proline at the +1 position. The proline-rich sequences play a role in association of Sos1 with the adapter protein Grb2, which binds to phosphorylated tyrosine residues on receptor tyrosine kinases. Alanine mutation of these phosphorylation sites in Sos1 demonstrated that feedback phosphorylation by ERK after serum stimulation inhibited the association of Sos1 with Grb2, suggesting a role for feedback phosphorylation in downregulation of signaling. Other studies have concluded that ERK phosphorylation of Sos1 acts not to decrease the Grb2/Sos1 complex itself, but to decrease its association with Shc and the EGF receptor (Rozakis-Adcock et al., 1995). One or both of these mechanisms may be occurring in any given cell type, but both mechanisms support a role for ERK phosphorylation of Sos1 to decrease signals coming from upstream of ERK, allowing for a downregulation of signals that induce ERK activation. Interestingly, Sos2, which has a higher affinity for Grb2 than does Sos1 (Yang, Van Aelst, & Bar-Sagi, 1995), only contains one potential ERK phosphorylation site, and the Grb2/Sos2 interaction was not affected by serum stimulation (Corbalan-Garcia et al., 1996).

5.6 Signaling to Transcription Factors for Cell Proliferation

Addition of mitogenic factors to quiescent cells induces cell cycle entry and G1 progression, leading to DNA replication and mitosis. Many of these factors result in the translocation to or activation of ERK in the nucleus (Chen et al., 1992; Gonzalez et al., 1993; Lenormand et al., 1993), which is required for cell cycle entry and progression through G1 (Brunet et al., 1999). Once in the nucleus, ERK promotes the transcriptional expression of immediate early genes (IEGs), which are genes whose expression is upregulated within the first 30–60 min after growth factor stimulation of quiescent cells. Many of the IEGs that ERK upregulates are transcription factors that are required to promote cell cycle progression, driving the transcription of genes such as *cyclin D1*. Upon entering the nucleus after growth factor or serum stimulation of cells, ERK phosphorylates ETS (E26 transformation-specific) family members such as Elk1 (Cruzalegui, Cano, & Treisman, 1999; Gille et al., 1995; Price, Rogers, & Treisman, 1995; Whitmarsh, Shore, Sharrocks, & Davis, 1995), which interacts with the serum response factor and integrates mitogen signaling into transcriptional responses, resulting in the transcription of IEGs such as *c-fos*, *fra*, *myc*, and *egr-1* (Murphy et al., 2002). Depending on the cell type and mitogen added, a transient or sustained activation of ERK has been observed, both of which result in induction of IEG transcription, but only sustained ERK activation results in cell proliferation in fibroblasts (Balmanno & Cook, 1999; Cook & McCormick, 1996; Roovers, Davey, Zhu, Bottazzi, & Assoian, 1999; Vouret-Craviari, Van Obberghen-Schilling, Scimeca, Van Obberghen, & Pouyssegur, 1993; Weber, Raben, Phillips, & Baldassare, 1997). The graded mammalian ERK activation in response to cellular stimuli is converted to a switch-like mechanism to control *c-fos* induction during the IEG response (Mackeigan, Murphy, Dimitri, & Blenis, 2005), in part through regulation of ERK nuclear translocation (Shindo et al., 2016). With only transient ERK activation, c-Fos and other IEG protein products are synthesized, but are unstable and quickly become degraded (Murphy et al., 2002). However, with mitogens that induce sustained ERK activation through G1, ERK, and p90RSK, an ERK substrate itself (Sturgill, Ray, Erikson, & Maller, 1988), phosphorylate the C-terminus of newly synthesized c-Fos protein, enhancing c-Fos stability and exposing the DEF domain, which further enhances ERK interaction with c-Fos and c-Fos-mediated activity (Murphy et al., 2002). Stabilized c-Fos is then able to induce genes required for G1 progression. Other immediate early transcription factors, such as Myc and Fra, are similarly phosphorylated to enhance their stability and their ability to induce G1-phase genes (Murphy et al., 2002).

ERK can also inhibit negative regulators of cell cycle progression to enhance proliferation and tumorigenesis. Interestingly, ERK signaling through the AP-1 transcriptional complex, both through transcriptional upregulation and phosphorylation, stimulates downregulation of antiproliferative genes that inhibit G1 progression, including *sox6*, *jun-d*, *gadd45a*, and *tob1*, among others (Yamamoto et al., 2006). Sustained activation of ERK is required throughout G1 to repress these genes, as pharmacological ERK inhibition in mid-G1 results in reexpression of these antiproliferative genes, inhibiting G1 progression and proliferation. While suppressing *tob1* (transducer of erbB2.1) transcription, the Tob1 protein is also a direct ERK target of phosphorylation on Ser152, Ser154, and Ser164 and phosphorylation inhibits Tob1 antiproliferative function (Maekawa et al., 2002; Suzuki et al., 2002), a requirement for the induction of proliferation and transformation by Ras (Suzuki et al., 2002). Two ways that Tob inhibits proliferation are by acting as a transcriptional corepressor for *cyclin D1* expression and by recruiting the CAF1–CCR4 deadenylation complex to mRNA, although this latter function is ERK phosphorylation independent (Ezzeddine, Chen, & Shyu, 2012). Another ERK target, the FOXO3a transcription factor, promotes cell cycle arrest by repression of *cyclin D* (Schmidt et al., 2002) and activation of the cyclin-dependent kinase inhibitor p27Kip1 (Dijkers et al., 2000), as well as by activating apoptosis through upregulation of *bim* and *fasL* (Yang, Xia, & Hu, 2006). Phosphorylation of FOXO3a on serines 294, 344, and 425 by ERK promotes FOXO3a association with the E3 ubiquitin ligase Mdm2, prompting FOXO3a ubiquitination and degradation, thereby removing the growth inhibitory effect of FOXO3a and supporting cell survival and tumorigenesis (Yang et al., 2008).

ERK can also inhibit other negative regulators of epithelial cell proliferation, such as signaling from the transforming growth factor beta (TGFβ) receptors. TGFβ is growth inhibitory to normal epithelial cells; however, oncogenic Ras can often bypass TGFβ-mediated growth inhibition (Houck, Michalopoulos, & Strom, 1989; Longstreet, Miller, & Howe, 1992; Valverius et al., 1989). One of the main pathways that the serine–threonine kinase TGFβ receptors send antiproliferative signals is through the SMAD family of transcription factors. TGFβ receptors phosphorylate the R-SMADs (receptor-SMADs) to induce their dimerization and heterotrimerization with the co-SMAD SMAD4, and the complex then translocates to the nucleus to regulate gene transcription. Oncogenic

Ras, and to a lesser extent EGF, induces ERK phosphorylation of R-SMADs in their linker region, inhibiting R-SMAD nuclear accumulation (Kretzschmar, Doody, Timokhina, & Massague, 1999). Alanine mutation of these phosphorylation sites restores TGFβ growth inhibitory responses in Ras-transformed cells, demonstrating a direct mechanism for Ras signaling through ERK in the inhibition of TGFβ cellular effects.

Positive regulators of cell survival are also directly affected by ERK phosphorylation. The Bcl-2 family protein myeloid leukemia cell 1 (Mcl-1) enhances cell survival and contains a PEST sequence that makes it susceptible to ubiquitination and proteosomal degradation. Mcl-1 is phosphorylated by both ERK (Domina, Smith, & Craig, 2000; Domina, Vrana, Gregory, Hann, & Craig, 2004; Nifoussi et al., 2012) and JNK (Morel, Carlson, White, & Davis, 2009) on Thr163 in the PEST sequence and phosphorylation by JNK in response to UV light serves as a priming phosphorylation for GSK3b, enhancing Mcl-1 degradation by the proteosome (Morel et al., 2009). However, in cancer cells that have suppressed this degradation pathway, Thr163 phosphorylation by ERK enhances Mcl-1 stability and cellular resistance to chemotherapeutics such as Ara-C, etoposide, vinblastine, and cisplatin (Nifoussi et al., 2012). An isoform of Bim, a proapoptotic member of the Bcl2 family, is also a substrate of ERK. There are three splice variants of Bim, with the longest, Bim(EL), acting as a substrate for ERK upon serum withdrawal or activation of Raf-1 (Weston et al., 2003). Serum withdrawal in CC139 lung fibroblasts caused upregulation of *bim(EL)*, which played a role in the induction of apoptosis. Activation of a ΔRaf-1/estrogen receptor chimera with estrogen reduced *bim* mRNA levels. Additionally, Bim(EL) protein was shown to contain a DEF-type docking domain for ERK and to coimmunoprecipitate with ERK (Ley et al., 2004; Ley, Hadfield, Howes, & Cook, 2005). ERK phosphorylates Bim(EL) and promotes its ubiquitination and proteasome-dependent degradation, thereby inhibiting its proapoptotic function.

5.7 ERK Signaling to Focal Adhesions

FAK plays an important role in integrating cell adhesion signals from integrins to affect signaling from growth factors. As mentioned earlier, cell adhesion is an important determinant in regulating the cellular response to growth factors in normal cells and can be overcome by activated FAK in cells that have lost cell adhesion, allowing them to activate ERK under anchorage-independent conditions. Hunger-Glasser and coworkers showed that in response to a variety of cell stimuli, such as bombesin, lysophosphatidic acid, PDGF, and EGF, but not insulin, in Swiss 3T3 cells FAK is phosphorylated on Ser910 by ERK (Hunger-Glaser, Fan, Perez-Salazar, & Rozengurt, 2004; Hunger-Glaser, Salazar, Sinnett-Smith, & Rozengurt, 2003). Thus, both factors that signal through receptor tyrosine kinases and G-protein-coupled receptors signal to FAK through ERK in this manner. Ser910 phosphorylation in response to PDGF and EGF was regulated independently of the activating Tyr397 phosphorylation site of FAK in response to these growth factors, which was downstream of PI3 kinase (Hunger-Glaser et al., 2004). Mutation of Ser910 to alanine increased the binding of FAK to the focal adhesion docking protein paxillin, suggesting that ERK phosphorylation of this site may downregulate adhesion signaling in response to growth factor stimulation (Hunger-Glaser et al., 2003). Paxillin itself is also a target for ERK phosphorylation. Stimulation of murine inner medullary collecting duct (mIMCD-3) epithelial cells with hepatocyte growth factor (HGF), an inducer of cell migration in many cell types, induced an ERK-dependent band shift in paxillin on western blots, suggesting phosphorylation (Liu, Yu, Nickel, Thomas, & Cantley, 2002), and ERK was shown to induce ERK phosphorylation of paxillin at Ser83 (Ishibe, Joly, Liu, & Cantley, 2004; Liu et al., 2002). Pretreatment with the MEK inhibitor U0126 before HGF stimulation inhibited ERK and FAK activation. Inhibiting paxillin phosphorylation by ERK or the interaction between paxillin and ERK not only inhibited FAK and Rac activation but also inhibited cell spreading and migration. These results demonstrate that a major mediator of HGF signaling from the cMet receptor at the cell surface to induce cell migration occurs by signaling of ERK to paxillin in focal adhesions to help bring about FAK and Rac activation as well as focal adhesion turnover. Another study found that in response to a variety of extracellular stimuli, ERK also phosphorylates paxillin on Ser130 and that this phosphorylation acts as a priming event for paxillin phosphorylation on Ser126 (Cai, Li, Vrana, & Schaller, 2006). Phosphorylation of these two sites on paxillin was required for cell spreading, as a double-alanine mutant inhibited spreading in paxillin null cells and cell spreading induced by lipopolysaccharide in RAW264.7 cells. Moreover, this paxillin mutant slowed neurite outgrowth in PC12 cells in response to nerve growth factor. This study shows additional regulation of ERK in cell spreading through phosphorylation of the focal adhesion protein paxillin.

5.8 ERK Regulation of RNA Processing

Pre-mRNA produced by gene transcription undergoes processing through constitutive and alternative splicing to produce the mature transcripts that are exported to the cytoplasm and translated into protein. Alternative splicing is a major mechanism of generating proteins with differing function from the same gene, and approximately 95% of all genes have their message undergo RNA splicing (Pan, Shai, Lee, Frey, & Blencowe, 2008; Wang et al., 2008). Splice-site mutations contribute to approximately 10%–15% of the total number of somatic mutations known in cancer and up to 50% of human genetic diseases (Matlin, Clark, & Smith, 2005). Alterations in alternative mRNA-splicing factor expression have also been shown in numerous cancers and can strongly influence the apoptotic response (Schwerk & Schulze-Osthoff, 2005). Matter, Herrlich, and Konig (2002) were the first to show that ERK phosphorylation of splicing factors could affect alternative mRNA splicing. Ras signaling induces the inclusion of exon v5 of CD44 into mature CD44 transcripts. The RNA-binding protein Sam68 (Src-associated in mitosis 68 kDa protein), a member of the STAR (signal transduction activator of RNA) family of RNA-binding proteins (Lukong & Richard, 2003), was shown to be required for exon v5 inclusion and required ERK phosphorylation of Sam68 on Ser58, Thr71, and Thr84 for this effect (Matter et al., 2002). Overall, ERK was found to regulate the phosphorylation of eight residues on Sam68. A key component of cancer cell metastasis is the epithelial to mesenchymal transition (EMT), which allows cells to lose their cell–cell contacts, to evade anoikis (apoptosis caused by the loss of cell anchorage), and to increase their migratory and invasive potential (Polyak & Weinberg, 2009; Thiery & Sleeman, 2006). Sam68 regulates EMT in colon cancer cells through regulation of an additional splicing regulator, the SF2/ASF protooncogene (Valacca et al., 2010). Epithelial-derived signals stimulate ERK phosphorylation of Sam68 and induce changes in the protein levels of SF2/ASF through an alternative splicing-activated nonsense-mediated mRNA decay mechanism. As part of this mechanism, Sam68 was shown to bind directly to the 3′ untranslated region (UTR) of SF2/ASF and inhibit splicing of a 3′ UTR intron. Mutation of the eight ERK phosphorylation sites on Sam68 affected the ability of Sam68 to modulate SF2/ASF alternative splicing (Valacca et al., 2010). Additionally, Sam68 has been shown to translocate to the cytoplasm in spermatocytes in an ERK-dependent manner, likely involving its direct phosphorylation by ERK, to associate with polysomes, and posttranscriptionally to regulate a subset of mRNAs involved in spermatogenesis (Paronetto et al., 2006). This serves as an example of how ERK phosphorylation of a single substrate may not only affect alternative splicing but also protein translation.

As mentioned earlier, we recently showed the phosphorylation of the alternative splicing factor SPF45 by ERK on Thr71 and Ser222. ERK phosphorylation affected SPF45-dependent exclusion of exon 6 of the pre-mRNA for the *fas* death receptor, a known SPF45 target (Corsini et al., 2007), in a minigene assay in cells (Al-Ayoubi et al., 2012). Exon 6 encodes the transmembrane domain of Fas, and its exclusion generates a soluble decoy receptor that inhibits Fas/FasL-mediated apoptosis (Cheng et al., 1994). ERK phosphorylation of SPF45 also regulated the incorporation of extra domain A (EDA) region into mature fibronectin mRNA transcripts, regulating cell adhesion to a fibronectin extracellular matrix (Al-Ayoubi et al., 2012). EDA inclusion occurs less in adult tissues, but is mainly seen in embryonic tissue, fibrotic liver, and wound healing (Blaustein, Pelisch, & Srebrow, 2007), and we have shown that SPF45 overexpression enhances cell migration (Liu et al., 2013). We originally identified SPF45 as an ERK-associated substrate in suspended ovarian cancer cells using a gatekeeper ERK2 mutant and were able to demonstrate that ERK specifically phosphorylated SPF45 in suspended cells, an action that was inhibited with the MEK inhibitor U0126 (Al-Ayoubi et al., 2012). Overexpression of SPF45 or a phosphomimetic mutant of SPF45 inhibited cell proliferation, while cells expressing a double-alanine mutant of SPF45 grew at the same rate as vector control cells, suggesting that ERK phosphorylation of SPF45 slows proliferation. While antiproliferation is initially counterintuitive in the context of cancer, ovarian cancer multicellular spheroids in intraperitoneal ascites fluid survive in part by slowing their proliferation and upregulating fibronection expression to evade anoikis (Auersperg, Maines-Bandiera, & Dyck, 1997; Casey et al., 2001), both of which were observed with SPF45 overexpression and ERK phosphorylation.

5.9 ERK Regulation of Protein Synthesis

External factors such as nutrients and growth factors regulate the physiological state of the cell and control the production of the factors required for protein synthesis. Increased protein synthesis is an integral part of cellular growth and proliferation and is regulated at many levels, including inactivation of factors that repress the protein synthesis machinery and an increase in the production of ribosomal components and transfer RNAs (tRNAs). The initiation of the protein synthesis machinery is controlled in part by mTOR (mammalian target of rapamycin) (Gingras, Raught, & Sonenberg, 2001), resulting in phosphorylation of 4EBP1 (4E-binding protein 1) and derepression of translation initiation (Brunn et al., 1997). mTOR activity is upregulated by Rheb (Ras homology enriched in brain), a Ras GTPase superfamily member that acts to promote cell growth (Stocker et al., 2003). Rheb activity is inhibited by binding of the TSC1/TSC2 (tuberous sclerosis) complex, as TSC2 is a GTPase-activating protein for Rheb (Garami et al., 2003; Inoki, Li, Xu, & Guan, 2003; Tee, Manning, Roux, Cantley, & Blenis, 2003). Activated ERK binds to TSC2 and phosphorylates it on Ser664, stimulating dissociation of the TSC1/TSC2 complex and removal of its repression of Rheb, resulting in increased cell proliferation and transformation (Ma, Chen, Erdjument-Bromage, Tempst, & Pandolfi, 2005; Saucedo et al., 2003). Moreover, phosphorylation of TSC2 on S664 by ERK has been shown to be a marker for ERK-mediated mTOR activation in human tumors (Ma et al., 2007).

Accompanying an increase in protein synthesis in cells stimulated by mitogens for enhanced G1 progression, there is a parallel need for increased production of ribosomes and tRNAs to facilitate the new synthesis of proteins required for an increase in cell size and cell division. Like transcription of mRNA, increased transcription of ribosomal RNA (rRNA) and tRNA have been shown to be enhanced by external cues that promote cell growth and proliferation. An increase in rRNA transcription is required for ribosome production and is driven by RNA polymerase I, whose protein levels are relatively unchanged after mitogen stimulation (Zhao, Yuan, Frodin, & Grummt, 2003). Stimulation of serum-starved cells with growth factors such as EGF, basic fibroblast growth factor, and the phorbol ester TPA enhances the transcription of 45S rRNA through an ERK-dependent pathway, and sustained ERK activation is required to maintain rRNA transcription (Stefanovsky et al., 2001; Zhao et al., 2003). Two different transcription factors, both of which are ERK targets, have been shown to regulate the induction of pre-rRNA synthesis in response to these mitogens. TIF-1A (transcription initiation factor 1A) binds to the promoter of the 45S rRNA in a complex with RNA polymerase I and other transcription factors, including SL-1, UBF (upstream-binding factor), and Myc (Oskarsson & Trumpp, 2005). TIF-1A is phosphorylated by ERK on serines 633 and 649 upon mitogen stimulation and cooperates with these transcription factors to initiate rRNA synthesis (Zhao et al., 2003). Of these two sites, Ser649 shows greater importance; however, both are required for maximal transcriptional activation and prevention of their phosphorylation by mutation or pharmacological inhibitors of the ERK pathway inhibit rRNA transcription. The architectural transcription factor UBF binds to both the 45S rRNA promoter and throughout the 45S gene and regulates both the initiation and elongation steps of rRNA transcription. Its primary role appears to be in elongation by affecting DNA bending and the remodeling of rRNA gene chromatin (Stefanovsky, Langlois, Bazett-Jones, Pelletier, & Moss, 2006; Stefanovsky et al., 2001). After activation with EGF, ERK phosphorylates UBF on two evolutionarily conserved threonines, 117 and 201, which lie in the DNA-binding HMG (high-mobility group) boxes 1 and 2. ERK phosphorylation of UBF on these sites negatively regulates the association of UBF with linear DNA. Reversible phosphorylation of UBF by ERK allows UBF to regulate structural elements throughout the rRNA gene, enhancing the ability of RNA polymerase I to elongate rRNA transcripts (Stefanovsky, Langlois, Bazett-Jones, et al., 2006; Stefanovsky, Langlois, Gagnon-Kugler, Rothblum, & Moss, 2006; Stefanovsky et al., 2001). In addition, tRNA synthesis, which is driven by RNA polymerase III, itself an IEG, is regulated by ERK as well. The BRF1 (B-related factor 1) subunit of RNA polymerase III is an ERK-binding partner and substrate (Felton-Edkins et al., 2003). Inhibition of ERK activation decreases association of BRF1 and RNA pol III with tRNA(Leu) genes. Mutation of BRF1 to prevent ERK binding or phosphorylation inhibits the ability of RNA polymerase III to induce tRNA transcription, demonstrating a critical role for ERK in generating tRNAs for the enhancement of protein synthesis. Overall, these examples demonstrate that ERK regulates the process of cell proliferation at multiple levels through direct phosphorylation of its protein targets.

6. CONCLUDING REMARKS

Since its initial discovery, we have gained a greater understanding of how the Ras to ERK pathway is regulated, both in terms of normal regulation during development and cellular homeostasis, and in human disease, particularly in multiple forms of cancer. Many downstream substrates of ERK have been identified, giving insight into the specific mechanisms of ERK biological function. Further study of the pathway will aid in the development of novel therapies that can inhibit ERK activation or biological action and overcome the resistance problems associated with many of the current therapies that target the Ras to ERK pathway.

ACKNOWLEDGMENTS
This work was supported by Grant number 5R01CA187342 from the National Cancer Institute.

REFERENCES
Adachi, M., Fukuda, M., & Nishida, E. (1999). Two co-existing mechanisms for nuclear import of MAP kinase: Passive diffusion of a monomer and active transport of a dimer. *The EMBO Journal, 18*(19), 5347–5358.
Adachi, M., Fukuda, M., & Nishida, E. (2000). Nuclear export of MAP kinase (ERK) involves a MAP kinase kinase (MEK)-dependent active transport mechanism. *The Journal of Cell Biology, 148*(5), 849–856.
Aebersold, D. M., Shaul, Y. D., Yung, Y., Yarom, N., Yao, Z., Hanoch, T., et al. (2004). Extracellular signal-regulated kinase 1c (ERK1c), a novel 42-kilodalton ERK, demonstrates unique modes of regulation, localization, and function. *Molecular and Cellular Biology, 24*(22), 10000–10015.
Ahn, N. G., & Krebs, E. G. (1990). Evidence for an epidermal growth factor-stimulated protein kinase cascade in Swiss 3T3 cells. Activation of serine peptide kinase activity by myelin basic protein kinases in vitro. *The Journal of Biological Chemistry, 265*(20), 11495–11501.
Ahn, N. G., Seger, R., Bratlien, R. L., Diltz, C. D., Tonks, N. K., & Krebs, E. G. (1991). Multiple components in an epidermal growth factor-stimulated protein kinase cascade. In vitro activation of a myelin basic protein/microtubule-associated protein 2 kinase. *The Journal of Biological Chemistry, 266*(7), 4220–4227.
Ahn, N. G., Weiel, J. E., Chan, C. P., & Krebs, E. G. (1990). Identification of multiple epidermal growth factor-stimulated protein serine/threonine kinases from Swiss 3T3 cells. *The Journal of Biological Chemistry, 265*(20), 11487–11494.
Al-Ayoubi, A., Tarcsafalvi, A., Zheng, H., Sakati, W., & Eblen, S. T. (2008). ERK activation and nuclear signaling induced by the loss of cell/matrix adhesion stimulates anchorage-independent growth of ovarian cancer cells. *Journal of Cellular Biochemistry, 105*(3), 875–884.
Al-Ayoubi, A. M., Zheng, H., Liu, Y., Bai, T., & Eblen, S. T. (2012). Mitogen-activated protein kinase phosphorylation of splicing factor 45 (SPF45) regulates SPF45 alternative splicing site utilization, proliferation, and cell adhesion. *Molecular and Cellular Biology, 32*(14), 2880–2893.
Alessi, D. R., Saito, Y., Campbell, D. G., Cohen, P., Sithanandam, G., Rapp, U., et al. (1994). Identification of the sites in MAP kinase kinase-1 phosphorylated by p74raf-1. *The EMBO Journal, 13*(7), 1610–1619.
Ardighieri, L., Zeppernick, F., Hannibal, C. G., Vang, R., Cope, L., Junge, J., et al. (2014). Mutational analysis of BRAF and KRAS in ovarian serous borderline (atypical proliferative) tumours and associated peritoneal implants. *The Journal of Pathology, 232*(1), 16–22.
Auersperg, N., Edelson, M. I., Mok, S. C., Johnson, S. W., & Hamilton, T. C. (1998). The biology of ovarian cancer. *Seminars in Oncology, 25*(3), 281–304.
Auersperg, N., Maines-Bandiera, S. L., & Dyck, H. G. (1997). Ovarian carcinogenesis and the biology of ovarian surface epithelium. *Journal of Cellular Physiology, 173*(2), 261–265.
Balmanno, K., & Cook, S. J. (1999). Sustained MAP kinase activation is required for the expression of cyclin D1, p21Cip1 and a subset of AP-1 proteins in CCL39 cells. *Oncogene, 18*(20), 3085–3097.
Bardwell, A. J., Frankson, E., & Bardwell, L. (2009). Selectivity of docking sites in MAPK kinases. *The Journal of Biological Chemistry, 284*(19), 13165–13173.

Blaustein, M., Pelisch, F., & Srebrow, A. (2007). Signals, pathways and splicing regulation. *The International Journal of Biochemistry & Cell Biology, 39*(11), 2031–2048.

Boulton, T. G., Gregory, J. S., & Cobb, M. H. (1991). Purification and properties of extracellular signal-regulated kinase 1, an insulin-stimulated microtubule-associated protein 2 kinase. *Biochemistry, 30*(1), 278–286.

Boulton, T. G., Gregory, J. S., Jong, S. M., Wang, L. H., Ellis, L., & Cobb, M. H. (1990). Evidence for insulin-dependent activation of S6 and microtubule-associated protein-2 kinases via a human insulin receptor/v-ros hybrid. *The Journal of Biological Chemistry, 265*(5), 2713–2719.

Boulton, T. G., Nye, S. H., Robbins, D. J., Ip, N. Y., Radziejewska, E., Morgenbesser, S. D., et al. (1991). ERKs: A family of protein-serine/threonine kinases that are activated and tyrosine phosphorylated in response to insulin and NGF. *Cell, 65*(4), 663–675.

Boulton, T. G., Yancopoulos, G. D., Gregory, J. S., Slaughter, C., Moomaw, C., Hsu, J., et al. (1990). An insulin-stimulated protein kinase similar to yeast kinases involved in cell cycle control. *Science, 249*(4964), 64–67.

Brummer, T., Naegele, H., Reth, M., & Misawa, Y. (2003). Identification of novel ERK-mediated feedback phosphorylation sites at the C-terminus of B-Raf. *Oncogene, 22*(55), 8823–8834.

Brummer, T., Shaw, P. E., Reth, M., & Misawa, Y. (2002). Inducible gene deletion reveals different roles for B-Raf and Raf-1 in B-cell antigen receptor signalling. *The EMBO Journal, 21*(21), 5611–5622.

Brunet, A., Pages, G., & Pouyssegur, J. (1994). Growth factor-stimulated MAP kinase induces rapid retrophosphorylation and inhibition of MAP kinase kinase (MEK1). *FEBS Letters, 346*(2–3), 299–303.

Brunet, A., Roux, D., Lenormand, P., Dowd, S., Keyse, S., & Pouyssegur, J. (1999). Nuclear translocation of p42/p44 mitogen-activated protein kinase is required for growth factor-induced gene expression and cell cycle entry. *The EMBO Journal, 18*(3), 664–674.

Brunn, G. J., Hudson, C. C., Sekulic, A., Williams, J. M., Hosoi, H., Houghton, P. J., et al. (1997). Phosphorylation of the translational repressor PHAS-I by the mammalian target of rapamycin. *Science, 277*(5322), 99–101.

Bryant, K. L., Mancias, J. D., Kimmelman, A. C., & Der, C. J. (2014). KRAS: Feeding pancreatic cancer proliferation. *Trends in Biochemical Sciences, 39*(2), 91–100.

Burack, W. R., & Shaw, A. S. (2005). Live cell imaging of ERK and MEK: Simple binding equilibrium explains the regulated nucleocytoplasmic distribution of ERK. *The Journal of Biological Chemistry, 280*(5), 3832–3837.

Burkhard, K. A., Chen, F., & Shapiro, P. (2011). Quantitative analysis of ERK2 interactions with substrate proteins: Roles for kinase docking domains and activity in determining binding affinity. *The Journal of Biological Chemistry, 286*(4), 2477–2485.

Busca, R., Pouyssegur, J., & Lenormand, P. (2016). ERK1 and ERK2 map kinases: Specific roles or functional redundancy? *Frontiers in Cell and Development Biology, 4*, 53.

Cai, X., Li, M., Vrana, J., & Schaller, M. D. (2006). Glycogen synthase kinase 3- and extracellular signal-regulated kinase-dependent phosphorylation of paxillin regulates cytoskeletal rearrangement. *Molecular and Cellular Biology, 26*(7), 2857–2868.

Canagarajah, B. J., Khokhlatchev, A., Cobb, M. H., & Goldsmith, E. J. (1997). Activation mechanism of the MAP kinase ERK2 by dual phosphorylation. *Cell, 90*(5), 859–869.

Cargnello, M., & Roux, P. P. (2011). Activation and function of the MAPKs and their substrates, the MAPK-activated protein kinases. *Microbiology and Molecular Biology Reviews, 75*(1), 50–83.

Carlson, S. M., Chouinard, C. R., Labadorf, A., Lam, C. J., Schmelzle, K., Fraenkel, E., et al. (2011). Large-scale discovery of ERK2 substrates identifies ERK-mediated transcriptional regulation by ETV3. *Science Signaling, 4*(196), rs11.

Casar, B., Pinto, A., & Crespo, P. (2008). Essential role of ERK dimers in the activation of cytoplasmic but not nuclear substrates by ERK-scaffold complexes. *Molecular Cell, 31*(5), 708–721.

Casey, R. C., Burleson, K. M., Skubitz, K. M., Pambuccian, S. E., Oegema, T. R., Jr., Ruff, L. E., et al. (2001). Beta 1-integrins regulate the formation and adhesion of ovarian carcinoma multicellular spheroids. *The American Journal of Pathology, 159*(6), 2071–2080.

Catling, A. D., Schaeffer, H. J., Reuter, C. W., Reddy, G. R., & Weber, M. J. (1995). A proline-rich sequence unique to MEK1 and MEK2 is required for raf binding and regulates MEK function. *Molecular and Cellular Biology, 15*(10), 5214–5225.

Chaudhary, A., King, W. G., Mattaliano, M. D., Frost, J. A., Diaz, B., Morrison, D. K., et al. (2000). Phosphatidylinositol 3-kinase regulates Raf1 through Pak phosphorylation of serine 338. *Current Biology, 10*(9), 551–554.

Chen, R. H., Sarnecki, C., & Blenis, J. (1992). Nuclear localization and regulation of erk- and rsk-encoded protein kinases. *Molecular and Cellular Biology, 12*(3), 915–927.

Cheng, J., Zhou, T., Liu, C., Shapiro, J. P., Brauer, M. J., Kiefer, M. C., et al. (1994). Protection from Fas-mediated apoptosis by a soluble form of the Fas molecule. *Science, 263*(5154), 1759–1762.

Cherniack, A. D., Klarlund, J. K., & Czech, M. P. (1994). Phosphorylation of the Ras nucleotide exchange factor son of sevenless by mitogen-activated protein kinase. *The Journal of Biological Chemistry, 269*(7), 4717–4720.

Chuderland, D., Konson, A., & Seger, R. (2008). Identification and characterization of a general nuclear translocation signal in signaling proteins. *Molecular Cell, 31*(6), 850–861.

Colicelli, J. (2004). Human RAS superfamily proteins and related GTPases. *Science's STKE, 2004*(250), RE13.

Cook, S. J., & McCormick, F. (1996). Kinetic and biochemical correlation between sustained p44ERK1 (44 kDa extracellular signal-regulated kinase 1) activation and lysophosphatidic acid-stimulated DNA synthesis in Rat-1 cells. *The Biochemical Journal, 320*(Pt. 1), 237–245.

Cooper, J., Nakamura, K. D., Hunter, T., & Weber, M. J. (1983). Phosphotyrosine-containing proteins and expression of transformation parameters in cells infected with partial transformation mutants of Rous sarcoma virus. *Journal of Virology, 46*(1), 15–28.

Cooper, J. A., Sefton, B. M., & Hunter, T. (1984). Diverse mitogenic agents induce the phosphorylation of two related 42,000-dalton proteins on tyrosine in quiescent chick cells. *Molecular and Cellular Biology, 4*(1), 30–37.

Corbalan-Garcia, S., Yang, S. S., Degenhardt, K. R., & Bar-Sagi, D. (1996). Identification of the mitogen-activated protein kinase phosphorylation sites on human Sos1 that regulate interaction with Grb2. *Molecular and Cellular Biology, 16*(10), 5674–5682.

Corsini, L., Bonna, S., Basquin, J., Hothorn, M., Scheffzek, K., Valcarcel, J., et al. (2007). U2AF-homology motif interactions are required for alternative splicing regulation by SPF45. *Nature Structural & Molecular Biology, 14*(7), 620–629.

Costa, M., Marchi, M., Cardarelli, F., Roy, A., Beltram, F., Maffei, L., et al. (2006). Dynamic regulation of ERK2 nuclear translocation and mobility in living cells. *Journal of Cell Science, 119*(Pt. 23), 4952–4963.

Cruzalegui, F. H., Cano, E., & Treisman, R. (1999). ERK activation induces phosphorylation of Elk-1 at multiple S/T-P motifs to high stoichiometry. *Oncogene, 18*(56), 7948–7957.

Davidson, B., Givant-Horwitz, V., Lazarovici, P., Risberg, B., Nesland, J. M., Trope, C. G., et al. (2003). Matrix metalloproteinases (MMP), EMMPRIN (extracellular matrix metalloproteinase inducer) and mitogen-activated protein kinases (MAPK): Co-expression in metastatic serous ovarian carcinoma. *Clinical & Experimental Metastasis, 20*(7), 621–631.

Davies, H., Bignell, G. R., Cox, C., Stephens, P., Edkins, S., Clegg, S., et al. (2002). Mutations of the BRAF gene in human cancer. *Nature, 417*(6892), 949–954.

Dhillon, A. S., & Kolch, W. (2002). Untying the regulation of the Raf-1 kinase. *Archives of Biochemistry and Biophysics, 404*(1), 3–9.

Dijkers, P. F., Medema, R. H., Pals, C., Banerji, L., Thomas, N. S., Lam, E. W., et al. (2000). Forkhead transcription factor FKHR-L1 modulates cytokine-dependent transcriptional regulation of p27(KIP1). *Molecular and Cellular Biology, 20*(24), 9138–9148.

Dimitri, C. A., Dowdle, W., MacKeigan, J. P., Blenis, J., & Murphy, L. O. (2005). Spatially separate docking sites on ERK2 regulate distinct signaling events in vivo. *Current Biology, 15*(14), 1319–1324.

Domina, A. M., Smith, J. H., & Craig, R. W. (2000). Myeloid cell leukemia 1 is phosphorylated through two distinct pathways, one associated with extracellular signal-regulated kinase activation and the other with G2/M accumulation or protein phosphatase 1/2A inhibition. *The Journal of Biological Chemistry, 275*(28), 21688–21694.

Domina, A. M., Vrana, J. A., Gregory, M. A., Hann, S. R., & Craig, R. W. (2004). MCL1 is phosphorylated in the PEST region and stabilized upon ERK activation in viable cells, and at additional sites with cytotoxic okadaic acid or taxol. *Oncogene, 23*(31), 5301–5315.

Dougherty, M. K., Muller, J., Ritt, D. A., Zhou, M., Zhou, X. Z., Copeland, T. D., et al. (2005). Regulation of Raf-1 by direct feedback phosphorylation. *Molecular Cell, 17*(2), 215–224.

Eblen, S. T., Kumar, N. V., Shah, K., Henderson, M. J., Watts, C. K., Shokat, K. M., et al. (2003). Identification of novel ERK2 substrates through use of an engineered kinase and ATP analogs. *The Journal of Biological Chemistry, 278*(17), 14926–14935.

Eblen, S. T., Kumar, N. V., & Weber, M. J. (2007). Using genetically engineered kinases to screen for novel protein kinase substrates: Phosphorylation of kinase-associated substrates. *CSH Protocols, 2007*. pdb prot4638.

Eblen, S. T., Slack, J. K., Weber, M. J., & Catling, A. D. (2002). Rac-PAK signaling stimulates extracellular signal-regulated kinase (ERK) activation by regulating formation of MEK1-ERK complexes. *Molecular and Cellular Biology, 22*(17), 6023–6033.

Eblen, S. T., Slack-Davis, J. K., Tarcsafalvi, A., Parsons, J. T., Weber, M. J., & Catling, A. D. (2004). Mitogen-activated protein kinase feedback phosphorylation regulates MEK1 complex formation and activation during cellular adhesion. *Molecular and Cellular Biology, 24*(6), 2308–2317.

Esteban, L. M., Vicario-Abejon, C., Fernandez-Salguero, P., Fernandez-Medarde, A., Swaminathan, N., Yienger, K., et al. (2001). Targeted genomic disruption of H-ras and N-ras, individually or in combination, reveals the dispensability of both loci for mouse growth and development. *Molecular and Cellular Biology, 21*(5), 1444–1452.

Ezzeddine, N., Chen, C. Y., & Shyu, A. B. (2012). Evidence providing new insights into TOB-promoted deadenylation and supporting a link between TOB's deadenylation-enhancing and antiproliferative activities. *Molecular and Cellular Biology, 32*(6), 1089–1098.

Fantz, D. A., Jacobs, D., Glossip, D., & Kornfeld, K. (2001). Docking sites on substrate proteins direct extracellular signal-regulated kinase to phosphorylate specific residues. *The Journal of Biological Chemistry, 276*(29), 27256–27265.

Felton-Edkins, Z. A., Fairley, J. A., Graham, E. L., Johnston, I. M., White, R. J., & Scott, P. H. (2003). The mitogen-activated protein (MAP) kinase ERK induces tRNA synthesis by phosphorylating TFIIIB. *The EMBO Journal, 22*(10), 2422–2432.

Fernandes, N., Bailey, D. E., Vanvranken, D. L., & Allbritton, N. L. (2007). Use of docking peptides to design modular substrates with high efficiency for mitogen-activated protein kinase extracellular signal-regulated kinase. *ACS Chemical Biology, 2*(10), 665–673.

Freeman, A. K., Ritt, D. A., & Morrison, D. K. (2013). The importance of Raf dimerization in cell signaling. *Small GTPases, 4*(3), 180–185.

Fremin, C., Saba-El-Leil, M. K., Levesque, K., Ang, S. L., & Meloche, S. (2015). Functional redundancy of ERK1 and ERK2 MAP kinases during development. *Cell Reports, 12*(6), 913–921.

Frost, J. A., Steen, H., Shapiro, P., Lewis, T., Ahn, N., Shaw, P. E., et al. (1997). Cross-cascade activation of ERKs and ternary complex factors by Rho family proteins. *The EMBO Journal, 16*(21), 6426–6438.

Fukuda, M., Asano, S., Nakamura, T., Adachi, M., Yoshida, M., Yanagida, M., et al. (1997). CRM1 is responsible for intracellular transport mediated by the nuclear export signal. *Nature, 390*(6657), 308–311.

Fukuda, M., Gotoh, I., Adachi, M., Gotoh, Y., & Nishida, E. (1997). A novel regulatory mechanism in the mitogen-activated protein (MAP) kinase cascade. Role of nuclear export signal of MAP kinase kinase. *The Journal of Biological Chemistry, 272*(51), 32642–32648.

Fukuda, M., Gotoh, I., Gotoh, Y., & Nishida, E. (1996). Cytoplasmic localization of mitogen-activated protein kinase kinase directed by its NH2-terminal, leucine-rich short amino acid sequence, which acts as a nuclear export signal. *The Journal of Biological Chemistry, 271*(33), 20024–20028.

Fukuda, M., Gotoh, Y., & Nishida, E. (1997). Interaction of MAP kinase with MAP kinase kinase: Its possible role in the control of nucleocytoplasmic transport of MAP kinase. *The EMBO Journal, 16*(8), 1901–1908.

Fukunaga, R., & Hunter, T. (1997). MNK1, a new MAP kinase-activated protein kinase, isolated by a novel expression screening method for identifying protein kinase substrates. *The EMBO Journal, 16*(8), 1921–1933.

Garami, A., Zwartkruis, F. J., Nobukuni, T., Joaquin, M., Roccio, M., Stocker, H., et al. (2003). Insulin activation of Rheb, a mediator of mTOR/S6K/4E-BP signaling, is inhibited by TSC1 and 2. *Molecular Cell, 11*(6), 1457–1466.

Gille, H., Kortenjann, M., Thomae, O., Moomaw, C., Slaughter, C., Cobb, M. H., et al. (1995). ERK phosphorylation potentiates Elk-1-mediated ternary complex formation and transactivation. *The EMBO Journal, 14*(5), 951–962.

Gingras, A. C., Raught, B., & Sonenberg, N. (2001). Regulation of translation initiation by FRAP/mTOR. *Genes & Development, 15*(7), 807–826.

Gonzalez, F. A., Raden, D. L., & Davis, R. J. (1991). Identification of substrate recognition determinants for human ERK1 and ERK2 protein kinases. *The Journal of Biological Chemistry, 266*(33), 22159–22163.

Gonzalez, F. A., Seth, A., Raden, D. L., Bowman, D. S., Fay, F. S., & Davis, R. J. (1993). Serum-induced translocation of mitogen-activated protein kinase to the cell surface ruffling membrane and the nucleus. *The Journal of Cell Biology, 122*(5), 1089–1101.

Hanahan, D., & Weinberg, R. A. (2000). The hallmarks of cancer. *Cell, 100*(1), 57–70.

Hanahan, D., & Weinberg, R. A. (2011). Hallmarks of cancer: The next generation. *Cell, 144*(5), 646–674.

Hancock, J. F., Paterson, H., & Marshall, C. J. (1990). A polybasic domain or palmitoylation is required in addition to the CAAX motif to localize p21ras to the plasma membrane. *Cell, 63*(1), 133–139.

Hatano, N., Mori, Y., Oh-hora, M., Kosugi, A., Fujikawa, T., Nakai, N., et al. (2003). Essential role for ERK2 mitogen-activated protein kinase in placental development. *Genes to Cells, 8*(11), 847–856.

Haystead, T. A., Sim, A. T., Carling, D., Honnor, R. C., Tsukitani, Y., Cohen, P., et al. (1989). Effects of the tumour promoter okadaic acid on intracellular protein phosphorylation and metabolism. *Nature, 337*(6202), 78–81.

Haystead, T. A., Weiel, J. E., Litchfield, D. W., Tsukitani, Y., Fischer, E. H., & Krebs, E. G. (1990). Okadaic acid mimics the action of insulin in stimulating protein kinase activity in isolated adipocytes. The role of protein phosphatase 2a in attenuation of the signal. *The Journal of Biological Chemistry, 265*(27), 16571–16580.

Herrero, A., Pinto, A., Colon-Bolea, P., Casar, B., Jones, M., Agudo-Ibanez, L., et al. (2015). Small molecule inhibition of ERK dimerization prevents tumorigenesis by RAS-ERK pathway oncogenes. *Cancer Cell*, *28*(2), 170–182.

Heublein, S., Grasse, K., Hessel, H., Burges, A., Lenhard, M., Engel, J., et al. (2013). KRAS, BRAF genotyping reveals genetic heterogeneity of ovarian borderline tumors and associated implants. *BMC Cancer*, *13*, 483.

Hotamisligil, G. S., & Davis, R. J. (2016). Cell signaling and stress responses. *Cold Spring Harbor Perspectives in Biology*, *8*(10).

Houck, K. A., Michalopoulos, G. K., & Strom, S. C. (1989). Introduction of a Ha-ras oncogene into rat liver epithelial cells and parenchymal hepatocytes confers resistance to the growth inhibitory effects of TGF-beta. *Oncogene*, *4*(1), 19–25.

Howe, A. K., & Juliano, R. L. (2000). Regulation of anchorage-dependent signal transduction by protein kinase A and p21-activated kinase. *Nature Cell Biology*, *2*(9), 593–600.

Hunger-Glaser, I., Fan, R. S., Perez-Salazar, E., & Rozengurt, E. (2004). PDGF and FGF induce focal adhesion kinase (FAK) phosphorylation at Ser-910: Dissociation from Tyr-397 phosphorylation and requirement for ERK activation. *Journal of Cellular Physiology*, *200*(2), 213–222.

Hunger-Glaser, I., Salazar, E. P., Sinnett-Smith, J., & Rozengurt, E. (2003). Bombesin, lysophosphatidic acid, and epidermal growth factor rapidly stimulate focal adhesion kinase phosphorylation at Ser-910: Requirement for ERK activation. *The Journal of Biological Chemistry*, *278*(25), 22631–22643.

Inoki, K., Li, Y., Xu, T., & Guan, K. L. (2003). Rheb GTPase is a direct target of TSC2 GAP activity and regulates mTOR signaling. *Genes & Development*, *17*(15), 1829–1834.

Ishibe, S., Joly, D., Liu, Z. X., & Cantley, L. G. (2004). Paxillin serves as an ERK-regulated scaffold for coordinating FAK and Rac activation in epithelial morphogenesis. *Molecular Cell*, *16*(2), 257–267.

Jaaro, H., Rubinfeld, H., Hanoch, T., & Seger, R. (1997). Nuclear translocation of mitogen-activated protein kinase kinase (MEK1) in response to mitogenic stimulation. *Proceedings of the National Academy of Sciences of the United States of America*, *94*(8), 3742–3747.

Jackson, J. H., Li, J. W., Buss, J. E., Der, C. J., & Cochrane, C. G. (1994). Polylysine domain of K-ras 4B protein is crucial for malignant transformation. *Proceedings of the National Academy of Sciences of the United States of America*, *91*(26), 12730–12734.

Jacobs, D., Glossip, D., Xing, H., Muslin, A. J., & Kornfeld, K. (1999). Multiple docking sites on substrate proteins form a modular system that mediates recognition by ERK MAP kinase. *Genes & Development*, *13*(2), 163–175.

Kaoud, T. S., Devkota, A. K., Harris, R., Rana, M. S., Abramczyk, O., Warthaka, M., et al. (2011). Activated ERK2 is a monomer in vitro with or without divalent cations and when complexed to the cytoplasmic scaffold PEA-15. *Biochemistry*, *50*(21), 4568–4578.

Karlsson, M., Mathers, J., Dickinson, R. J., Mandl, M., & Keyse, S. M. (2004). Both nuclear-cytoplasmic shuttling of the dual specificity phosphatase MKP-3 and its ability to anchor MAP kinase in the cytoplasm are mediated by a conserved nuclear export signal. *The Journal of Biological Chemistry*, *279*(40), 41882–41891.

Khokhlatchev, A. V., Canagarajah, B., Wilsbacher, J., Robinson, M., Atkinson, M., Goldsmith, E., et al. (1998). Phosphorylation of the MAP kinase ERK2 promotes its homodimerization and nuclear translocation. *Cell*, *93*(4), 605–615.

King, A. J., Sun, H., Diaz, B., Barnard, D., Miao, W., Bagrodia, S., et al. (1998). The protein kinase Pak3 positively regulates Raf-1 activity through phosphorylation of serine 338. *Nature*, *396*(6707), 180–183.

Koera, K., Nakamura, K., Nakao, K., Miyoshi, J., Toyoshima, K., Hatta, T., et al. (1997). K-ras is essential for the development of the mouse embryo. *Oncogene*, *15*(10), 1151–1159.

Kohno, M., & Pouyssegur, J. (1986). Alpha-thrombin-induced tyrosine phosphorylation of 43,000- and 41,000-Mr proteins is independent of cytoplasmic alkalinization in quiescent fibroblasts. *The Biochemical Journal, 238*(2), 451–457.

Kosako, H., & Motani, K. (2017). Global identification of ERK substrates by phosphoproteomics based on IMAC and 2D-DIGE. *Methods in Molecular Biology, 1487*, 137–149.

Kosako, H., Yamaguchi, N., Aranami, C., Ushiyama, M., Kose, S., Imamoto, N., et al. (2009). Phosphoproteomics reveals new ERK MAP kinase targets and links ERK to nucleoporin-mediated nuclear transport. *Nature Structural & Molecular Biology, 16*(10), 1026–1035.

Kretzschmar, M., Doody, J., Timokhina, I., & Massague, J. (1999). A mechanism of repression of TGFbeta/Smad signaling by oncogenic Ras. *Genes & Development, 13*(7), 804–816.

Kucharska, A., Rushworth, L. K., Staples, C., Morrice, N. A., & Keyse, S. M. (2009). Regulation of the inducible nuclear dual-specificity phosphatase DUSP5 by ERK MAPK. *Cellular Signalling, 21*(12), 1794–1805.

Kumar, N. V., Eblen, S. T., & Weber, M. J. (2004). Identifying specific kinase substrates through engineered kinases and ATP analogs. *Methods, 32*(4), 389–397.

Labidi-Galy, S. I., Papp, E., Hallberg, D., Niknafs, N., Adleff, V., Noe, M., et al. (2017). High grade serous ovarian carcinomas originate in the fallopian tube. *Nature Communications, 8*(1), 1093.

Lee, T., Hoofnagle, A. N., Kabuyama, Y., Stroud, J., Min, X., Goldsmith, E. J., et al. (2004). Docking motif interactions in MAP kinases revealed by hydrogen exchange mass spectrometry. *Molecular Cell, 14*(1), 43–55.

Lenormand, P., Brondello, J. M., Brunet, A., & Pouyssegur, J. (1998). Growth factor-induced p42/p44 MAPK nuclear translocation and retention requires both MAPK activation and neosynthesis of nuclear anchoring proteins. *The Journal of Cell Biology, 142*(3), 625–633.

Lenormand, P., Sardet, C., Pages, G., L'Allemain, G., Brunet, A., & Pouyssegur, J. (1993). Growth factors induce nuclear translocation of MAP kinases (p42mapk and p44mapk) but not of their activator MAP kinase kinase (p45mapkk) in fibroblasts. *The Journal of Cell Biology, 122*(5), 1079–1088.

Lewis, T. S., Hunt, J. B., Aveline, L. D., Jonscher, K. R., Louie, D. F., Yeh, J. M., et al. (2000). Identification of novel MAP kinase pathway signaling targets by functional proteomics and mass spectrometry. *Molecular Cell, 6*(6), 1343–1354.

Ley, R., Ewings, K. E., Hadfield, K., Howes, E., Balmanno, K., & Cook, S. J. (2004). Extracellular signal-regulated kinases 1/2 are serum-stimulated "Bim(EL) kinases" that bind to the BH3-only protein Bim(EL) causing its phosphorylation and turnover. *The Journal of Biological Chemistry, 279*(10), 8837–8847.

Ley, R., Hadfield, K., Howes, E., & Cook, S. J. (2005). Identification of a DEF-type docking domain for extracellular signal-regulated kinases 1/2 that directs phosphorylation and turnover of the BH3-only protein BimEL. *The Journal of Biological Chemistry, 280*(18), 17657–17663.

Lidke, D. S., Huang, F., Post, J. N., Rieger, B., Wilsbacher, J., Thomas, J. L., et al. (2010). ERK nuclear translocation is dimerization-independent but controlled by the rate of phosphorylation. *The Journal of Biological Chemistry, 285*(5), 3092–3102.

Lin, T. H., Chen, Q., Howe, A., & Juliano, R. L. (1997). Cell anchorage permits efficient signal transduction between ras and its downstream kinases. *The Journal of Biological Chemistry, 272*(14), 8849–8852.

Liu, Y., Conaway, L., Rutherford Bethard, J., Al-Ayoubi, A. M., Thompson Bradley, A., Zheng, H., et al. (2013). Phosphorylation of the alternative mRNA splicing factor 45 (SPF45) by Clk1 regulates its splice site utilization, cell migration and invasion. *Nucleic Acids Research, 41*, 4949–4962.

Liu, Z. X., Yu, C. F., Nickel, C., Thomas, S., & Cantley, L. G. (2002). Hepatocyte growth factor induces ERK-dependent paxillin phosphorylation and regulates paxillin-focal adhesion kinase association. *The Journal of Biological Chemistry, 277*(12), 10452–10458.

Longstreet, M., Miller, B., & Howe, P. H. (1992). Loss of transforming growth factor beta 1 (TGF-beta 1)-induced growth arrest and p34cdc2 regulation in ras-transfected epithelial cells. *Oncogene, 7*(8), 1549–1556.

Lukong, K. E., & Richard, S. (2003). Sam68, the KH domain-containing superSTAR. *Biochimica et Biophysica Acta, 1653*(2), 73–86.

Ma, L., Chen, Z., Erdjument-Bromage, H., Tempst, P., & Pandolfi, P. P. (2005). Phosphorylation and functional inactivation of TSC2 by Erk implications for tuberous sclerosis and cancer pathogenesis. *Cell, 121*(2), 179–193.

Ma, L., Teruya-Feldstein, J., Bonner, P., Bernardi, R., Franz, D. N., Witte, D., et al. (2007). Identification of S664 TSC2 phosphorylation as a marker for extracellular signal-regulated kinase mediated mTOR activation in tuberous sclerosis and human cancer. *Cancer Research, 67*(15), 7106–7112.

Mackeigan, J. P., Murphy, L. O., Dimitri, C. A., & Blenis, J. (2005). Graded mitogen-activated protein kinase activity precedes switch-like c-Fos induction in mammalian cells. *Molecular and Cellular Biology, 25*(11), 4676–4682.

Maekawa, M., Nishida, E., & Tanoue, T. (2002). Identification of the anti-proliferative protein Tob as a MAPK substrate. *The Journal of Biological Chemistry, 277*(40), 37783–37787.

Mandl, M., Slack, D. N., & Keyse, S. M. (2005). Specific inactivation and nuclear anchoring of extracellular signal-regulated kinase 2 by the inducible dual-specificity protein phosphatase DUSP5. *Molecular and Cellular Biology, 25*(5), 1830–1845.

Mansour, S. J., Resing, K. A., Candi, J. M., Hermann, A. S., Gloor, J. W., Herskind, K. R., et al. (1994). Mitogen-activated protein (MAP) kinase phosphorylation of MAP kinase kinase: Determination of phosphorylation sites by mass spectrometry and site-directed mutagenesis. *Journal of Biochemistry (Tokyo), 116*(2), 304–314.

Mason, C. S., Springer, C. J., Cooper, R. G., Superti-Furga, G., Marshall, C. J., & Marais, R. (1999). Serine and tyrosine phosphorylations cooperate in Raf-1, but not B-Raf activation. *The EMBO Journal, 18*(8), 2137–2148.

Matlin, A. J., Clark, F., & Smith, C. W. (2005). Understanding alternative splicing: Towards a cellular code. *Nature Reviews. Molecular Cell Biology, 6*(5), 386–398.

Matter, N., Herrlich, P., & Konig, H. (2002). Signal-dependent regulation of splicing via phosphorylation of Sam68. *Nature, 420*(6916), 691–695.

Mineo, C., Anderson, R. G., & White, M. A. (1997). Physical association with ras enhances activation of membrane-bound raf (RafCAAX). *The Journal of Biological Chemistry, 272*(16), 10345–10348.

Morel, C., Carlson, S. M., White, F. M., & Davis, R. J. (2009). Mcl-1 integrates the opposing actions of signaling pathways that mediate survival and apoptosis. *Molecular and Cellular Biology, 29*(14), 3845–3852.

Muratcioglu, S., Chavan, T. S., Freed, B. C., Jang, H., Khavrutskii, L., Freed, R. N., et al. (2015). GTP-dependent K-Ras dimerization. *Structure, 23*(7), 1325–1335.

Murphy, L. O., MacKeigan, J. P., & Blenis, J. (2004). A network of immediate early gene products propagates subtle differences in mitogen-activated protein kinase signal amplitude and duration. *Molecular and Cellular Biology, 24*(1), 144–153.

Murphy, L. O., Smith, S., Chen, R. H., Fingar, D. C., & Blenis, J. (2002). Molecular interpretation of ERK signal duration by immediate early gene products. *Nature Cell Biology, 4*(8), 556–564.

Nakamura, K. D., Martinez, R., & Weber, M. J. (1983). Tyrosine phosphorylation of specific proteins after mitogen stimulation of chicken embryo fibroblasts. *Molecular and Cellular Biology, 3*(3), 380–390.

Nifoussi, S. K., Vrana, J. A., Domina, A. M., De Biasio, A., Gui, J., Gregory, M. A., et al. (2012). Thr 163 phosphorylation causes Mcl-1 stabilization when degradation is independent of the adjacent GSK3-targeted phosphodegron, promoting drug resistance in cancer. *PLoS One, 7*(10), e47060.

Oskarsson, T., & Trumpp, A. (2005). The Myc trilogy: Lord of RNA polymerases. *Nature Cell Biology, 7*(3), 215–217.

Pages, G., Guerin, S., Grall, D., Bonino, F., Smith, A., Anjuere, F., et al. (1999). Defective thymocyte maturation in p44 MAP kinase (Erk 1) knockout mice. *Science, 286*(5443), 1374–1377.

Pan, C., Olsen, J. V., Daub, H., & Mann, M. (2009). Global effects of kinase inhibitors on signaling networks revealed by quantitative phosphoproteomics. *Molecular & Cellular Proteomics, 8*(12), 2796–2808.

Pan, Q., Shai, O., Lee, L. J., Frey, B. J., & Blencowe, B. J. (2008). Deep surveying of alternative splicing complexity in the human transcriptome by high-throughput sequencing. *Nature Genetics, 40*(12), 1413–1415.

Park, E. R., Eblen, S. T., & Catling, A. D. (2007). MEK1 activation by PAK: A novel mechanism. *Cellular Signalling, 19*(7), 1488–1496.

Paronetto, M. P., Zalfa, F., Botti, F., Geremia, R., Bagni, C., & Sette, C. (2006). The nuclear RNA-binding protein Sam68 translocates to the cytoplasm and associates with the polysomes in mouse spermatocytes. *Molecular Biology of the Cell, 17*(1), 14–24.

Payne, D. M., Rossomando, A. J., Martino, P., Erickson, A. K., Her, J. H., Shabanowitz, J., et al. (1991). Identification of the regulatory phosphorylation sites in pp42/mitogen-activated protein kinase (MAP kinase). *The EMBO Journal, 10*(4), 885–892.

Plotnikov, A., Chuderland, D., Karamansha, Y., Livnah, O., & Seger, R. (2011). Nuclear extracellular signal-regulated kinase 1 and 2 translocation is mediated by casein kinase 2 and accelerated by autophosphorylation. *Molecular and Cellular Biology, 31*(17), 3515–3530.

Plotnikov, A., Zehorai, E., Procaccia, S., & Seger, R. (2011). The MAPK cascades: Signaling components, nuclear roles and mechanisms of nuclear translocation. *Biochimica et Biophysica Acta, 1813*(9), 1619–1633.

Plowman, S. J., Williamson, D. J., O'Sullivan, M. J., Doig, J., Ritchie, A. M., Harrison, D. J., et al. (2003). While K-ras is essential for mouse development, expression of the K-ras 4A splice variant is dispensable. *Molecular and Cellular Biology, 23*(24), 9245–9250.

Polyak, K., & Weinberg, R. A. (2009). Transitions between epithelial and mesenchymal states: Acquisition of malignant and stem cell traits. *Nature Reviews. Cancer, 9*(4), 265–273.

Porfiri, E., & McCormick, F. (1996). Regulation of epidermal growth factor receptor signaling by phosphorylation of the ras exchange factor hSOS1. *The Journal of Biological Chemistry, 271*(10), 5871–5877.

Price, M. A., Rogers, A. E., & Treisman, R. (1995). Comparative analysis of the ternary complex factors Elk-1, SAP-1a and SAP-2 (ERP/NET). *The EMBO Journal, 14*(11), 2589–2601.

Rajakulendran, T., Sahmi, M., Lefrancois, M., Sicheri, F., & Therrien, M. (2009). A dimerization-dependent mechanism drives RAF catalytic activation. *Nature, 461*(7263), 542–545.

Rajalingam, K., Schreck, R., Rapp, U. R., & Albert, S. (2007). Ras oncogenes and their downstream targets. *Biochimica et Biophysica Acta, 1773*(8), 1177–1195.

Ranganathan, A., Yazicioglu, M. N., & Cobb, M. H. (2006). The nuclear localization of ERK2 occurs by mechanisms both independent of and dependent on energy. *The Journal of Biological Chemistry, 281*(23), 15645–15652.

Ray, L. B., & Sturgill, T. W. (1987). Rapid stimulation by insulin of a serine/threonine kinase in 3T3-L1 adipocytes that phosphorylates microtubule-associated protein 2 in vitro. *Proceedings of the National Academy of Sciences of the United States of America, 84*(6), 1502–1506.

Ray, L. B., & Sturgill, T. W. (1988). Characterization of insulin-stimulated microtubule-associated protein kinase. Rapid isolation and stabilization of a novel serine/threonine kinase from 3T3-L1 cells. *The Journal of Biological Chemistry, 263*(25), 12721–12727.

Ren, J. G., Li, Z., & Sacks, D. B. (2007). IQGAP1 modulates activation of B-Raf. *Proceedings of the National Academy of Sciences of the United States of America, 104*(25), 10465–10469.

Renshaw, M. W., Price, L. S., & Schwartz, M. A. (1999). Focal adhesion kinase mediates the integrin signaling requirement for growth factor activation of MAP kinase. *The Journal of Cell Biology, 147*(3), 611–618.

Renshaw, M. W., Ren, X. D., & Schwartz, M. A. (1997). Growth factor activation of MAP kinase requires cell adhesion. *The EMBO Journal, 16*(18), 5592–5599.

Ritt, D. A., Monson, D. M., Specht, S. I., & Morrison, D. K. (2010). Impact of feedback phosphorylation and Raf heterodimerization on normal and mutant B-Raf signaling. *Molecular and Cellular Biology, 30*(3), 806–819.

Roovers, K., Davey, G., Zhu, X., Bottazzi, M. E., & Assoian, R. K. (1999). Alpha5beta1 integrin controls cyclin D1 expression by sustaining mitogen-activated protein kinase activity in growth factor-treated cells. *Molecular Biology of the Cell, 10*(10), 3197–3204.

Roskoski, R., Jr. (2012). ERK1/2 MAP kinases: Structure, function, and regulation. *Pharmacological Research, 66*(2), 105–143.

Rozakis-Adcock, M., van der Geer, P., Mbamalu, G., & Pawson, T. (1995). MAP kinase phosphorylation of mSos1 promotes dissociation of mSos1-Shc and mSos1-EGF receptor complexes. *Oncogene, 11*(7), 1417–1426.

Saba-El-Leil, M. K., Vella, F. D., Vernay, B., Voisin, L., Chen, L., Labrecque, N., et al. (2003). An essential function of the mitogen-activated protein kinase Erk2 in mouse trophoblast development. *EMBO Reports, 4*(10), 964–968.

Saucedo, L. J., Gao, X., Chiarelli, D. A., Li, L., Pan, D., & Edgar, B. A. (2003). Rheb promotes cell growth as a component of the insulin/TOR signalling network. *Nature Cell Biology, 5*(6), 566–571.

Schaeffer, H. J., Catling, A. D., Eblen, S. T., Collier, L. S., Krauss, A., & Weber, M. J. (1998). MP1: A MEK binding partner that enhances enzymatic activation of the MAP kinase cascade. *Science, 281*(5383), 1668–1671.

Schaeffer, H. J., & Weber, M. J. (1999). Mitogen-activated protein kinases: Specific messages from ubiquitous messengers. *Molecular and Cellular Biology, 19*(4), 2435–2444.

Schevzov, G., Kee, A. J., Wang, B., Sequeira, V. B., Hook, J., Coombes, J. D., et al. (2015). Regulation of cell proliferation by ERK and signal-dependent nuclear translocation of ERK is dependent on Tm5NM1-containing actin filaments. *Molecular Biology of the Cell, 26*(13), 2475–2490.

Schmidt, M., Fernandez de Mattos, S., van der Horst, A., Klompmaker, R., Kops, G. J., Lam, E. W., et al. (2002). Cell cycle inhibition by FoxO forkhead transcription factors involves downregulation of cyclin D. *Molecular and Cellular Biology, 22*(22), 7842–7852.

Schwerk, C., & Schulze-Osthoff, K. (2005). Regulation of apoptosis by alternative pre-mRNA splicing. *Molecular Cell, 19*(1), 1–13.

Shah, K., Liu, Y., Deirmengian, C., & Shokat, K. M. (1997). Engineering unnatural nucleotide specificity for Rous sarcoma virus tyrosine kinase to uniquely label its direct substrates. *Proceedings of the National Academy of Sciences of the United States of America, 94*(8), 3565–3570.

Sharma, C., Vomastek, T., Tarcsafalvi, A., Catling, A. D., Schaeffer, H. J., Eblen, S. T., et al. (2005). MEK partner 1 (MP1): Regulation of oligomerization in MAP kinase signaling. *Journal of Cellular Biochemistry, 94*(4), 708–719.

Sheridan, D. L., Kong, Y., Parker, S. A., Dalby, K. N., & Turk, B. E. (2008). Substrate discrimination among mitogen-activated protein kinases through distinct docking sequence motifs. *The Journal of Biological Chemistry, 283*(28), 19511–19520.

Shindo, Y., Iwamoto, K., Mouri, K., Hibino, K., Tomita, M., Kosako, H., et al. (2016). Conversion of graded phosphorylation into switch-like nuclear translocation via autoregulatory mechanisms in ERK signalling. *Nature Communications*, 7, 10485.

Singer, G., Oldt, R., 3rd, Cohen, Y., Wang, B. G., Sidransky, D., Kurman, R. J., et al. (2003). Mutations in BRAF and KRAS characterize the development of low-grade ovarian serous carcinoma. *Journal of the National Cancer Institute*, 95(6), 484–486.

Slack-Davis, J. K., Eblen, S. T., Zecevic, M., Boerner, S. A., Tarcsafalvi, A., Diaz, H. B., et al. (2003). PAK1 phosphorylation of MEK1 regulates fibronectin-stimulated MAPK activation. *The Journal of Cell Biology*, 162(2), 281–291.

Stefanovsky, V. Y., Langlois, F., Bazett-Jones, D., Pelletier, G., & Moss, T. (2006). ERK modulates DNA bending and enhancesome structure by phosphorylating HMG1-boxes 1 and 2 of the RNA polymerase I transcription factor UBF. *Biochemistry*, 45(11), 3626–3634.

Stefanovsky, V., Langlois, F., Gagnon-Kugler, T., Rothblum, L. I., & Moss, T. (2006). Growth factor signaling regulates elongation of RNA polymerase I transcription in mammals via UBF phosphorylation and r-chromatin remodeling. *Molecular Cell*, 21(5), 629–639.

Stefanovsky, V. Y., Pelletier, G., Hannan, R., Gagnon-Kugler, T., Rothblum, L. I., & Moss, T. (2001). An immediate response of ribosomal transcription to growth factor stimulation in mammals is mediated by ERK phosphorylation of UBF. *Molecular Cell*, 8(5), 1063–1073.

Stocker, H., Radimerski, T., Schindelholz, B., Wittwer, F., Belawat, P., Daram, P., et al. (2003). Rheb is an essential regulator of S6K in controlling cell growth in Drosophila. *Nature Cell Biology*, 5(6), 559–565.

Stokoe, D., Macdonald, S. G., Cadwallader, K., Symons, M., & Hancock, J. F. (1994). Activation of Raf as a result of recruitment to the plasma membrane. *Science*, 264(5164), 1463–1467.

Sturgill, T. W., Ray, L. B., Erikson, E., & Maller, J. L. (1988). Insulin-stimulated MAP-2 kinase phosphorylates and activates ribosomal protein S6 kinase II. *Nature*, 334(6184), 715–718.

Suzuki, T., K-Tsuzuku, J., Ajima, R., Nakamura, T., Yoshida, Y., & Yamamoto, T. (2002). Phosphorylation of three regulatory serines of Tob by Erk1 and Erk2 is required for Ras-mediated cell proliferation and transformation. *Genes & Development*, 16(11), 1356–1370.

Tanoue, T., Adachi, M., Moriguchi, T., & Nishida, E. (2000). A conserved docking motif in MAP kinases common to substrates, activators and regulators. *Nature Cell Biology*, 2(2), 110–116.

Tee, A. R., Manning, B. D., Roux, P. P., Cantley, L. C., & Blenis, J. (2003). Tuberous sclerosis complex gene products, Tuberin and Hamartin, control mTOR signaling by acting as a GTPase-activating protein complex toward Rheb. *Current Biology*, 13(15), 1259–1268.

Teis, D., Wunderlich, W., & Huber, L. A. (2002). Localization of the MP1-MAPK scaffold complex to endosomes is mediated by p14 and required for signal transduction. *Developmental Cell*, 3(6), 803–814.

Thiery, J. P., & Sleeman, J. P. (2006). Complex networks orchestrate epithelial-mesenchymal transitions. *Nature Reviews. Molecular Cell Biology*, 7(2), 131–142.

Tsang, Y. T., Deavers, M. T., Sun, C. C., Kwan, S. Y., Kuo, E., Malpica, A., et al. (2013). KRAS (but not BRAF) mutations in ovarian serous borderline tumour are associated with recurrent low-grade serous carcinoma. *The Journal of Pathology*, 231(4), 449–456.

Turski, M. L., Vidwans, S. J., Janku, F., Garrido-Laguna, I., Munoz, J., Schwab, R., et al. (2016). Genomically driven tumors and actionability across histologies: BRAF-mutant cancers as a paradigm. *Molecular Cancer Therapeutics*, 15(4), 533–547.

Unal, E. B., Uhlitz, F., & Bluthgen, N. (2017). A compendium of ERK targets. *FEBS Letters*, *591*(17), 2607–2615.

Valacca, C., Bonomi, S., Buratti, E., Pedrotti, S., Baralle, F. E., Sette, C., et al. (2010). Sam68 regulates EMT through alternative splicing-activated nonsense-mediated mRNA decay of the SF2/ASF proto-oncogene. *The Journal of Cell Biology*, *191*(1), 87–99.

Valverius, E. M., Walker-Jones, D., Bates, S. E., Stampfer, M. R., Clark, R., McCormick, F., et al. (1989). Production of and responsiveness to transforming growth factor-beta in normal and oncogene-transformed human mammary epithelial cells. *Cancer Research*, *49*(22), 6269–6274.

Vetterkind, S., Poythress, R. H., Lin, Q. Q., & Morgan, K. G. (2013). Hierarchical scaffolding of an ERK1/2 activation pathway. *Cell Communication and Signaling: CCS*, *11*, 65.

Volmat, V., Camps, M., Arkinstall, S., Pouyssegur, J., & Lenormand, P. (2001). The nucleus, a site for signal termination by sequestration and inactivation of p42/p44 MAP kinases. *Journal of Cell Science*, *114*(Pt. 19), 3433–3443.

Vomastek, T., Iwanicki, M. P., Burack, W. R., Tiwari, D., Kumar, D., Parsons, J. T., et al. (2008). Extracellular signal-regulated kinase 2 (ERK2) phosphorylation sites and docking domain on the nuclear pore complex protein Tpr cooperatively regulate ERK2-Tpr interaction. *Molecular and Cellular Biology*, *28*(22), 6954–6966.

Vouret-Craviari, V., Van Obberghen-Schilling, E., Scimeca, J. C., Van Obberghen, E., & Pouyssegur, J. (1993). Differential activation of p44mapk (ERK1) by alpha-thrombin and thrombin-receptor peptide agonist. *The Biochemical Journal*, *289*(Pt. 1), 209–214.

Wang, E. T., Sandberg, R., Luo, S., Khrebtukova, I., Zhang, L., Mayr, C., et al. (2008). Alternative isoform regulation in human tissue transcriptomes. *Nature*, *456*(7221), 470–476.

Waskiewicz, A. J., Flynn, A., Proud, C. G., & Cooper, J. A. (1997). Mitogen-activated protein kinases activate the serine/threonine kinases Mnk1 and Mnk2. *The EMBO Journal*, *16*(8), 1909–1920.

Weber, J. D., Raben, D. M., Phillips, P. J., & Baldassare, J. J. (1997). Sustained activation of extracellular-signal-regulated kinase 1 (ERK1) is required for the continued expression of cyclin D1 in G1 phase. *The Biochemical Journal*, *326*(Pt. 1), 61–68.

Weber, C. K., Slupsky, J. R., Kalmes, H. A., & Rapp, U. R. (2001). Active Ras induces heterodimerization of cRaf and BRaf. *Cancer Research*, *61*(9), 3595–3598.

Wennerberg, K., Rossman, K. L., & Der, C. J. (2005). The Ras superfamily at a glance. *Journal of Cell Science*, *118*(Pt. 5), 843–846.

Weston, C. R., Balmanno, K., Chalmers, C., Hadfield, K., Molton, S. A., Ley, R., et al. (2003). Activation of ERK1/2 by deltaRaf-1:ER* represses Bim expression independently of the JNK or PI3K pathways. *Oncogene*, *22*(9), 1281–1293.

Weston, C. R., & Davis, R. J. (2007). The JNK signal transduction pathway. *Current Opinion in Cell Biology*, *19*(2), 142–149.

Whitmarsh, A. J., Shore, P., Sharrocks, A. D., & Davis, R. J. (1995). Integration of MAP kinase signal transduction pathways at the serum response element. *Science*, *269*(5222), 403–407.

Wong, K. K., Tsang, Y. T., Deavers, M. T., Mok, S. C., Zu, Z., Sun, C., et al. (2010). BRAF mutation is rare in advanced-stage low-grade ovarian serous carcinomas. *The American Journal of Pathology*, *177*(4), 1611–1617.

Wunderlich, W., Fialka, I., Teis, D., Alpi, A., Pfeifer, A., Parton, R. G., et al. (2001). A novel 14-kilodalton protein interacts with the mitogen-activated protein kinase scaffold mp1 on a late endosomal/lysosomal compartment. *The Journal of Cell Biology*, *152*(4), 765–776.

Yamamoto, T., Ebisuya, M., Ashida, F., Okamoto, K., Yonehara, S., & Nishida, E. (2006). Continuous ERK activation downregulates antiproliferative genes throughout G1 phase to allow cell-cycle progression. *Current Biology*, *16*(12), 1171–1182.

Yang, S. S., Van Aelst, L., & Bar-Sagi, D. (1995). Differential interactions of human Sos1 and Sos2 with Grb2. *The Journal of Biological Chemistry*, *270*(31), 18212–18215.

Yang, J. Y., Xia, W., & Hu, M. C. (2006). Ionizing radiation activates expression of FOXO3a, Fas ligand, and Bim, and induces cell apoptosis. *International Journal of Oncology*, *29*(3), 643–648.

Yang, J. Y., Zong, C. S., Xia, W., Yamaguchi, H., Ding, Q., Xie, X., et al. (2008). ERK promotes tumorigenesis by inhibiting FOXO3a via MDM2-mediated degradation. *Nature Cell Biology*, *10*(2), 138–148.

Yao, Y., Li, W., Wu, J., Germann, U. A., Su, M. S., Kuida, K., et al. (2003). Extracellular signal-regulated kinase 2 is necessary for mesoderm differentiation. *Proceedings of the National Academy of Sciences of the United States of America*, *100*(22), 12759–12764.

Yoon, S., & Seger, R. (2006). The extracellular signal-regulated kinase: Multiple substrates regulate diverse cellular functions. *Growth Factors*, *24*(1), 21–44.

Yung, Y., Yao, Z., Hanoch, T., & Seger, R. (2000). ERK1b, a 46-kDa ERK isoform that is differentially regulated by MEK. *The Journal of Biological Chemistry*, *275*(21), 15799–15808.

Zeppernick, F., Ardighieri, L., Hannibal, C. G., Vang, R., Junge, J., Kjaer, S. K., et al. (2014). BRAF mutation is associated with a specific cell type with features suggestive of senescence in ovarian serous borderline (atypical proliferative) tumors. *The American Journal of Surgical Pathology*, *38*(12), 1603–1611.

Zhao, J., Yuan, X., Frodin, M., & Grummt, I. (2003). ERK-dependent phosphorylation of the transcription initiation factor TIF-IA is required for RNA polymerase I transcription and cell growth. *Molecular Cell*, *11*(2), 405–413.

Zheng, H., Al-Ayoubi, A., & Eblen, S. T. (2010). Identification of novel substrates of MAP Kinase cascades using bioengineered kinases that uniquely utilize analogs of ATP to phosphorylate substrates. *Methods in Molecular Biology*, *661*, 167–183.

CHAPTER FIVE

Role of MDA-7/IL-24 a Multifunction Protein in Human Diseases

Mitchell E. Menezes*,[2], Praveen Bhoopathi*,[2], Anjan K. Pradhan*,[2], Luni Emdad*,[†,‡,2], Swadesh K. Das*,[†,‡], Chunqing Guo*, Xiang-Yang Wang*,[†,‡], Devanand Sarkar*,[†,‡], Paul B. Fisher*,[†,‡,1]

*Virginia Commonwealth University, School of Medicine, Richmond, VA, United States
†VCU Institute of Molecular Medicine, Virginia Commonwealth University, School of Medicine, Richmond, VA, United States
‡VCU Massey Cancer Center, Virginia Commonwealth University, School of Medicine, Richmond, VA, United States
[1]Corresponding author: e-mail address: paul.fisher@vcuhealth.org

Contents

1. Introduction — 144
2. Characteristic Features of MDA-7/IL-24 — 145
   2.1 Identification of MDA-7/IL-24 — 145
   2.2 Structure of MDA-7/IL-24 — 146
   2.3 Splice Variants/Isoforms of MDA-7/IL-24 — 147
   2.4 Deletions, Modifications, and Enhancing Stability of MDA-7/IL-24 — 149
   2.5 Receptors of MDA-7/IL-24 — 150
3. Physiological Role of MDA-7/IL-24 — 150
   3.1 Naturally Occurring Cellular Source of MDA-7/IL-24 — 150
   3.2 MDA-7/IL-24 Function Under Physiological Conditions — 151
4. Functional Role of MDA-7/IL-24 in Cancer — 152
   4.1 Stem Cells and Differentiation — 152
   4.2 Apoptosis — 153
   4.3 Autophagy — 156
   4.4 Angiogenesis — 157
   4.5 Invasion and Metastasis — 158
   4.6 Synergistic Effects — 158
   4.7 Bystander Activity — 159
5. Role of MDA-7/IL-24 in Other Diseases — 162
   5.1 Inflammation — 162
   5.2 Inflammatory Bowel Disease (IBD) — 163
   5.3 Psoriasis — 164
   5.4 Cardiovascular Disease — 165

---

[2] Contributed equally to this review: M.E.M., P.B., A.K.P., and L.E.

Advances in Cancer Research, Volume 138
ISSN 0065-230X
https://doi.org/10.1016/bs.acr.2018.02.005

5.5 Rheumatoid Arthritis (RA)	167
5.6 Tuberculosis	168
5.7 Influenza Virus Replication	169
6. Immunological Effects of MDA-7/IL-24	169
7. Conclusions and Future Perspectives	171
Acknowledgments	172
Conflict of Interest	172
References	172

Abstract

Subtraction hybridization identified genes displaying differential expression as metastatic human melanoma cells terminally differentiated and lost tumorigenic properties by treatment with recombinant fibroblast interferon and mezerein. This approach permitted cloning of multiple genes displaying enhanced expression when melanoma cells terminally differentiated, called melanoma differentiation associated (*mda*) genes. One *mda* gene, *mda-7*, has risen to the top of the list based on its relevance to cancer and now inflammation and other pathological states, which based on presence of a secretory sequence, chromosomal location, and an IL-10 signature motif has been named interleukin-24 (MDA-7/IL-24). Discovered in the early 1990s, MDA-7/IL-24 has proven to be a potent, near ubiquitous cancer suppressor gene capable of inducing cancer cell death through apoptosis and toxic autophagy in cancer cells in vitro and in preclinical animal models in vivo. In addition, MDA-7/IL-24 embodied profound anticancer activity in a Phase I/II clinical trial following direct injection with an adenovirus (Ad.*mda-7*; INGN-241) in tumors in patients with advanced cancers. In multiple independent studies, MDA-7/IL-24 has been implicated in many pathological states involving inflammation and may play a role in inflammatory bowel disease, psoriasis, cardiovascular disease, rheumatoid arthritis, tuberculosis, and viral infection. This review provides an up-to-date review on the multifunctional gene *mda-7/IL-24*, which may hold potential for the therapy of not only cancer, but also other pathological states.

1. INTRODUCTION

Melanoma differentiation associated gene-7 (MDA-7), also known as interleukin-24 (IL-24), is a secreted cytokine and a member of the IL-10 gene family. Although MDA-7/IL-24 was discovered several decades ago, new discoveries of the role that MDA-7/IL-24 plays in normal physiology as well as in multiple human pathologies are still unfolding. So far, researchers have confirmed that MDA-7/IL-24 is not only involved in normal immune function and wound healing, but it also has several additional beneficial effects in a variety of human diseases. As examples, MDA-7/IL-24 functions as an anticancer gene in multiple diverse cancers including melanoma (Lebedeva et al., 2002; Sarkar et al., 2008), prostate cancer (Greco et al., 2010; Lebedeva, Sarkar, et al., 2003; Lebedeva, Su, Sarkar, & Fisher, 2003), breast cancer (Bhutia et al., 2013; Menezes et al., 2015; Pradhan et al., 2017; Sarkar et al., 2005), osteosarcoma (Zhuo et al., 2017), neuroblastoma (Bhoopathi et al., 2016), pancreatic cancer (Sarkar, Quinn, Shen, Dent, et al., 2015), renal carcinoma (Park et al., 2009), leukemia (Rahmani et al., 2010), lung cancer (Lv, Su, Liang, Hu, & Yuan, 2016; Shapiro et al., 2016), esophageal squamous cell carcinoma (Ma, Jin, et al., 2016), and hepatocellular carcinoma (Wang, Ye, Zhong, Xiang, & Yang, 2007). MDA-7/IL-24 provides protection against autoimmune diseases and bacterial infections (Leng, Pan, Tao, & Ye, 2011; Ma et al., 2009). MDA-7/IL-24 is also relevant in inflammation (Pasparakis, Haase, & Nestle, 2014), rheumatoid arthritis (RA) (Kragstrup et al., 2008), and cardiovascular diseases (Vargas-Alarcon et al., 2014). In this review, we discuss in detail the roles of MDA-7/IL-24 in both normal physiology as well as the various disease states mentioned earlier. We begin with a discussion of the characteristic features of MDA-7/IL-24 that allows this molecule to play a key role in normal cellular function as well as contributing to a variety of disease states.

2. CHARACTERISTIC FEATURES OF MDA-7/IL-24

We begin with an overview of the initial cloning of the MDA-7/IL-24 gene, followed by its structure, isoforms, and modifications that have helped enhance MDA-7/IL-24 potency. We then discuss the receptors that MDA-7/IL-24 utilizes for cellular signaling.

2.1 Identification of MDA-7/IL-24

As the name suggests, MDA-7 was initially identified and cloned from terminally differentiating human melanoma cells in the Fisher laboratory by Jiang in 1993 and reported in detail in 1995 (Jiang & Fisher, 1993; Jiang, Lin, Su, Goldstein, & Fisher, 1995). HO-1 human metastatic melanoma cells were treated with a combination of recombinant human fibroblast interferon (IFN-beta) and mezerein (MEZ) to induce terminal differentiation and suppression of growth and tumorigenic abilities. Next subtraction hybridization of cDNA libraries was performed to assess genes that were differentially expressed in melanoma cells before and after terminal differentiation (Jiang & Fisher, 1993). MDA-7 was identified as one of the transcripts that was induced in terminally differentiating melanoma cells (Jiang et al., 1995; Jiang & Fisher, 1993). In subsequent years, MDA-7 was found to have tumor suppressive abilities against several different cancer indications, while leaving normal counterparts unharmed (Jiang et al., 1996; Su et al., 1998). In 2001, Huang and colleagues in the Fisher laboratory identified the genomic structure and chromosomal localization of MDA-7 (Huang et al., 2001). They determined that MDA-7 was located in a region of the chromosome that contained a cluster of genes associated with the IL-10 cytokine family (Huang et al., 2001). MDA-7 also had an IL-10 signature sequence and was specifically expressed in tissues associated with the immune system including the spleen, thymus, and peripheral blood leukocytes (Huang et al., 2001). Given the conserved chromosomal location, presence of a putative secretory motif, an IL-10 signature sequence and the expression profile of MDA-7, the Human Gene Organization (HUGO) designated this gene as IL-24 (Sarkar et al., 2002a). Additionally in 2002, Caudell and colleagues provided evidence that MDA-7/IL-24 had functional immunostimulatory attributes justifying its designation as an interleukin (Caudell et al., 2002).

2.2 Structure of MDA-7/IL-24

Located on chromosome 1q32–33 in humans, MDA-7/IL-24 is a secreted cytokine that belongs to the IL-10 gene family (Caudell et al., 2002; Huang et al., 2001). MDA-7/IL-24 contains seven exons and six introns. The cDNA of MDA-7/IL-24 is 1718 base pairs and the protein encodes 206-amino acids (Huang et al., 2001). Being a secreted cytokine, MDA-7/IL-24 has a 49-amino acid N-terminal hydrophobic signal peptide that allows for protein secretion (Fig. 1). Sauane and colleagues in the Fisher laboratory utilized the Prosite database to analyze the peptide sequence of MDA-7/IL-24 and identified three putative *N*-glycosylation sites at amino acids 85, 99, and 126 (Sauane, Gopalkrishnan, Sarkar, et al., 2003). In addition, an IL-10 signature motif was identified from amino acids 101–121; three protein kinase C consensus phosphorylation sites were identified at amino acids 88, 133, and 161; and three casein kinase II consensus phosphorylation sites were identified at amino acids 101, 111, and 161 using this database (Sauane, Gopalkrishnan, Sarkar, et al., 2003). The predicted tertiary structure of MDA-7/IL-24 is that of a compact globular molecule comprised of four strongly helical regions interspersed by loops of unpredicted structure (Sauane, Gopalkrishnan, Sarkar, et al., 2003). MDA-7/IL-24 can also form N-linked glycosylated dimers through intermolecular disulfide bonds and these dimers are functionally active (Mumm, Ekmekcioglu, Poindexter, Chada, & Grimm, 2006).

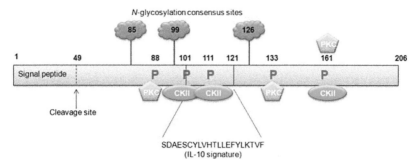

Fig. 1 Schematic representation of MDA-7/IL-24 protein with predicted and established domains and protein modification sites indicated. Cleavage of the 49-amino acid signal peptide allows for secretion of the MDA-7/IL-24 protein. The IL-10 signature sequence is located between amino acids 101 and 121. N-glycosylation can occur at amino acids 85, 99, and 126. Protein kinase C consensus phosphorylation sites are present at amino acids 88, 133, and 161. Casein kinase II (CKII) consensus phosphorylation sites are present at amino acids 101, 111, and 161. Numbers indicate amino acids. Not drawn to scale. *Figure reproduced from Menezes, M. E., Bhatia, S., Bhoopathi, P., Das, S. K., Emdad, L., Dasgupta, S., et al. (2014). MDA-7/IL-24: Multifunctional cancer killing cytokine. Advances in Experimental Medicine and Biology, 818, 127–153.*

2.3 Splice Variants/Isoforms of MDA-7/IL-24

Allen and colleagues identified a splice variant of MDA-7/IL-24, *mda-7s*, which lacked exons 3 and 5 (Allen et al., 2004). They observed that MDA-7S could heterodimerize with full-length MDA-7/IL-24 but noted that this interaction did not affect the apoptotic abilities of MDA-7/IL-24 in melanoma cells. Since the expression of *mda-7s* was reduced or absent in melanoma as compared to normal melanocytes, the authors also suggested an association between loss of *mda-7s* and metastatic melanoma (Allen et al., 2004). This same group also identified and published a short study on the presence of two splice variants that lacked exons 3 and 5, respectively, that were expressed in normal human melanocytes but not in metastatic melanoma (Allen et al., 2005). Filippov and colleagues identified splice isoforms of MDA-7/IL-24 while studying the effects of a ubiquitous splicing factor SRp55, that is upregulated by DNA damage in the absence of p53 and whose inactivation enhanced DNA damage resistance in a p53-dependent manner (Filippov, Schmidt, Filippova, & Duerksen-Hughes, 2008). U2OS human osteosarcoma cells treated with siRNA to SRp55 were assessed using a splice-specific microarray analysis to identify the relevance of SRp55 on the splicing patterns of genes involved in apoptosis. At least four isoforms of MDA-7 were identified, out of which one isoform (that lacks exons

2 and 3) was sensitive to splicing by SRp55 and silencing SRp55 splicing activity caused an increase in this isoform. In a follow-up study, Whitaker and colleagues identified and characterized five alternatively spliced isoforms of MDA-7/IL-24 (Whitaker, Filippov, Filippova, Guerrero-Juarez, & Duerksen-Hughes, 2011). Overall, they observed six differentially spliced transcripts of MDA-7/IL-24 in addition to the full-length transcript (Fig. 2). The splice variants identified were: *mda-7/IL-24δ3,5*—lacking exons 3 and 5 (described and characterized previously by Allen et al., 2004); *mda-7/IL-24δ5*—lacking exon 5; *mda-7/IL-24δ2,3*—lacking exons 2 and 3;

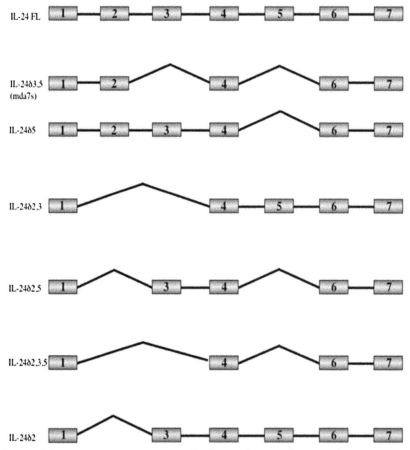

Fig. 2 Schematic representation of the splice isoforms of MDA-7/IL-24. *Figure reproduced from Whitaker, E. L., Filippov, V., Filippova, M., Guerrero-Juarez, C. F., Duerksen-Hughes, P. J. (2011). Splice variants of mda-7/IL-24 differentially affect survival and induce apoptosis in U2OS cells.* Cytokine, 56, 272–281.

mda-7/IL-24δ2,5—lacking exons 2 and 5; *mda-7/IL-24δ2,3,5*—lacking exons 2, 3, and 5; and *mda-7/IL-24δ2*—lacking exon 2. All seven exons were present in the full-length transcript. An important point to note is that the expression and distribution of *mda-7/IL-24* isoforms might vary based on different cell types. Full-length MDA-7/IL-24 as well as spliced isoforms δ5, δ2,3,5, δ2,5, and δ2 were capable of reducing U2OS cell viability with no effect on the viability of noncancerous immortalized NOK cells (Whitaker et al., 2011). Interestingly, apoptosis was higher in U2OS cells expressing mda-7/IL-24δ2,3,5 than cells expressing full-length MDA-7/IL-24.

2.4 Deletions, Modifications, and Enhancing Stability of MDA-7/IL-24

In an effort to identify the molecular basis of tumor cell selectivity of MDA-7/IL-24, Gupta and colleagues in the Fisher laboratory constructed several amino terminal deletion mutants of MDA-7/IL-24 and labeled them M1–M6 (Gupta, Walter, et al., 2006). The signal peptide was deleted in M1; α-helical domain A was disrupted in M2; α-helical domain B was disrupted in M3; α-helical domains C, D, E, and F were present in M4; α-helical domains D, E, and F were present in M5; and α-helical domains E and F were present in M6. As would be expected, deletion of the signal peptide (M1) did not disrupt the tumor inhibitory effects of MDA-7/IL-24. Interestingly, however, all the other deletions, except M4, caused a loss of tumor inhibitory effects. M4 showed tumor suppressive effects in HeLa and DU-145 cells but did not affect normal prostate epithelial P69 cells and was capable of inducing cancer cell-specific apoptosis (Gupta, Walter, et al., 2006).

MDA-7/IL-24 protein gets ubiquitinated and degraded via the 26S proteasome. In order to determine the exact site of ubiquitination, Tian and colleagues mutated each of the 10 lysine sites within the MDA-7/IL-24 protein and converted them to arginine (Tian et al., 2012). They suggested lysine 123 as the critical internal lysine involved in MDA-7/IL-24 ubiquitination. They report that further conversion of lysine 123 to arginine enhanced MDA-7/IL-24 protein stability as well as tumor suppressive abilities (Tian et al., 2012). Based on the original sequence of the MDA-7/IL-24 protein, as reported in the database, the lysine site at 122 is the key site, which when we changed to arginine resulted in enhanced stability of the MDA-7/IL-24 protein (M. Menezes et al., unpublished data).

2.5 Receptors of MDA-7/IL-24

IL-10 cytokine family members signal through receptor dimers that consist of an R1 type receptor (with a long cytoplasmic domain) and an R2 type receptor (with a short cytoplasmic domain). The IL-10 cytokine family of receptors has three R1 and two R2 subunits. The R1 subunits are IL-10R1, IL-20R1, and IL-22R1 and the R2 subunits are IL-20R2 and IL-10R2. In order to identify the receptors of MDA-7/IL-24, Wang and colleagues utilized a biochemical approach using IL-24 affinity tagged to the secreted human placental alkaline phosphatase (IL-24-AP) (Wang, Tan, Zhang, Kotenko, & Liang, 2002). They observed that MDA-7/IL-24 utilized two heterodimeric receptors, IL-22R1/IL-20R2 and IL20-R1/IL-20R2, to activate downstream signaling (Wang et al., 2002). Dumoutier and colleagues utilized ligand-dependent STAT (signal transducer and activator of transcription) activation as readout for receptor activation and independently identified these same receptors (Dumoutier, Leemans, Lejeune, Kotenko, & Renauld, 2001). More recent studies by Dash and colleagues in the Fisher laboratory demonstrated that MDA-7/IL-24 can also signal and induce growth suppression and apoptosis in a cancer-selective manner using the IL-20R1/IL22-R1 heterodimeric receptors (Dash et al., 2014; Pradhan et al., 2017). However, the mechanism through which these two R1 receptor dimers to promote signaling after interacting with MDA-7/IL-24 remains to be determined.

3. PHYSIOLOGICAL ROLE OF MDA-7/IL-24

Extensive studies were performed to understand the role of MDA-7/IL-24 in cancer, however, our understanding of the physiological role of MDA-7/IL-24 is fairly limited. In the following section, we discuss the cellular source of MDA-7/IL-24 and its functions in normal physiology.

3.1 Naturally Occurring Cellular Source of MDA-7/IL-24

MDA-7/IL-24 can be produced by immune cells (myeloid cells and lymphoid cells and monocytes) in response to treatment with lipopolysaccharides or specific cytokines (Buzas, Oppenheim, & Zack Howard, 2011). Physiological levels of MDA-7/IL-24 are induced in Th2 lymphocytes by stimulation with phorbol myristate acetate and ionomycin and in T cells, especially CD4+ naïve and memory cells activated by anti-CD3 monoclonal antibody (Sahoo et al., 2011; Schaefer, Venkataraman, & Schindler, 2001).

In monocytes, MDA-7/IL-24 is induced by antigenic stimulation with lipopolysaccharide, concanavalin A, and cytokines (Caudell et al., 2002; Wolk et al., 2004). B-cell receptor signaling also triggers MDA-7/IL-24 expression in B-lymphocytes (Maarof et al., 2010). Nonlymphoid cells can also produce physiological levels of MDA-7/IL-24 in response to cytokines secreted by immune cells (Persaud et al., 2016). Several in vitro and in vivo studies established that epithelial cells when stimulated with cytokines can secret MDA-7/IL-24 (Buzas et al., 2011; Persaud et al., 2016; Whitaker, Filippov, & Duerksen-Hughes, 2012). Additionally, IL-1 can stimulate MDA-7/IL-24 expression in keratinocytes and human colon cells (Andoh et al., 2009). Basal expression of MDA-7/IL-24 at physiological levels is found in melanocytes and expression gradually decreases as the melanocytes begin to transform into metastatic melanoma (Ekmekcioglu et al., 2001; Ellerhorst et al., 2002; Jiang et al., 1995).

3.2 MDA-7/IL-24 Function Under Physiological Conditions

MDA-7/IL-24 is produced by various immune cells and exerts a range of immune functions (Persaud et al., 2016). At lower physiological concentrations, MDA-7/IL-24 mainly functions as a cytokine. MDA-7/IL-24, when secreted, interacts with distinct sets of receptors including IL-20R1/IL-20R2, IL-22R1/IL-20R2, or IL-22R1/IL-20R1 receptor complexes (Dash et al., 2014; Dumoutier et al., 2001; Wang & Liang, 2005; Wang et al., 2002). Most immune cells lack the cognate pairs of receptors and chiefly express IL-20R2. One study by Caudell and colleagues assessed the secretion profile of peripheral blood mononuclear cells treated with MDA-7/IL-24 protein, which showed increased secretion of immune modulatory cytokines such as IL-6, IL-1β, IFN-γ, TNFα, IL-12, and GM-CSF (Caudell et al., 2002). The enhanced secretion of IFN-γ in turn upregulates IL-22R1 expression in keratinocytes, which facilitates formation of IL-22R1/IL-20R2 receptor pairs and induces innate immunity responses (Wolk et al., 2004). In addition to immune functions, MDA-7/IL-24 also induces several additional changes in normal skin cells. He and colleagues developed transgenic mice, which overexpress MDA-7/IL-24 specifically in skin (He & Liang, 2010). This genetically modified mouse is embryonic lethal and exhibits epidermal hyperplasia and abnormal keratinocyte differentiation. In contrast, treatment of human keratinocytes with MDA-7/IL-24 in a wound-healing model results in suppression of keratinocyte proliferation, suggesting a potential therapeutic role of this cytokine in proliferating skin lesions (Liang et al., 2011; Poindexter et al., 2010). MDA-7/IL-24 also impedes B-cell maturation to plasma cells by regulating several transcription factors, which are important for plasma cell differentiation (Maarof et al., 2010). Additionally, MDA-7/IL-24 plays a diverse role in proinflammatory, infectious, and autoimmune skin diseases, which is discussed in further detail below (Persaud et al., 2016).

Apart from these immune and dermatologic functions, several studies have also reported other biological functions of MDA-7/IL-24 in vascular diseases and inflammatory bowel disease (IBD) (Persaud et al., 2016). MDA-7/IL-24 is also expressed in normal cultured fetal membranes, suggesting a potential role in normal pregnancy (Nace, Fortunato, Maul, & Menon, 2010).

4. FUNCTIONAL ROLE OF MDA-7/IL-24 IN CANCER

The role of *mda-7/IL-24* has been extensively studied in cancer. In this section, we describe briefly some of the important findings.

4.1 Stem Cells and Differentiation

Tumors are comprised of heterogeneous cell populations with diverse biological properties. Cancer stem cells are immortal cells within tumors which display the property of self-renewal. They can divide and differentiate to give rise to a heterogeneous cell population, in which subsets of cells can form distant tumors (Talukdar, Emdad, Das, Sarkar, & Fisher, 2016). Stem cells detach from the primary tumor, migrate, and generate tumors at distant sites. Cancer stem cells can relapse and metastasize making the need for specific therapies against them essential (Eyler & Rich, 2008; Talukdar et al., 2016). They are also resistant to conventional therapies and divide more rapidly (Morrison, Morris, & Steel, 2013).

mda-7/IL-24 inhibits the growth of breast cancer stem cells. Specifically, infection of Ad.*mda-7* decreased proliferation of breast cancer initiating cells without harming normal stem cells (Bhutia, Das, et al., 2013). Overexpression of *mda-7/IL-24* induces apoptosis and endoplasmic reticulum (ER) stress in sorted stem cell populations of breast cancer cells, which is similar to what is observed in unsorted breast cancer cells (Bhutia, Das, et al., 2013). Overexpression of *mda-7/IL-24* also decreases the self-renewal capabilities of cancer stem cells. *mda-7/IL-24* suppresses β-catenin/Wnt signaling (Chada et al., 2005; Sieger et al., 2004) and regulates the proliferation of stem cells. The Wnt/β catenin pathway is one of the key signaling pathways that promote self-renewal of stem cells (Xu et al., 2016). Wnt proteins interact with Frizzled and LRP receptors to signal β-catenin to activate Wnt target genes (MacDonald & He, 2012). It can also signal through ROR/RYK receptors as an alternative pathway (Green, Nusse, & van Amerongen, 2014). In cancer, these are dynamically expressed and this causes an imbalance in the proliferation and differentiation of cancer stem cells. Alteration of the β-catenin signaling pathway increases the survival of stem cells. This suggests that *mda-7/IL-24*-mediated blockage in proliferation of stem cells is facilitated through the β-catenin pathway. In a subcutaneous human tumor xenograft nude mouse model, injection of Ad.*mda-7* inhibited the growth of subcutaneous tumors. Tumor growth inhibition is associated with inhibition in cellular proliferation and angiogenesis (Bhutia, Das, et al., 2013).

Overexpression of *mda-7/IL-24* by an adenoviral system increased the expression of tumor suppressors including PTEN, E-cadherin, GSK-3β, and APC and downregulated protooncogenes involved in β-catenin and PI3K signaling (Gupta, Su, et al., 2006). β-Catenin translocates to the plasma membrane from the nucleus upon *mda-7/IL-24* treatment, which reduces the transcriptional activity of TCF/LEF (Mhashilkar et al., 2003). This upregulates the expression of E-cadherin–β-catenin adhesion in a cancer-selective manner. In lung and breast cancer, *mda-7/IL-24* regulates cell–cell adhesion by modulating these signaling cascades (Mhashilkar et al., 2003). These effects are not common in normal cells and are specific for cancer cells.

Ad.*mda-7* downregulates the tendency of breast cancer cells to form mammospheres and also inhibits the formation of distant tumors (Bhutia, Das, et al., 2013). *MDA-7/IL-24* regulates the PI3K/Akt pathway, decreases β-catenin phosphorylation, and proteosomal degradation pathways (Bhutia, Das, et al., 2013; Mhashilkar et al., 2003). Stem cells also display overexpression of Akt, Bcl-2, and Bcl-xL (Wang & Scadden, 2015). *mda-7/IL-24* can induce apoptosis by downregulating Akt, Bcl-2, and Bcl-xL as described earlier (Fig. 3).

4.2 Apoptosis

Programmed cell death or apoptosis plays a pivotal therapeutic role in cancer drug sensitivity (Naik, Karrim, & Hanahan, 1996). One of the hallmarks of cancer is apoptosis (Hanahan & Weinberg, 2000). It involves a series of signaling events that are disrupted in cancer. Cancer cells bypass the apoptotic signaling pathway and evade this mechanism of cell death (Fernald & Kurokawa, 2013). Much of the research focusing on cancer therapeutics

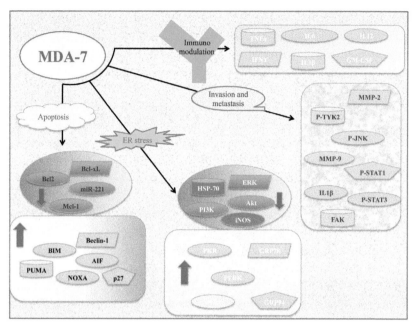

Fig. 3 Schematic representation of the pathways regulated by MDA-7/IL-24. MDA-7/IL-24 regulates both pro- and antiapoptotic molecules to induce tumor-specific cell death. This involves a series of signaling events including downregulation of Mcl-1 and Bcl-xL and activation of tumor suppressors, i.e., SARI, PUMA, AIF, PERK, and others as shown in the figure. Also the cytokine induces ER stress and regulates a number of genes/proteins to block invasion and metastasis. MDA-7/IL-24 also modulates the immune pathways by deregulating a number of cytokines, which in turn activates the immune system to induce cytotoxic cell death.

involves the ability of the therapy to induce apoptosis specifically in cancer cells (Lebedeva, Sarkar, et al., 2003). Understanding the mechanism by which cancer cells evade the general apoptotic pathways is critical to develop new therapies against cancer.

mda-7/IL-24 regulates ER stress and the mitochondrial apoptotic pathway (Fisher, 2005; Gopalkrishnan, Sauane, & Fisher, 2004; Lebedeva, Sauane, et al., 2005; Lebedeva, Su, et al., 2005; Lebedeva, Su, Sarkar, Kitada, et al., 2003; Sauane et al., 2008; Sieger et al., 2004). Overexpression of *mda-7/IL-24* has been shown to induce apoptosis in different cancer cells without any harmful effect to normal cells (reviewed in Fisher, 2005). This cancer cell-specific death is both time- and dose-dependent. The p38MAPK or mitogen protein kinase pathway is altered due to overexpression of *mda-7/IL-24* (Sarkar et al., 2002b). SB203580, an inhibitor of the p38MAPK pathway, inhibits Ad.mda-7-induced apoptosis. This induces growth arrest and DNA damage genes leading to cell cycle arrest and cell death (Sarkar et al., 2002b). AIF-mediated apoptosis by *mda-7/IL-24* has recently been demonstrated to occur uniquely in neuroblastoma (Bhoopathi et al., 2016). A recent study from our group showed that *mda-7/IL-24* regulates a subset of microRNAs (Pradhan et al., 2017). One microRNA, miR-221, was downregulated following treatment with *mda-7/IL-24*. miR-221 targets PUMA, a proapoptotic gene, blocking apoptosis (Pradhan et al., 2017). *mda-7/IL-24* downregulates miR-221, which in turn upregulates PUMA inducing cell death (Pradhan et al., 2017). *mda-7/IL-24* downregulates the expression of antiapoptotic proteins Mcl-1, Bcl-xL, and Bcl-2, while inducing proapoptotic proteins such as Bid, Bim, Bax, and Bak (Menezes et al., 2014). In so doing, *mda-7/IL-24* increases the Bax/Bcl-2 ratio (Pei et al., 2012). Previous studies also demonstrated a role of PERK in *mda-7/IL-24*-mediated cell death (Park et al., 2008).

Enhanced expression of *mda-7/IL-24* induces the production of reactive oxygen species (ROS), which regulates multiple signaling cascades deregulating the mitochondrial integrity and cell death (Dent et al., 2010; Lebedeva, Su, et al., 2005; Lebedeva, Su, Sarkar, Kitada, et al., 2003). The role of ROS in *mda-7/IL-24*-mediated cell death is well established. N-acetyl cysteine (Nace et al., 2010) inhibits cell death mediated by *mda-7/IL-24* (Lebedeva, Sauane, et al., 2005). Simultaneously, ROS inducers enhance cell death mediated by *mda-7/IL-24* (Lebedeva, Su, et al., 2005; Sauane et al., 2008). These results confirm the role of ROS and mitochondrial membrane potential as an important component in cell death promoted in cancer cells by the cytokine MDA-7/IL-24.

mda-7/IL-24 also upregulates SARI, a tumor suppressor, which is cancer specific (Dash et al., 2014). Ectopic expression of *mda-7/IL-24* induces SARI mRNA and protein in a broad panel of cancer cells (Dash et al., 2014). SARI expression is required for the antitumor effects of *mda-7/IL-24*. Recombinant MDA-7/IL-24 protein also induces SARI expression through binding to its cognate receptors, IL-22R1/IL-20R2 or IL20-R1/IL-20R2 (Dash et al., 2014).

The FasL signaling pathway is another pathway activated by Ad.*mda-7*, which results in cancer cell-selective apoptosis (Gopalan et al., 2005). Ad.*mda-7* induces activation of the transcription factors c-Jun and activating transcription factor 2 which then activates downstream target, FasL-Fas (Gopalan et al., 2005). siRNA targeting Fas decreased *mda-7/IL-24*-induced cell death in ovarian cancer cells (Gopalan et al., 2005).

This work reveals a role of *mda-7/IL-24* in regulating the Fas–FasL signaling cascade to induce cancer cell death.

mda-7/IL-24 upregulates PKR (serine/threonine protein kinase) in non-small cell lung cancer, which is independent of p53 expression (Mhashilkar et al., 2003). The regulation of PKR by *mda-7/IL-24* is posttranscriptional (Gupta, Su, et al., 2006). Exogenous recombinant *mda-7/IL-24* also induces PKR and *mda-7/IL-24* interacts with PKR in cancer cells (Pataer et al., 2005).

Apoptosis mediated by *mda-7/IL-24* is independent of p53 mutations and functions (Gupta, Su, et al., 2006; Su et al., 2003). It is established that *mda-7/IL-24* induces apoptosis in diverse breast cancer cells, i.e., MCF7 (p53-wt), MDA-MB-231 (mutant p53), MDA-MB-453 (mutant p53), and T47D (mutant p53) (Chada et al., 2006). Based on the different genetic backgrounds, these results indicate that cell death induction by *mda-7/IL-24* is also independent of ER/PR/HER2 status in breast cancer cells. In these contexts, *mda-7*/IL-24-induced apoptosis is distinct from other identified tumor suppressors.

Secreted MDA-7/IL-24 protein binds to its cognate receptor pairs and induces phosphorylation and nuclear translocation of STAT3 (Chada, Mhashilkar, et al., 2004). This receptor interaction induces BAX protein leading to cell death (Gupta, Su, et al., 2006). This process is STAT3-independent as other interleukins (IL-10, IL-19, IL-20, and IL-22) also activate STAT3 without promoting cell death (Mosser & Zhang, 2008). *mda-7/IL-24* binds IL-20/IL-22 receptor complexes resulting in activation of the JAK/STAT cascade. Studies have shown that *mda-7/IL-24* induces apoptosis of cancer cells independent of the JAK/STAT pathway (Sauane, Gopalkrishnan, Lebedeva, et al., 2003). Specifically, inhibitors of JAK/STAT pathway do not inhibit apoptosis mediated by *mda-7/IL-24* (Sauane, Gopalkrishnan, Lebedeva, et al., 2003). These results demonstrate that *mda-7/IL-24* functions independent of tyrosine kinase activation.

4.3 Autophagy

Autophagy is the process of degradation of organelles located in the cytoplasm. This process is complex due to its differential context dependent role. "Is autophagy good or bad for life and cancer?" is a difficult question to answer (Bhutia, Mukhopadhyay, et al., 2013). Sometimes it is protective, helping cancer cells to survive adverse conditions but it can also be toxic toward cancer cells (Bhutia, Mukhopadhyay, et al., 2013; Liu &

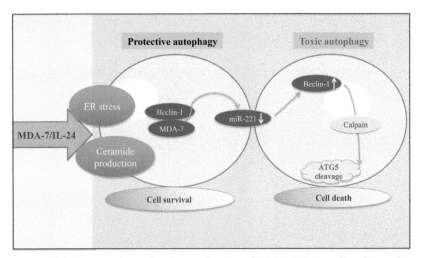

Fig. 4 Model depicting the molecular mechanism of MDA-7/IL-24-mediated autophagy induction. MDA-7/IL-24 regulates autophagy mediated through ER stress and ceramide production. Also MDA-7/IL-24 downregulates miR-221, which in turn upregulates Beclin-1 to induce toxic autophagy leading to cell death. The transition of protective to toxic autophagy is mediated by the cleavage of ATG5 by Calpain.

Debnath, 2016) (Fig. 4). Small molecules that can control autophagy may in certain contexts provide therapeutic benefit. *mda-7/IL-24* induces autophagy, which is mediated by PERK (Park et al., 2008) and Beclin-1 (Bhutia et al., 2010). *mda-7/IL-24* regulates a subset of microRNAs, including the oncogenic microRNA, miR-221 (Pradhan et al., 2017). Beclin-1 was identified as a new transcriptional target of miR-221 (Pradhan et al., 2017). *mda-7/IL-24* downregulates miR-221, which in turn induces beclin-1, leading to autophagy (Pradhan et al., 2017). Cleavage of LC3, a marker of autophagy, is also observed. In renal and ovarian cancers, CD95 is an important regulatory molecule in the induction of autophagy mediated by *mda-7/IL-24* (Park et al., 2009).

4.4 Angiogenesis

Cancer and metastatic spread depends on an adequate supply of nutrients and oxygen to cells (Welch & Fisher, 2016). Additionally, removal of waste products also requires new blood and lymph vessels. The process of formation of new blood vessels is called angiogenesis, which represents another hallmark of cancer (Hanahan & Weinberg, 2011). Angiogenesis is regulated by a number of activator and inhibitor molecules. Although not as effective as anticipated when used as a single agent, angiogenesis inhibitors combined with other therapeutic agents are showing promise in the treatment of various cancers.

Overexpression of *mda-7/IL-24* in human umbilical vascular endothelial cells inhibits endothelial cell differentiation (Dash et al., 2010; Wang, Wang, Chen, & Lv, 2016). Similarly, treatment of tumor xenografts with *mda-7/IL-24* reduces expression of angiogenesis markers (Bhutia et al., 2012). Vascular endothelial growth factor and basic fibroblast growth factor, which induce angiogenesis, are inhibited by MDA-7/IL-24 protein (Nishikawa, Ramesh, Munshi, Chada, & Meyn, 2004). The PI3K/Akt pathway is another signaling cascade known to regulate angiogenesis (Karar & Maity, 2011), and *mda-7/IL-24* downregulates phospho Akt and can therefore negatively modulate angiogenesis (Dash et al., 2010).

4.5 Invasion and Metastasis

mda-7/IL-24 has been shown to impede the migration of cancer cells (Ramesh et al., 2004). Also, overexpression of *mda-7/IL-24* results in a decrease in the in vitro invasion of an array of different cancer cell types (Ramesh et al., 2004). Lung cancer cells showed an inhibition in migration and invasion by modulating a number of signaling cascades following overexpression of *mda-7/IL-24* (Panneerselvam et al., 2015). Focal adhesion kinase (FAK) and matrix metalloproteinases (MMPs) play a critical role in migration and invasion of cells (Hauck, Hsia, & Schlaepfer, 2002; Lin et al., 2000). *mda-7/IL-24* downregulates FAK and MMP-2/MMP-9 protein, which indirectly inhibits migration and invasion of cancer cells (Menezes et al., 2014; Ramesh et al., 2004). *mda-7/IL-24* has been shown to promote potent antiinvasive activity in lung cancer cells, cervical cancer cells, and liver cancer cells (Emdad et al., 2009; Lebedeva et al., 2007). *mda-7/IL-24* regulates a number of molecules related to metastasis, i.e., cyclin-B1, TGF-β, Survinin, Twist, ICAM-1, and CD44 (Huo, Li, Zhu, Bao, & An, 2013). Also, E-cadherin, NF-κB, and PERK are regulated by *mda-7/IL-24* (Panneerselvam, Munshi, & Ramesh, 2013). *mda-7/IL-24*-mediated inhibition in invasion and metastasis is both receptor-dependent and receptor-independent (Menezes et al., 2014).

4.6 Synergistic Effects

Cancer is a complex process that is mediated by multiple genetic and epigenetic changes that impact directly and indirectly on a number of pivotal signaling pathways involved in cell growth, survival, resistance to apoptosis, and additional physiologically relevant processes (Hanahan & Weinberg, 2000, 2011). Considering this complexity, it is not surprising that a single targeting molecule fails to provide complete therapy resulting in a cure in most cancers. Conversely, a combinatorial approach using multiple target-selective agents directed toward specific signaling abnormalities in defined cancers has shown promise in cancer therapy.

Based on Phase I/II clinical studies, *mda-7/IL-24* has been shown to have a therapeutic role in cancer (Cunningham et al., 2005; Fisher et al., 2003, 2007; Tong et al., 2005). Also, preclinical studies have confirmed synergistic therapeutic responses when *mda-7/IL-24* is combined with existing therapies, including radiation, chemotherapy, antibody-based therapies, small molecule, and immunotherapies (Table 1). The mechanisms underlying this synergy include the regulation of similar pathways as well as different pathways by *mda-7/IL-24*. This is tabulated in Table 1.

4.7 Bystander Activity

Evidence of bystander activity of *mda-7/IL-24* (Su et al., 2005) was shown in vivo in animal studies, where tumor cells were injected in both flanks of nude mice (Pradhan et al., 2017; Sarkar et al., 2007, 2008, 2005; Su et al., 2005). A tumor on one flank was treated while the tumor on the other flank was left untreated. Tumor measurements showed a decrease in tumor size in the treated as well as the untreated tumor. The inhibitory action on distant tumors can be explained by the antitumor "bystander" activity of the secreted *mda-7/IL-24* cytokine and its ability to induce apoptosis and promote production of MDA-7/IL-24 through dimeric receptor pairs in the untreated tumor (Menezes et al., 2014; Sauane et al., 2008). Additionally, in a syngeneic model this distant antitumor effect can also be explained by the activation of immune pathways, i.e., cytotoxic T cells and NK cells by administration of *mda-7/IL-24* (Gao et al., 2008; Menezes et al., 2015; Miyahara et al., 2006). Overexpression of *mda-7/IL-24* gene results in production of MDA-7/IL-24 protein which is secreted as a glycosylated protein (Dash et al., 2010; Fuson et al., 2009; Sauane et al., 2006). Infection of Ad. *mda-7*, which is dependent on coxsackie and adenovirus viral receptors on cells, or treatment with GST-MDA-7 is not dependent on the IL-22R1/IL-20R2 or IL20-R1/IL-20R2 receptors (Dent et al., 2010; Sauane et al., 2004). In contrast, to provoke a signaling and biological effect, secreted MDA-7/IL-24 requires a complete set of dimeric cell surface receptors

Table 1 Combinatorial Enhancement of Therapy by Combining *mda-7/IL-24* With Other Therapeutic Modalities

Therapeutic Agent	Cancer Types	References
Trastuzumab	Breast cancer	McKenzie et al., (2004)
Bevacizumab	Lung cancer	Inoue et al. (2007)
Erlotinib	Melanoma	Deng, Kwon, Ekmekcioglu, Poindexter, and Grimm (2011)
Gefitinib	NSCLC	Emdad, Lebedeva, Su, Gupta, et al. (2007)
Temozolamide	Glioblastoma	Hamed, Yacoub, Park, Eulitt, Dash, et al. (2010)
Tarceva	NSCLC	Gupta et al. (2008)
Arsenic trioxide	Renal carcinoma	Yacoub et al. (2003)
Cisplatin	Liver, colorectal	Wu et al. (2009)
Sabutoclax (Mcl-1 inhibitor)	Prostate	Dash et al. (2011)
Sabutoclax (Mcl-1 inhibitor)	Colorectal	Azab et al. (2012)
BI-97D6 (Mcl-1 inhibitor)	Prostate	Sarkar, Quinn, Shen, Dash, et al. (2015)
Grp170	Prostate	Gao et al. (2008)
Radiation	Prostate	Su et al. (2006)
BI-69A11	Colon	Pal et al. (2014)
5-FU	Esophageal	Ma et al. (2014)
HSP90 inhibitors	Pancreatic	Zhang et al. (2013)
HDAC inhibitors	Renal carcinoma	Hamed, Das, et al. (2013)
HDAC inhibitors	Glioblastoma	Hamed, Yacoub, et al. (2013)
5-FU, doxorubicin	Colon	Xu et al. (2013)
Sorafenib	Renal carcinoma	Eulitt et al. (2010)
Doxorubicin	Hepatocellular carcinoma	Wang et al. (2010)
Dichloroacetate	Hepatocellular carcinoma	Xiao et al. (2010)
osu-03012	Glioblastoma	Hamed, Yacoub, Park, Eulitt, Sarkar, et al. (2010)

Table 1 Combinatorial Enhancement of Therapy by Combining *mda-7/IL-24* With Other Therapeutic Modalities—cont'd

Therapeutic Agent	Cancer Types	References
Perillyl alcohol	Pancreatic cancer	Lebedeva et al. (2008)
CDDP, Epirubicin, VCR	B cell lymphoma	Ma, Zhao, et al. (2016)
Doxorubicin	Colorectal	Emdad, Lebedeva, Su, Sarkar, et al. (2007)
Vitamin E succinate	Ovarian cancer	Shanker et al. (2007)
Geldanamycin	Lung cancer	Pataer et al. (2007)
Radiation	Ovarian	Emdad et al. (2006)
Celecoxib	Breast cancer	Suh et al. (2005)
Sulindac	Lung cancer	Oida et al. (2005)

(Dash et al., 2014; Dumoutier et al., 2001; Wang et al., 2002). Secreted MDA-7/IL-24 binds to the dimeric receptor pair and induces cancer cell death (Dash et al., 2014; Menezes et al., 2014). By the use of IL-20R1/IL-20R2 antibodies, it has been demonstrated that mda-7/IL-24-mediated cell death is receptor dependent (Chada, Mhashilkar, et al., 2004). Zheng and colleagues described the role of IL-20R1/IL-20R2 receptor pair in *mda-7/IL-24*-mediated cell death and its independence of STAT3 phosphorylation (Zheng, Bocangel, et al., 2007). Biological activity of *mda-7/IL-24* was also shown to be independent of JAK/STAT signaling using inhibitors and various receptor mutant cells (Sauane, Gopalkrishnan, Lebedeva, et al., 2003).

Normal cells also promote "bystander" activity after exposure to *mda-7/IL-24*, which results in production and secretion of MDA-7/IL-24 without inducing toxicity or cell death. Infection of normal primary or immortal human cells, such primary human fetal astrocytes, FM-516 or P69, results in secretion of MDA-7/IL-24. Addition of supernatant from normal cells infected with Ad.*mda-7* to cancer cells results in suppression of their growth and induction of apoptosis. Since Ad.*mda-7* will result in MDA-7/IL-24 protein production in normal and cancer cells, this can result in a robust "bystander" effect that is observed both in preclinical animal models and in a Phase I/II clinical trial in patients with advanced cancers (Dash et al., 2010).

Activation of the immune system provides another important mechanism underlying the "bystander" activity of *mda-7/IL-24*. MDA-7/IL-24 induces IL-6, TNFα, IFN-γ, IL-1β, and IL-12, which are potent immunoregulatory molecules (Caudell et al., 2002; Deng et al., 2011; Menezes et al., 2014). Also, these immunoregulatory molecules can regulate APCs to present tumor antigens to trigger immune response (Caudell et al., 2002). In addition to immune-mediated effects, the "bystander" antitumor activity of MDA-7/IL-24 is also elicited through its direct proapoptotic and antiangiogenic activity (Dash et al., 2010).

5. ROLE OF MDA-7/IL-24 IN OTHER DISEASES

MDA-7/IL-24 has been extensively studied in cancer. In addition to its function as a tumor suppressor and apoptosis-toxic autophagy inducing cytokine in cancer, MDA-7/IL-24 has also been reported to play a significant role in inflammation, cardiovascular disease, autoimmune diseases, and viral replication.

5.1 Inflammation

The skin is the largest organ in the body and plays an essential role in promoting immunity and defense against pathogenic microorganisms. However, dysregulated immune reactions can cause chronic inflammatory skin diseases. Extensive crosstalk between the different cellular and microbial components of the skin regulates local immune responses to ensure efficient host defense, to maintain and restore homeostasis, and to prevent chronic disease. In this section, we briefly discuss recent findings that highlight a role of MDA-7/IL-24 in inflammation. IL-19 and MDA-7/IL-24 belong to the IL-20 subfamily and are known to be involved in host defense against bacteria and fungi, tissue remodeling, and wound healing (Fonseca-Camarillo, Furuzawa-Carballeda, Granados, & Yamamoto-Furusho, 2014). These groups of cytokines are involved in protecting the epithelial tissue from damage that is a consequence of bacterial and viral infections. MDA-7/IL-24 may be a member of a complex cascade of cytokines involved in inflammation as MDA-7/IL-24 can induce expression of many cytokines, including TNFα, IL-6, and IFN-γ (Wang & Liang, 2005). MDA-7/IL-24 and its receptor expression pattern support a major physiological function related to epidermal functions, such as wound healing, and abnormalities may be part of the cause of pathological skin conditions such as psoriasis.

5.2 Inflammatory Bowel Disease (IBD)

Chronic inflammation of all parts of the digestive tract may bring about IBD. This includes primarily ulcerative colitis (UC) and Crohn's disease (CD). The symptoms for both of these conditions include severe diarrhea, pain, fatigue, and weight loss. Genomic abnormalities and environmental factors can trigger IBD. Andoh and colleagues assessed the expression of MDA-7/IL-24 in inflamed mucosa of IBD patients and determined the molecular mechanism that resulted in MDA-7/IL-24 expression in colonic subepithelial myofibroblasts (Andoh et al., 2009). They demonstrated that MDA-7/IL-24 expression is enhanced in the inflamed mucosa of active IBD patients. Their data suggest that MDA-7/IL-24 targets epithelial cells and play antiinflammatory and protective roles in the intestinal mucosa. This elevated expression of MDA-7/IL-24 leads to increased Jak/Stat pathway signals leading to increased expression of different MUC genes in the mucosa. MUC genes are the primary component of the mucin barrier that divides the intestinal microbiota and the intestinal epithelium. MUC genes also play an important role in the pathogenesis of IBD. This study showed that MDA-7/IL-24 expression is elevated in inflamed mucosa of IBD patients compared to control patients. Work done by other researchers show that the IL-10 subfamily of cytokines is involved in immune regulation and inflammatory responses. To obtain an enhanced understanding of this group of cytokines for potential therapeutic applications, more focus is required on mechanism; some of them may in the future reduce adverse side effects and/or increase the efficacy typically observed in IL-10 therapy for IBD. In active IBD, MDA-7/IL-24 is synthesized by peripheral B cells, CD4+ T cells, CD8+ T cells, and monocytes. Overall, MDA-7/IL-24 can promote a suppressive inflammatory effect on colonic epithelial cells and mucosal inflammation in IBD.

Studies by Fonseca-Camarillo and colleagues explored the role of MDA-7/IL-24 in Mexican matzo patients with IBD (Fonseca-Camarillo et al., 2014). The authors studied a total of 113 patients that included 77 patients with UC and 36 patients with CD. This study also included 33 patients as control. They compared the gene expression profiles of IL-19 and MDA-7/IL-24 in these patients. The study found that IL-19 and MDA-7/IL-24 levels were elevated significantly with active IBD disease compared with inactive IBD at both a transcriptional and translational levels. Additionally, they showed that when compared with active UC and noninflammatory tissue an increase in IL-19 and MDA-7/IL-24 producing cells were observed in active CD. This study indicates that in patients with active IBD, circulating B cells and monocytes produce IL-19 and peripheral B cells, CD4+ T cells, CD8+ T cells and monocytes produce MDA-7/IL-24.

5.3 Psoriasis

Psoriasis is a common chronic inflammatory skin disease resulting from a complex interplay among the immune system, keratinocytes, susceptibility genes, and environmental factors with a prevalence of 2% in the Caucasian population. Kumari and colleagues observed the presence of MDA-7/IL-24 as well as IL-19 and IL-20 in psoriatic skin lesions (Kumari et al., 2013). Results from these studies showed that MDA-7/IL-24 was elevated in psoriatic skin compared to normal skin. It is also reported that MDA-7/IL-24 can induce different psoriasis-associated factors, which can promote inflammation and epidermal hyperplasia (Kumari et al., 2013).

The IL-10 family of cytokines including MDA-7/IL-24 has been implicated in the pathogenesis of psoriasis (Kunz et al., 2006; Leng et al., 2011; Romer et al., 2003; Weiss et al., 2004; Wolk et al., 2009). These reports also showed an increased expression of MDA-7/IL-24 in psoriatic skin compared to normal skin. MDA-7/IL-24 was mainly produced by keratinocytes, myeloid cells, and T cells (Conti et al., 2003; Kunz et al., 2006; Zheng, Danilenko, et al., 2007). High expression of MDA-7/IL-24 receptors are also found in keratinocytes and they signal by activating STAT3 (Dumoutier et al., 2001; Kunz et al., 2006; Parrish-Novak et al., 2002). STAT3 overexpression is also observed in psoriatic skin conditions and the expression of constitutively active STAT3 in epidermal keratinocytes also caused psoriasis-like skin inflammation in mice (Sano et al., 2005), which suggests an important role for epidermal STAT3 signaling in psoriasis (Kumari et al., 2013).

Kumari and colleagues report that epidermis-specific NF-κB inhibition increased MDA-7/IL-24 and STAT3 expression in keratinocytes in a TNFR1-dependent manner in psoriasis-like skin inflammation. In the psoriasis epidermis, MDA-7/IL-24 expression was elevated and inhibition of NF-κB increased MDA-7/IL-24 expression in TNF-stimulated human primary keratinocytes. This suggests the importance of this molecular pathway in human psoriasis. They also showed a new keratinocyte-intrinsic mechanism that linked TNFR1, NF-κB, ERK, MDA-7/IL-24, IL-22R1, and STAT3 signaling to disease initiation in psoriasis pathogenesis. The authors also show that skin inflammation requires both TNFR1 signaling in IKK2-deficient epidermal keratinocytes and also identified skin epithelial cells as the major cellular target of this model. This manuscript also demonstrates that in keratinocytes, TNFR1-induced, ROS-, and ERK-dependent expression of MDA-7/IL-24 is a key early event in skin inflammation. In the inflammatory process, epidermis-specific inhibition of NF-κB activates Stat3 and increases MDA-7/IL-24 expression in primary keratinocytes (Persaud et al., 2016). Taken together, the studies on MDA-7/IL-24 in psoriasis indicate a significant role in the expression of proinflammatory mediators thereby resulting in psoriatic skin lesions. The studies also provide evidence suggesting that MDA-7/IL-24 may play a key role in psoriasis initiation.

5.4 Cardiovascular Disease

Vascular calcification is a symptom of cardiovascular disease. Wang and colleagues showed that low concentration of H_2O_2 treatment induced abnormal proliferation of vascular endothelial cells and MDA-7/IL-24 inhibited this proliferation (Wang et al., 2016). They also showed that MDA-7/IL-24 could inhibit apoptosis by inhibiting ROS production in vascular endothelial cells. MDA-7/IL-24 is also involved in the downregulation of several genes that regulate cardiovascular disease. The authors concluded that MDA-7/IL-24 can provide a basic therapeutic strategy for treating vascular disease and cancer by inhibiting ROS production in vascular cells. Lower levels of MDA-7/IL-24 were observed in hypertensive rats compared to controls, and antihypertensive therapy increased MDA-7/IL-24 levels. Hypertension is also a hallmark of cardiovascular disease. MDA-7/IL-24 was identified as 1 of the 16 differentially regulated genes in spontaneously hypertensive rats. MDA-7/IL-24 also regulates the expression of inflammation- and hypertension-related genes in a H_2O_2-treated mouse vascular smooth muscle cell line, MOVAS. This study showed that MDA-7/IL-24 attenuates H_2O_2-induced activation of PI3K/Akt and Erk. Studies by Ki-Mo Lee and colleagues also suggests that MDA-7/IL-24 can inhibit ROS production by regulating mitochondrial ROS release mediated by PI3K/Akt and Erk pathway in H_2O_2-treated vascular smooth muscle cells (VSMCs) (Lee et al., 2012). This inhibition of ROS in VSMC leads to reduced cell growth and migration. Another study by Chen and colleagues indicated that adenovirus-mediated expression of MDA-7/IL-24 could inhibit pulmonary arterial smooth muscle cell line (PAC1-SMC) migration and proliferation, leading to reduced intimal hyperplasia

(Chen, Chada, Mhashilkar, & Miano, 2003). This study also emphasizes the role of MDA-7/IL-24 in cancer-specific cell death as the authors validated the inhibition of proliferation and induction of apoptosis in PAC1-SMCs (these cells have tumorigenic potential) compared to normal human coronary artery SMC and rat aortic SMC. Based on this data, MDA-7/IL-24 could be used as a therapeutic option for vascular proliferative disorders. Taken together, these studies suggest that MDA-7/IL-24 may be a novel therapeutic target for cardiovascular disease and/or hypertension.

Another study showed that MDA-7/IL-24 inhibits β-GP-induced VSMC calcification. Activation of the Wnt/β-catenin pathway by β-GP is inhibited by MDA-7/IL-24, which indicates that the inhibitory effect of MDA-7/IL-24 on VSMC calcification correlates with the inactivation of the Wnt/β-catenin pathway. This inhibition by MDA-7/IL-24 correlates with suppression of apoptosis, and the expression of osteoblast markers and calcification by downregulation of BMP-2 and the Wnt/β-catenin pathway. They also showed that β-GP increased the expression of calcification and osteoblastic markers in VSMCs (Persaud et al., 2016). This effect is specifically inhibited by MDA-7/IL-24 suggesting that MDA-7/IL-24 suppresses downstream molecules by inhibiting BMP-2 expression. The inhibitory effect MDA-7/IL-24 on VSMC calcification is mediated at least in part through antiapoptotic activity. The effect of MDA-7/IL-24 on VSMC calcification is similar to statins, which are hydroxy-3-methylglutaryl coenzyme A reductase inhibitors. These results explain the role of MDA-7/IL-24 in pathophysiology of vascular calcification.

Vargas-Alarcon and colleagues showed in a case–control association study that individuals with premature coronary artery disease (CAD), subclinical atherosclerosis (SA), and healthy controls who had several metabolic and cardiovascular risk factors were associated with MDA-7/IL-24 polymorphisms (Vargas-Alarcon et al., 2014). The authors used an informatics approach and showed that the rs1150253 and rs1150258 polymorphisms in MDA-7/IL-24 had a functional effect generating DNA binding sites for transcription factors. In Mexican populations, these polymorphisms can be used as risk factors for cardiovascular disease, hypertension, diabetes, and increased levels of lipids. The authors concluded that the association of MDA-7/IL-24 polymorphisms with metabolic parameters and cardiovascular risk factors was due to characteristic genetic background with important differences in Mexican populations compared to other populations.

Based on the available literature, MDA-7/IL-24 appears to play a distinct role in cardiovascular disease. MDA-7/IL-24 can promote the growth of

VSMCs by suppressing calcification and osteoblast marker expression, which is associated with atherosclerosis pathogenesis. MDA-7/IL-24 also may provide benefit in the treatment of vascular disorders since it selectively inhibits rat pulmonary arterial smooth muscle cell growth and migration. Polymorphisms in the MDA-7/IL-24 gene also correlate with cardiovascular and metabolic risk factors, further supporting a relationship between MDA-7/IL-24 and cardiovascular diseases.

5.5 Rheumatoid Arthritis (RA)

RA is an inflammatory autoimmune disease that can lead to progressive joint damage and disability. Cytokines including IL-1, IL-6, IL-8, IL-10, monocyte chemoattractant protein 1 (CCL2/MCP-1), and tumor necrosis factor (TNFα) play an important role in RA. A study in RA and spondyloarthropathy (SpA) patients with osteoarthritis (OA) patients as controls, analyzed the role of IL-20 and MDA-7/IL-24 by measuring levels of expression, cellular sources, and targets and effects on cytokine production. This study indicated increased levels of IL-20 and MDA-7/IL-24 in RA and SpA patients as compared with inflammatory disease controls and normal controls. They also found that MDA-7/IL-24 levels were almost 10-times greater in these samples as compared to IL-20 levels in synovial fluid, demonstrating the dominant role of MDA-7/IL-24 locally in the joints, because these two cytokines share the same receptors. This study also showed that IL-20R1 and IL-22R are expressed in granulocytes from the RA and SpA patients' synovial fluid. This indicates that these two cytokines could be involved in neutrophil chemotaxis in arthritis. This study also showed that IL-20 and MDA-7/IL-24 are not involved directly in TNFα and IL-6 production in arthritis, whereas increased expression of CCL2/MCP-1 in SFMC cultures was evident indicating a positive correlation in RA and SpA patients. Taken together, this study demonstrates the association of IL-20 and MDA-7/IL-24 to the synovium of RA and SpA. It also implicates the importance of IL-20 and MDA-7/IL-24 in endothelial cell function and recruitment of granulocytes and mononuclear cells to the synovial joint (Kragstrup et al., 2008).

Kragstrup and colleagues observed an increased plasma concentration of IL-20 and MDA-7/IL-24 in early RA patients as compared to normal healthy controls, and with conventional or antiinflammatory treatment these levels decreased (Kragstrup et al., 2008). Radiographic progression of the disease and the association of IL-20 and MDA-7/IL-24 suggest an involvement of these cytokines in bone destruction. These two cytokines link RA-associated autoantibodies and radiographic progression of IL-22R1. By showing the relationship between IL-20 and MDA-7/IL-24 and RA-associated immune complexes and osteoclasts stimulation via IL-22R1, the investigators demonstrate a correlation between the IL-20R axis and they also provide evidence for a relationship between the IL-20R axis and progression of structural damage. This study showed that targeting the IL-20R axis could be a viable treatment option for bone destruction in rheumatic disease. It also suggests that the dual inhibition of IL-20 and MDA-7/IL-24 or inhibition of IL-22R1 could be helpful in seropositive RA. These changes in this IL-20R axis provide a promising treatment modality for RA. These studies show a clear association of MDA-7/IL-24 and RA, though additional research is vital to fully understand the potential role of MDA-7/IL-24 in RA.

5.6 Tuberculosis

Tuberculosis (TB) is an infectious disease caused by *Mycobacterium tuberculosis* in humans. Although the lungs are the primary organs altered by TB infection, other parts of the body can also be affected. Wu and colleagues reported that active TB patients had decreased expression of MDA-7/IL-24 compared to individuals with latent TB infection. This observation led them to investigate the role of MDA-7/IL-24 in pulmonary TB patients. Since IFN-γ plays an important role in TB infection, and the levels of IFN-γ were similar to MDA-7/IL-24 levels in these patients, they investigated the role of MDA-7/IL-24 on IFN-γ expression. PBMCs isolated from these individuals were stimulated with *M. tuberculosis* early secreted Ag of 6 kDa (EAST-6) to determine the levels of gene expression. Exogenous MDA-7/IL-24 in the presence of EAST-6 stimulation in PBMCs increased IFN-γ levels and neutralizing MDA-7/IL-24 decreased IFN-γ. This upregulation of IFN-γ with exogenous MDA-7/IL-24 was due to increased levels of IL-12α, IL-12β, IL-23α, and IL-27. Taken together, these results show that MDA-7/IL-24 regulates IFN-γ in TB patients and targeting MDA-7/IL-24 might be a treatment option for these patients. Another study by Kumar and colleagues indicated significantly lower levels of MDA-7/IL-24 in TB patients (Kumar et al., 2015; Ma et al., 2011). Additional research is required in this area to decipher molecular mechanism of MDA-7/IL-24 in the pathophysiology of TB in patients.

5.7 Influenza Virus Replication

Influenza infection also known as flu, is associated with mild to severe symptoms including fever, headaches, runny nose, and fatigue. Weiss and colleagues studied the role of MDA-7/IL-24 in influenza A virus replication, as MDA-7/IL-24 is known to influence TLR3-mediated apoptosis and influenza virus can stimulate the TLR3 receptor (Weiss et al., 2015). In this study, the investigators demonstrated that the expression of MDA-7/IL-24 could decrease influenza A virus subtypes replication by inducing apoptosis. The reduction of viral replication by MDA-7/IL-24 could be independent of type I interferon. MDA-7/IL-24 could inhibit Mcl1 and induce caspase 3 cleavage due to initiation of TLR3-mediated apoptosis. This was further demonstrated by TLR3 knockdown or by treating cells with a Pan-Caspase inhibitor. Inhibition of antiapoptotic proteins Bcl-2, Bax, and Bcl-xL was also observed following MDA-7/IL-24 expression. They established that Mcl1 is the key factor in MDA-7/IL-24-mediated inhibition of influenza A virus replication. They also showed that MDA-7/IL-24 expressed by influenza A virus vector does not have any toxicity in mice. Another study by Seong and colleagues also showed that MDA-7/IL-24 expression decreased influenza viral replication (Seong, Choi, & Shin, 2016). MDA-7/IL-24 decreased the transcript level of the viral nucleoprotein (NP) gene following influenza virus infection as compared to viral infection alone, confirming an inhibitory role of MDA-7/IL-24 in viral replication. Furthermore, an MDA-7/IL-24 expressing recombinant adenovirus did not induce toxicity as compared to a wild-type adenovirus, suggesting that MDA-7/IL-24 can specifically target virus-infected cells. Taken together, these studies suggest that MDA-7/IL-24 exerts potent inhibitory activity of influenza viral replication and can be used as a promising novel approach to suppress viral infections (Seong et al., 2016; Weiss et al., 2015).

6. IMMUNOLOGICAL EFFECTS OF MDA-7/IL-24

The role of MDA-7/IL-24 in normal physiology and disease pathology is quite diverse and depends principally on the source of production/secretion and the target tissue. As a cytokine, MDA-7/IL-24 exerts immune-modulatory functions in diverse autoimmune, infectious, and immunopathological diseases including RA, psoriasis, IBDs and others, as discussed in detail earlier (also reviewed in Persaud et al., 2016).

MDA-7/IL-24 also plays a prominent role in host defense by inducing innate immune response in epithelial tissue during infection and inflammation by induction of chemokines and recruitment/activation of leukocytes (Jin, Choi, Chun, & Noh, 2014; Tamai et al., 2012).

Apart from these immune-modulatory roles in diverse biological diseases, MDA-7/IL-24 also exerts a profound immune stimulatory effect in the context of cancer. Forced expression of MDA-7/IL-24 induces IFN-γ and IL-6 secretion from melanoma cells and displays potent antitumor functions (Caudell et al., 2002; Chada, Sutton, et al., 2004). Transduction of MDA-7/IL-24 via an adenoviral vector resulted in a significant increase in the CD3+ and CD8+ population, thereby facilitating immune activation and antitumor immunity. In one recent study, Ma et al. evaluated the efficacy of MDA-7/IL-24 in inhibiting colon cancer progression in murine models with an intact immune system and explored the immune-modulatory role of MDA-7/IL-24 in colon cancer progression (Ma, Ren, et al., 2016). The investigators found that MDA-7/IL-24 promoted CD4 + CD8+ T cells to secrete IFN-γ and facilitated the cytotoxicity of CD8 + T cells. In another recent study, Menezes and colleagues in the Fisher laboratory assessed the relevance of immune response in MDA-7/IL-24-mediated tumor suppression in a transgenic murine mouse model of breast cancer with an intact immune system (Menezes et al., 2015). The investigators found that intratumoral injection of Ad.5-*CTV* (replication competent cancer-selective adenovirus expressing MDA-7/IL-24; a cancer terminator virus) resulted in a marked increased IFN-γ expression and intratumoral CD8+ T cell infiltration. Interestingly, a significant increase in infiltrating CD8+ T cells, along with increased IFN-γ and granzyme B expression was also observed in nontreated tumors derived from MMTV-PyMT transgenic mice that received Ad.5-*CTV* suggesting that MDA-7/IL-24 is capable of inducing a systemic immune response in an intact immune microenvironment (Menezes et al., 2015). Another study by the Wang and Fisher laboratories evaluated the therapeutic efficacy of Ad.*mda*-7 in combination with an ER resident chaperone grp170 (Ad. sgrp170) in a prostate cancer model (Gao, Sun, Chen, Subjeck, & Wang, 2009; Gao et al., 2008). The investigators demonstrated that the combination treatment of MDA-7/IL-24 and grp170 was more effective in inhibiting TRAMP-C2 prostate tumor growth as compared to a single agent. The combination treatment resulted in increased IFN-γ production and cytolytic activity suggesting an antigen and tumor-specific T-cell response. Interestingly, the combination treatment was able to reduce distant tumor burden suggesting induction of profound "bystander" systemic antitumor immunity (Gao et al., 2009, 2008). Additionally, a vaccine effect was evident with subsequent tumor challenge experiments associated with a significant increase in the CD3+ and CD8+ cell populations. All of these studies in diverse cancer models strongly support an anticancer immune modulatory role of MDA-7/IL-24.

Evidence of immune activation was also evident in a Phase I/II clinical trial of Ad.*mda*-7 (INGN-241) in patients with advanced cancers (Cunningham et al., 2005; Dash et al., 2010; Fisher et al., 2007; Sarkar et al., 2007; Tong et al., 2005). A majority of the patients receiving intermediate- or high-dose injections of Ad.*mda*-7 (INGN-241) showed a marked increase in CD3+ and CD8+ T cells at day 15 following injection as well as transient increases in circulating cytokines, such as IL-6, IL-10, and TNFα (Cunningham et al., 2005; Tong et al., 2005). A few patients showed elevated levels of GM-CSF and IL-2 as well. These immune and cytokine profiles following injection of Ad.*mda*-7 (INGN-241) in patients mimic a TH-1 type immune response and strongly support an immune stimulatory function of MDA-7/IL-24 in eliciting an antitumor response.

7. CONCLUSIONS AND FUTURE PERSPECTIVES

As described in this review, MDA-7/IL-24 plays significant roles in a number of different human diseases. When initially identified, MDA-7/IL-24 was primarily recognized for its role as a tumor suppressor in cancer. However, as more information regarding the role of MDA-7/IL-24 became available our understanding of its relevance in other diseases has also increased. A detailed understanding of the molecular mechanisms defining the function of MDA-7/IL-24 has helped develop several preclinical therapeutic options as well as therapeutic targets against cancer. As mentioned previously, MDA-7/IL-24 has already been tested in clinical trials for cancer and a Phase I clinical trial with MDA-7/IL-24 (INGN 241) showed promising results (Cunningham et al., 2005; Tong et al., 2005). Currently, research is focused on developing novel approaches to enhance MDA-7/IL-24 potency and tumor-specific delivery. The search for new molecules and compounds that can enhance or stabilize MDA-7/IL-24 protein are also ongoing. Finally, combination therapies that would enhance MDA-7/IL-24-mediated tumor cell killing and prevent tumor growth and metastasis are also being identified and tested preclinically (Menezes et al., 2014). As new MDA-7/IL-24 therapeutic options are developed in one disease indication, they will also be valuable against other human diseases with

MDA-7/IL-24 involvement. Further information gained regarding the role of MDA-7/IL-24 in diseases where MDA-7/IL-24 is overexpressed will allow researchers and clinicians to develop newer approaches to manage these conditions. Given the currently known functions of MDA-7/IL-24, it is likely that MDA-7/IL-24 will also be implicated in other disease indications. Such information will be critical for understanding the multifaceted role of MDA-7/IL-24 in human physiology.

ACKNOWLEDGMENTS

The research reported in this review was supported in part by National Institutes of Health, National Cancer Institute grants R01 CA097318, R01 CA108520, P01 CA104177, P30 CA16059, and P50 CA058236; DOD grant W81XWH-14-1-0409; the Samuel Waxman Cancer Research Foundation; and the National Foundation for Cancer Research (NFCR). Support from the Human and Molecular Genetics Enhancement Fund was provided to S.K.D. and L.E. P.B.F. is the holder of the Thelma Newmeyer Corman Chair in Cancer Research in the VCU Massey Cancer Center. D.S. is the holder of the Harrison Endowed Chair.

CONFLICT OF INTEREST

P.B.F. is a cofounder and owns stock in Cancer Targeting Systems (CTS). Virginia Commonwealth University, Johns Hopkins University, and Columbia University own stock in CTS.

REFERENCES

Allen, M., Pratscher, B., Krepler, C., Frei, K., Schofer, C., Pehamberger, H., et al. (2005). Alternative splicing of IL-24 in melanocytes by deletion of exons 3 and 5. *International Journal of Immunogenetics*, *32*, 375–378.

Allen, M., Pratscher, B., Roka, F., Krepler, C., Wacheck, V., Schofer, C., et al. (2004). Loss of novel mda-7 splice variant (mda-7s) expression is associated with metastatic melanoma. *The Journal of Investigative Dermatology*, *123*, 583–588.

Andoh, A., Shioya, M., Nishida, A., Bamba, S., Tsujikawa, T., Kim-Mitsuyama, S., et al. (2009). Expression of IL-24, an activator of the JAK1/STAT3/SOCS3 cascade, is enhanced in inflammatory bowel disease. *Journal of Immunology*, *183*, 687–695.

Azab, B., Dash, R., Das, S. K., Bhutia, S. K., Shen, X. N., Quinn, B. A., et al. (2012). Enhanced delivery of mda-7/IL-24 using a serotype chimeric adenovirus (Ad.5/3) in combination with the apogossypol derivative BI-97C1 (Sabutoclax) improves therapeutic efficacy in low CAR colorectal cancer cells. *Journal of Cellular Physiology*, *227*, 2145–2153.

Bhoopathi, P., Lee, N., Pradhan, A. K., Shen, X. N., Das, S. K., Sarkar, D., et al. (2016). mda-7/IL-24 induces cell death in neuroblastoma through a novel mechanism involving AIF and ATM. *Cancer Research*, *76*, 3572–3582.

Bhutia, S. K., Das, S. K., Azab, B., Menezes, M. E., Dent, P., Wang, X. Y., et al. (2013). Targeting breast cancer-initiating/stem cells with melanoma differentiation-associated gene-7/interleukin-24. *International Journal of Cancer*, *133*, 2726–2736.

Bhutia, S. K., Das, S. K., Kegelman, T. P., Azab, B., Dash, R., Su, Z. Z., et al. (2012). mda-7/IL-24 differentially regulates soluble and nuclear clusterin in prostate cancer. *Journal of Cellular Physiology*, *227*, 1805–1813.

Bhutia, S. K., Dash, R., Das, S. K., Azab, B., Su, Z. Z., Lee, S. G., et al. (2010). Mechanism of autophagy to apoptosis switch triggered in prostate cancer cells by antitumor cytokine melanoma differentiation-associated gene 7/interleukin-24. *Cancer Research, 70*, 3667–3676.

Bhutia, S. K., Mukhopadhyay, S., Sinha, N., Das, D. N., Panda, P. K., Patra, S. K., et al. (2013). Autophagy: Cancer's friend or foe? *Advances in Cancer Research, 118*, 61–95.

Buzas, K., Oppenheim, J. J., & Zack Howard, O. M. (2011). Myeloid cells migrate in response to IL-24. *Cytokine, 55*, 429–434.

Caudell, E. G., Mumm, J. B., Poindexter, N., Ekmekcioglu, S., Mhashilkar, A. M., Yang, X. H., et al. (2002). The protein product of the tumor suppressor gene, melanoma differentiation-associated gene 7, exhibits immunostimulatory activity and is designated IL-24. *Journal of Immunology, 168*, 6041–6046.

Chada, S., Bocangel, D., Ramesh, R., Grimm, E. A., Mumm, J. B., Mhashilkar, A. M., et al. (2005). mda-7/IL24 kills pancreatic cancer cells by inhibition of the Wnt/PI3K signaling pathways: Identification of IL-20 receptor-mediated bystander activity against pancreatic cancer. *Molecular Therapy, 11*, 724–733.

Chada, S., Mhashilkar, A. M., Liu, Y., Nishikawa, T., Bocangel, D., Zheng, M., et al. (2006). mda-7 gene transfer sensitizes breast carcinoma cells to chemotherapy, biologic therapies and radiotherapy: Correlation with expression of bcl-2 family members. *Cancer Gene Therapy, 13*, 490–502.

Chada, S., Mhashilkar, A. M., Ramesh, R., Mumm, J. B., Sutton, R. B., Bocangel, D., et al. (2004). Bystander activity of Ad-mda7: Human MDA-7 protein kills melanoma cells via an IL-20 receptor-dependent but STAT3-independent mechanism. *Molecular Therapy, 10*, 1085–1095.

Chada, S., Sutton, R. B., Ekmekcioglu, S., Ellerhorst, J., Mumm, J. B., Leitner, W. W., et al. (2004). MDA-7/IL-24 is a unique cytokine—Tumor suppressor in the IL-10 family. *International Immunopharmacology, 4*, 649–667.

Chen, J., Chada, S., Mhashilkar, A., & Miano, J. M. (2003). Tumor suppressor MDA-7/IL-24 selectively inhibits vascular smooth muscle cell growth and migration. *Molecular Therapy, 8*, 220–229.

Conti, P., Kempuraj, D., Frydas, S., Kandere, K., Boucher, W., Letourneau, R., et al. (2003). IL-10 subfamily members: IL-19, IL-20, IL-22, IL-24 and IL-26. *Immunology Letters, 88*, 171–174.

Cunningham, C. C., Chada, S., Merritt, J. A., Tong, A., Senzer, N., Zhang, Y., et al. (2005). Clinical and local biological effects of an intratumoral injection of mda-7 (IL24; INGN 241) in patients with advanced carcinoma: A phase I study. *Molecular Therapy, 11*, 149–159.

Dash, R., Azab, B., Quinn, B. A., Shen, X., Wang, X. Y., Das, S. K., et al. (2011). Apogossypol derivative BI-97C1 (Sabutoclax) targeting Mcl-1 sensitizes prostate cancer cells to mda-7/IL-24-mediated toxicity. *Proceedings of the National Academy of Sciences of the United States of America, 108*, 8785–8790.

Dash, R., Bhoopathi, P., Das, S. K., Sarkar, S., Emdad, L., Dasgupta, S., et al. (2014). Novel mechanism of MDA-7/IL-24 cancer-specific apoptosis through SARI induction. *Cancer Research, 74*, 563–574.

Dash, R., Bhutia, S. K., Azab, B., Su, Z. Z., Quinn, B. A., Kegelmen, T. P., et al. (2010). mda-7/IL-24: A unique member of the IL-10 gene family promoting cancer-targeted toxicity. *Cytokine & Growth Factor Reviews, 21*, 381–391.

Deng, W. G., Kwon, J., Ekmekcioglu, S., Poindexter, N. J., & Grimm, E. A. (2011). IL-24 gene transfer sensitizes melanoma cells to erlotinib through modulation of the Apaf-1 and Akt signaling pathways. *Melanoma Research, 21*, 44–56.

Dent, P., Yacoub, A., Hamed, H. A., Park, M. A., Dash, R., Bhutia, S. K., et al. (2010). MDA-7/IL-24 as a cancer therapeutic: From bench to bedside. *Anti-Cancer Drugs, 21*, 725–731.

Dumoutier, L., Leemans, C., Lejeune, D., Kotenko, S. V., & Renauld, J. C. (2001). Cutting edge: STAT activation by IL-19, IL-20 and mda-7 through IL-20 receptor complexes of two types. *Journal of Immunology, 167*, 3545–3549.

Ekmekcioglu, S., Ellerhorst, J., Mhashilkar, A. M., Sahin, A. A., Read, C. M., Prieto, V. G., et al. (2001). Down-regulated melanoma differentiation associated gene (mda-7) expression in human melanomas. *International Journal of Cancer, 94*, 54–59.

Ellerhorst, J. A., Prieto, V. G., Ekmekcioglu, S., Broemeling, L., Yekell, S., Chada, S., et al. (2002). Loss of MDA-7 expression with progression of melanoma. *Journal of Clinical Oncology, 20*, 1069–1074.

Emdad, L., Lebedeva, I. V., Su, Z. Z., Gupta, P., Sarkar, D., Settleman, J., et al. (2007). Combinatorial treatment of non-small-cell lung cancers with gefitinib and Ad.mda-7 enhances apoptosis-induction and reverses resistance to a single therapy. *Journal of Cellular Physiology, 210*, 549–559.

Emdad, L., Lebedeva, I. V., Su, Z. Z., Gupta, P., Sauane, M., Dash, R., et al. (2009). Historical perspective and recent insights into our understanding of the molecular and biochemical basis of the antitumor properties of mda-7/IL-24. *Cancer Biology & Therapy, 8*, 391–400.

Emdad, L., Lebedeva, I. V., Su, Z. Z., Sarkar, D., Dent, P., Curiel, D. T., et al. (2007). Melanoma differentiation associated gene-7/interleukin-24 reverses multidrug resistance in human colorectal cancer cells. *Molecular Cancer Therapeutics, 6*, 2985–2994.

Emdad, L., Sarkar, D., Lebedeva, I. V., Su, Z. Z., Gupta, P., Mahasreshti, P. J., et al. (2006). Ionizing radiation enhances adenoviral vector expressing mda-7/IL-24-mediated apoptosis in human ovarian cancer. *Journal of Cellular Physiology, 208*, 298–306.

Eulitt, P. J., Park, M. A., Hossein, H., Cruikshanks, N., Yang, C., Dmitriev, I. P., et al. (2010). Enhancing mda-7/IL-24 therapy in renal carcinoma cells by inhibiting multiple protective signaling pathways using sorafenib and by Ad.5/3 gene delivery. *Cancer Biology & Therapy, 10*, 1290–1305.

Eyler, C. E., & Rich, J. N. (2008). Survival of the fittest: Cancer stem cells in therapeutic resistance and angiogenesis. *Journal of Clinical Oncology, 26*, 2839–2845.

Fernald, K., & Kurokawa, M. (2013). Evading apoptosis in cancer. *Trends in Cell Biology, 23*, 620–633.

Filippov, V., Schmidt, E. L., Filippova, M., & Duerksen-Hughes, P. J. (2008). Splicing and splice factor SRp55 participate in the response to DNA damage by changing isoform ratios of target genes. *Gene, 420*, 34–41.

Fisher, P. B. (2005). Is mda-7/IL-24 a "magic bullet" for cancer? *Cancer Research, 65*, 10128–10138.

Fisher, P. B., Gopalkrishnan, R. V., Chada, S., Ramesh, R., Grimm, E. A., Rosenfeld, M. R., et al. (2003). mda-7/IL-24, a novel cancer selective apoptosis inducing cytokine gene: From the laboratory into the clinic. *Cancer Biology & Therapy, 2*, S23–37.

Fisher, P. B., Sarkar, D., Lebedeva, I. V., Emdad, L., Gupta, P., Sauane, M., et al. (2007). Melanoma differentiation associated gene-7/interleukin-24 (mda-7/IL-24): Novel gene therapeutic for metastatic melanoma. *Toxicology and Applied Pharmacology, 224*, 300–307.

Fonseca-Camarillo, G., Furuzawa-Carballeda, J., Granados, J., & Yamamoto-Furusho, J. K. (2014). Expression of interleukin (IL)-19 and IL-24 in inflammatory bowel disease patients: A cross-sectional study. *Clinical and Experimental Immunology, 177*, 64–75.

Fuson, K. L., Zheng, M., Craxton, M., Pataer, A., Ramesh, R., Chada, S., et al. (2009). Structural mapping of post-translational modifications in human interleukin-24: Role of N-linked glycosylation and disulfide bonds in secretion and activity. *Journal of Biological Chemistry, 284*, 30526–30533.

Gao, P., Sun, X., Chen, X., Subjeck, J., & Wang, X. Y. (2009). Secretion of stress protein grp170 promotes immune-mediated inhibition of murine prostate tumor. *Cancer Immunology, Immunotherapy, 58*, 1319–1328.

Gao, P., Sun, X., Chen, X., Wang, Y., Foster, B. A., Subjeck, J., et al. (2008). Secretable chaperone Grp170 enhances therapeutic activity of a novel tumor suppressor, mda-7/IL-24. *Cancer Research, 68,* 3890–3898.

Gopalan, B., Litvak, A., Sharma, S., Mhashilkar, A. M., Chada, S., & Ramesh, R. (2005). Activation of the Fas-FasL signaling pathway by MDA-7/IL-24 kills human ovarian cancer cells. *Cancer Research, 65,* 3017–3024.

Gopalkrishnan, R. V., Sauane, M., & Fisher, P. B. (2004). Cytokine and tumor cell apoptosis inducing activity of mda-7/IL-24. *International Immunopharmacology, 4,* 635–647.

Greco, A., Di Benedetto, A., Howard, C. M., Kelly, S., Nande, R., Dementieva, Y., et al. (2010). Eradication of therapy-resistant human prostate tumors using an ultrasound-guided site-specific cancer terminator virus delivery approach. *Molecular Therapy, 18,* 295–306.

Green, J., Nusse, R., & van Amerongen, R. (2014). The role of Ryk and Ror receptor tyrosine kinases in Wnt signal transduction. *Cold Spring Harbor Perspectives in Biology, 6,* a009175.

Gupta, P., Emdad, L., Lebedeva, I. V., Sarkar, D., Dent, P., Curiel, D. T., et al. (2008). Targeted combinatorial therapy of non-small cell lung carcinoma using a GST-fusion protein of full-length or truncated MDA-7/IL-24 with Tarceva. *Journal of Cellular Physiology, 215,* 827–836.

Gupta, P., Su, Z. Z., Lebedeva, I. V., Sarkar, D., Sauane, M., Emdad, L., et al. (2006). mda-7/IL-24: Multifunctional cancer-specific apoptosis-inducing cytokine. *Pharmacology and Therapeutics, 111,* 596–628.

Gupta, P., Walter, M. R., Su, Z. Z., Lebedeva, I. V., Emdad, L., Randolph, A., et al. (2006). BiP/GRP78 is an intracellular target for MDA-7/IL-24 induction of cancer-specific apoptosis. *Cancer Research, 66,* 8182–8191.

Hamed, H. A., Das, S. K., Sokhi, U. K., Park, M. A., Cruickshanks, N., Archer, K., et al. (2013). Combining histone deacetylase inhibitors with MDA-7/IL-24 enhances killing of renal carcinoma cells. *Cancer Biology & Therapy, 14,* 1039–1049.

Hamed, H. A., Yacoub, A., Park, M. A., Archer, K., Das, S. K., Sarkar, D., et al. (2013). Histone deacetylase inhibitors interact with melanoma differentiation associated-7/interleukin-24 to kill primary human glioblastoma cells. *Molecular Pharmacology, 84,* 171–181.

Hamed, H. A., Yacoub, A., Park, M. A., Eulitt, P. J., Dash, R., Sarkar, D., et al. (2010). Inhibition of multiple protective signaling pathways and Ad.5/3 delivery enhances mda-7/IL-24 therapy of malignant glioma. *Molecular Therapy, 18,* 1130–1142.

Hamed, H. A., Yacoub, A., Park, M. A., Eulitt, P., Sarkar, D., Dmitriev, I. P., et al. (2010). OSU-03012 enhances Ad.7-induced GBM cell killing via ER stress and autophagy and by decreasing expression of mitochondrial protective proteins. *Cancer Biology & Therapy, 9,* 526–536.

Hanahan, D., & Weinberg, R. A. (2000). The hallmarks of cancer. *Cell, 100,* 57–70.

Hanahan, D., & Weinberg, R. A. (2011). Hallmarks of cancer: The next generation. *Cell, 144,* 646–674.

Hauck, C. R., Hsia, D. A., & Schlaepfer, D. D. (2002). The focal adhesion kinase—A regulator of cell migration and invasion. *IUBMB Life, 53,* 115–119.

He, M., & Liang, P. (2010). IL-24 transgenic mice: In vivo evidence of overlapping functions for IL-20, IL-22, and IL-24 in the epidermis. *Journal of Immunology, 184,* 1793–1798.

Huang, E. Y., Madireddi, M. T., Gopalkrishnan, R. V., Leszczyniecka, M., Su, Z., Lebedeva, I. V., et al. (2001). Genomic structure, chromosomal localization and expression profile of a novel melanoma differentiation associated (mda-7) gene with cancer specific growth suppressing and apoptosis inducing properties. *Oncogene, 20,* 7051–7063.

Huo, W., Li, Z. M., Zhu, X. M., Bao, Y. M., & An, L. J. (2013). MDA-7/IL-24 suppresses tumor adhesion and invasive potential in hepatocellular carcinoma cell lines. *Oncology Reports, 30,* 986–992.

Inoue, S., Hartman, A., Branch, C. D., Bucana, C. D., Bekele, B. N., Stephens, L. C., et al. (2007). mda-7 In combination with bevacizumab treatment produces a synergistic and complete inhibitory effect on lung tumor xenograft. *Molecular Therapy, 15*, 287–294.

Jiang, H., & Fisher, P. B. (1993). Use of a sensitive and efficient subtraction hybridization protocol for the identification of genes differentially regulated during the induction of differentiation in human melanoma cells. *Molecular and Cellular Differentiation, 1*(3), 285–299.

Jiang, H., Lin, J. J., Su, Z. Z., Goldstein, N. I., & Fisher, P. B. (1995). Subtraction hybridization identifies a novel melanoma differentiation associated gene, mda-7, modulated during human melanoma differentiation, growth and progression. *Oncogene, 11*, 2477–2486.

Jiang, H., Su, Z. Z., Lin, J. J., Goldstein, N. I., Young, C. S., & Fisher, P. B. (1996). The melanoma differentiation associated gene mda-7 suppresses cancer cell growth. *Proceedings of the National Academy of Sciences of the United States of America, 93*, 9160–9165.

Jin, S. H., Choi, D., Chun, Y. J., & Noh, M. (2014). Keratinocyte-derived IL-24 plays a role in the positive feedback regulation of epidermal inflammation in response to environmental and endogenous toxic stressors. *Toxicology and Applied Pharmacology, 280*, 199–206.

Karar, J., & Maity, A. (2011). PI3K/AKT/mTOR pathway in angiogenesis. *Frontiers in Molecular Neuroscience, 4*, 51.

Kragstrup, T. W., Otkjaer, K., Holm, C., Jorgensen, A., Hokland, M., Iversen, L., et al. (2008). The expression of IL-20 and IL-24 and their shared receptors are increased in rheumatoid arthritis and spondyloarthropathy. *Cytokine, 41*, 16–23.

Kumar, N. P., Banurekha, V. V., Nair, D., Kumaran, P., Dolla, C. K., & Babu, S. (2015). Type 2 diabetes—Tuberculosis co-morbidity is associated with diminished circulating levels of IL-20 subfamily of cytokines. *Tuberculosis (Edinburgh, Scotland), 95*, 707–712.

Kumari, S., Bonnet, M. C., Ulvmar, M. H., Wolk, K., Karagianni, N., Witte, E., et al. (2013). Tumor necrosis factor receptor signaling in keratinocytes triggers interleukin-24-dependent psoriasis-like skin inflammation in mice. *Immunity, 39*, 899–911.

Kunz, S., Wolk, K., Witte, E., Witte, K., Doecke, W. D., Volk, H. D., et al. (2006). Interleukin (IL)-19, IL-20 and IL-24 are produced by and act on keratinocytes and are distinct from classical ILs. *Experimental Dermatology, 15*, 991–1004.

Lebedeva, I. V., Emdad, L., Su, Z. Z., Gupta, P., Sauane, M., Sarkar, D., et al. (2007). mda-7/IL-24, novel anticancer cytokine: Focus on bystander antitumor, radiosensitization and antiangiogenic properties and overview of the phase I clinical experience (Review). *International Journal of Oncology, 31*, 985–1007.

Lebedeva, I. V., Sarkar, D., Su, Z. Z., Kitada, S., Dent, P., Stein, C. A., et al. (2003). Bcl-2 and Bcl-x(L) differentially protect human prostate cancer cells from induction of apoptosis by melanoma differentiation associated gene-7, mda-7/IL-24. *Oncogene, 22*, 8758–8773.

Lebedeva, I. V., Sauane, M., Gopalkrishnan, R. V., Sarkar, D., Su, Z. Z., Gupta, P., et al. (2005). mda-7/IL-24: Exploiting cancer's Achilles' heel. *Molecular Therapy, 11*, 4–18.

Lebedeva, I. V., Su, Z. Z., Chang, Y., Kitada, S., Reed, J. C., & Fisher, P. B. (2002). The cancer growth suppressing gene mda-7 induces apoptosis selectively in human melanoma cells. *Oncogene, 21*, 708–718.

Lebedeva, I. V., Su, Z. Z., Sarkar, D., & Fisher, P. B. (2003). Restoring apoptosis as a strategy for cancer gene therapy: Focus on p53 and mda-7. *Seminars in Cancer Biology, 13*, 169–178.

Lebedeva, I. V., Su, Z. Z., Sarkar, D., Gopalkrishnan, R. V., Waxman, S., Yacoub, A., et al. (2005). Induction of reactive oxygen species renders mutant and wild-type K-ras pancreatic carcinoma cells susceptible to Ad.mda-7-induced apoptosis. *Oncogene, 24*, 585–596.

Lebedeva, I. V., Su, Z. Z., Sarkar, D., Kitada, S., Dent, P., Waxman, S., et al. (2003). Melanoma differentiation associated gene-7, mda-7/interleukin-24, induces apoptosis in prostate cancer cells by promoting mitochondrial dysfunction and inducing reactive oxygen species. *Cancer Research, 63*, 8138–8144.

Lebedeva, I. V., Su, Z. Z., Vozhilla, N., Chatman, L., Sarkar, D., Dent, P., et al. (2008). Mechanism of in vitro pancreatic cancer cell growth inhibition by melanoma differentiation-associated gene-7/interleukin-24 and perillyl alcohol. *Cancer Research, 68*, 7439–7447.

Lee, K. M., Kang, H. A., Park, M., Lee, H. Y., Song, M. J., Ko, K., et al. (2012). Interleukin-24 suppresses the growth of vascular smooth muscle cells by inhibiting H(2)O(2)-induced reactive oxygen species production. *Pharmacology, 90*, 332–341.

Leng, R. X., Pan, H. F., Tao, J. H., & Ye, D. Q. (2011). IL-19, IL-20 and IL-24: Potential therapeutic targets for autoimmune diseases. *Expert Opinion on Therapeutic Targets, 15*, 119–126.

Liang, J., Huang, R. L., Huang, Q., Peng, Z., Zhang, P. H., & Wu, Z. X. (2011). Adenovirus-mediated human interleukin 24 (MDA-7/IL-24) selectively suppresses proliferation and induces apoptosis in keloid fibroblasts. *Annals of Plastic Surgery, 66*, 660–666.

Lin, S. W., Lee, M. T., Ke, F. C., Lee, P. P., Huang, C. J., Ip, M. M., et al. (2000). TGFbeta1 stimulates the secretion of matrix metalloproteinase 2 (MMP2) and the invasive behavior in human ovarian cancer cells, which is suppressed by MMP inhibitor BB3103. *Clinical and Experimental Metastasis, 18*, 493–499.

Liu, J., & Debnath, J. (2016). The evolving, multifaceted roles of autophagy in cancer. *Advances in Cancer Research, 130*, 1–53.

Lv, C., Su, Q., Liang, Y., Hu, J., & Yuan, S. (2016). Oncolytic vaccine virus harbouring the IL-24 gene suppresses the growth of lung cancer by inducing apoptosis. *Biochemical and Biophysical Research Communications, 476*, 21–28.

Ma, Y., Chen, H., Wang, Q., Luo, F., Yan, J., & Zhang, X. L. (2009). IL-24 protects against Salmonella typhimurium infection by stimulating early neutrophil Th1 cytokine production, which in turn activates CD8+ T cells. *European Journal of Immunology, 39*, 3357–3368.

Ma, Y., Chen, H. D., Wang, Y., Wang, Q., Li, Y., Zhao, Y., et al. (2011). Interleukin 24 as a novel potential cytokine immunotherapy for the treatment of Mycobacterium tuberculosis infection. *Microbes and Infection, 13*, 1099–1110.

Ma, Q., Jin, B., Zhang, Y., Shi, Y., Zhang, C., Luo, D., et al. (2016). Secreted recombinant human IL-24 protein inhibits the proliferation of esophageal squamous cell carcinoma Eca-109 cells in vitro and in vivo. *Oncology Reports, 35*, 2681–2690.

Ma, G., Kawamura, K., Shan, Y., Okamoto, S., Li, Q., Namba, M., et al. (2014). Combination of adenoviruses expressing melanoma differentiation-associated gene-7 and chemotherapeutic agents produces enhanced cytotoxicity on esophageal carcinoma. *Cancer Gene Therapy, 21*, 31–37.

Ma, Y. F., Ren, Y., Wu, C. J., Zhao, X. H., Xu, H., Wu, D. Z., et al. (2016). Interleukin (IL)-24 transforms the tumor microenvironment and induces anticancer immunity in a murine model of colon cancer. *Molecular Immunology, 75*, 11–20.

Ma, M., Zhao, L., Sun, G., Zhang, C., Liu, L., Du, Y., et al. (2016). Mda-7/IL-24 enhances sensitivity of B cell lymphoma to chemotherapy drugs. *Oncology Reports, 35*, 3122–3130.

Maarof, G., Bouchet-Delbos, L., Gary-Gouy, H., Durand-Gasselin, I., Krzysiek, R., & Dalloul, A. (2010). Interleukin-24 inhibits the plasma cell differentiation program in human germinal center B cells. *Blood, 115*, 1718–1726.

MacDonald, B. T., & He, X. (2012). Frizzled and LRP5/6 receptors for Wnt/beta-catenin signaling. *Cold Spring Harbor Perspectives in Biology, 4*, a007880.

McKenzie, T., Liu, Y., Fanale, M., Swisher, S. G., Chada, S., & Hunt, K. K. (2004). Combination therapy of Ad-mda7 and trastuzumab increases cell death in Her-2/neu-overexpressing breast cancer cells. *Surgery, 136*, 437–442.

Menezes, M. E., Bhatia, S., Bhoopathi, P., Das, S. K., Emdad, L., Dasgupta, S., et al. (2014). MDA-7/IL-24: Multifunctional cancer killing cytokine. *Advances in Experimental Medicine and Biology, 818*, 127–153.

Menezes, M. E., Shen, X. N., Das, S. K., Emdad, L., Guo, C., Yuan, F., et al. (2015). MDA-7/IL-24 functions as a tumor suppressor gene in vivo in transgenic mouse models of breast cancer. *Oncotarget, 6*, 36928–36942.

Mhashilkar, A. M., Stewart, A. L., Sieger, K., Yang, H. Y., Khimani, A. H., Ito, I., et al. (2003). MDA-7 negatively regulates the beta-catenin and PI3K signaling pathways in breast and lung tumor cells. *Molecular Therapy, 8*, 207–219.

Miyahara, R., Banerjee, S., Kawano, K., Efferson, C., Tsuda, N., Miyahara, Y., et al. (2006). Melanoma differentiation-associated gene-7 (mda-7)/interleukin (IL)-24 induces anticancer immunity in a syngeneic murine model. *Cancer Gene Therapy, 13*, 753–761.

Morrison, B. J., Morris, J. C., & Steel, J. C. (2013). Lung cancer-initiating cells: A novel target for cancer therapy. *Targeted Oncology, 8*, 159–172.

Mosser, D. M., & Zhang, X. (2008). Interleukin-10: New perspectives on an old cytokine. *Immunological Reviews, 226*, 205–218.

Mumm, J. B., Ekmekcioglu, S., Poindexter, N. J., Chada, S., & Grimm, E. A. (2006). Soluble human MDA-7/IL-24: Characterization of the molecular form(s) inhibiting tumor growth and stimulating monocytes. *Journal of Interferon & Cytokine Research, 26*, 877–886.

Nace, J., Fortunato, S. J., Maul, H., & Menon, R. (2010). The expression pattern of two novel cytokines (IL-24 and IL-29) in human fetal membranes. *Journal of Perinatal Medicine, 38*, 665–670.

Naik, P., Karrim, J., & Hanahan, D. (1996). The rise and fall of apoptosis during multistage tumorigenesis: Down-modulation contributes to tumor progression from angiogenic progenitors. *Genes and Development, 10*, 2105–2116.

Nishikawa, T., Ramesh, R., Munshi, A., Chada, S., & Meyn, R. E. (2004). Adenovirus-mediated mda-7 (IL24) gene therapy suppresses angiogenesis and sensitizes NSCLC xenograft tumors to radiation. *Molecular Therapy, 9*, 818–828.

Oida, Y., Gopalan, B., Miyahara, R., Inoue, S., Branch, C. D., Mhashilkar, A. M., et al. (2005). Sulindac enhances adenoviral vector expressing mda-7/IL-24-mediated apoptosis in human lung cancer. *Molecular Cancer Therapeutics, 4*, 291–304.

Pal, I., Sarkar, S., Rajput, S., Dey, K. K., Chakraborty, S., Dash, R., et al. (2014). BI-69A11 enhances susceptibility of colon cancer cells to mda-7/IL-24-induced growth inhibition by targeting Akt. *British Journal of Cancer, 111*, 101–111.

Panneerselvam, J., Jin, J., Shanker, M., Lauderdale, J., Bates, J., Wang, Q., et al. (2015). IL-24 inhibits lung cancer cell migration and invasion by disrupting the SDF-1/CXCR4 signaling axis. *PLoS One, 10*, e0122439.

Panneerselvam, J., Munshi, A., & Ramesh, R. (2013). Molecular targets and signaling pathways regulated by interleukin (IL)-24 in mediating its antitumor activities. *Journal of Molecular Signaling, 8*, 15.

Park, M. A., Walker, T., Martin, A. P., Allegood, J., Vozhilla, N., Emdad, L., et al. (2009). MDA-7/IL-24-induced cell killing in malignant renal carcinoma cells occurs by a ceramide/CD95/PERK-dependent mechanism. *Molecular Cancer Therapeutics, 8*, 1280–1291.

Park, M. A., Yacoub, A., Sarkar, D., Emdad, L., Rahmani, M., Spiegel, S., et al. (2008). PERK-dependent regulation of MDA-7/IL-24-induced autophagy in primary human glioma cells. *Autophagy, 4*, 513–515.

Parrish-Novak, J., Xu, W., Brender, T., Yao, L., Jones, C., West, J., et al. (2002). Interleukins 19, 20, and 24 signal through two distinct receptor complexes. Differences in receptor–ligand interactions mediate unique biological functions. *Journal of Biological Chemistry, 277*, 47517–47523.

Pasparakis, M., Haase, I., & Nestle, F. O. (2014). Mechanisms regulating skin immunity and inflammation. *Nature Reviews. Immunology, 14*, 289–301.

Pataer, A., Bocangel, D., Chada, S., Roth, J. A., Hunt, K. K., & Swisher, S. G. (2007). Enhancement of adenoviral MDA-7-mediated cell killing in human lung cancer cells by geldanamycin and its 17-allyl- amino-17-demethoxy analogue. *Cancer Gene Therapy*, *14*, 12–18.

Pataer, A., Vorburger, S. A., Chada, S., Balachandran, S., Barber, G. N., Roth, J. A., et al. (2005). Melanoma differentiation-associated gene-7 protein physically associates with the double-stranded RNA-activated protein kinase PKR. *Molecular Therapy*, *11*, 717–723.

Pei, D. S., Yang, Z. X., Zhang, B. F., Yin, X. X., Li, L. T., Li, H. Z., et al. (2012). Enhanced apoptosis-inducing function of MDA-7/IL-24 RGD mutant via the increased adhesion to tumor cells. *Journal of Interferon & Cytokine Research*, *32*, 66–73.

Persaud, L., De Jesus, D., Brannigan, O., Richiez-Paredes, M., Huaman, J., Alvarado, G., et al. (2016). Mechanism of action and applications of interleukin 24 in immunotherapy. *International Journal of Molecular Sciences*, *17*, E869.

Poindexter, N. J., Williams, R. R., Powis, G., Jen, E., Caudle, A. S., Chada, S., et al. (2010). IL-24 is expressed during wound repair and inhibits TGFalpha-induced migration and proliferation of keratinocytes. *Experimental Dermatology*, *19*, 714–722.

Pradhan, A. K., Talukdar, S., Bhoopathi, P., Shen, X. N., Emdad, L., Das, S. K., et al. (2017). mda-7/IL-24 mediates cancer cell-specific death via regulation of miR-221 and the beclin-1 axis. *Cancer Research*, *77*, 949–959.

Rahmani, M., Mayo, M., Dash, R., Sokhi, U. K., Dmitriev, I. P., Sarkar, D., et al. (2010). Melanoma differentiation associated gene-7/interleukin-24 potently induces apoptosis in human myeloid leukemia cells through a process regulated by endoplasmic reticulum stress. *Molecular Pharmacology*, *78*, 1096–1104.

Ramesh, R., Ito, I., Gopalan, B., Saito, Y., Mhashilkar, A. M., & Chada, S. (2004). Ectopic production of MDA-7/IL-24 inhibits invasion and migration of human lung cancer cells. *Molecular Therapy*, *9*, 510–518.

Romer, J., Hasselager, E., Norby, P. L., Steiniche, T., Thorn Clausen, J., & Kragballe, K. (2003). Epidermal overexpression of interleukin-19 and -20 mRNA in psoriatic skin disappears after short-term treatment with cyclosporine a or calcipotriol. *Journal of Investigative Dermatology*, *121*, 1306–1311.

Sahoo, A., Lee, C. G., Jash, A., Son, J. S., Kim, G., Kwon, H. K., et al. (2011). Stat6 and c-Jun mediate Th2 cell-specific IL-24 gene expression. *Journal of Immunology*, *186*, 4098–4109.

Sano, S., Chan, K. S., Carbajal, S., Clifford, J., Peavey, M., Kiguchi, K., et al. (2005). Stat3 links activated keratinocytes and immunocytes required for development of psoriasis in a novel transgenic mouse model. *Nature Medicine*, *11*, 43–49.

Sarkar, D., Lebedeva, I. V., Gupta, P., Emdad, L., Sauane, M., Dent, P., et al. (2007). Melanoma differentiation associated gene-7 (mda-7)/IL-24: A 'magic bullet' for cancer therapy? *Expert Opinion on Biological Therapy*, *7*, 577–586.

Sarkar, S., Quinn, B. A., Shen, X. N., Dash, R., Das, S. K., Emdad, L., et al. (2015). Therapy of prostate cancer using a novel cancer terminator virus and a small molecule BH-3 mimetic. *Oncotarget*, *6*, 10712–10727.

Sarkar, S., Quinn, B. A., Shen, X., Dent, P., Das, S. K., Emdad, L., et al. (2015). Reversing translational suppression and induction of toxicity in pancreatic cancer cells using a chemoprevention gene therapy approach. *Molecular Pharmacology*, *87*, 286–295.

Sarkar, D., Su, Z. Z., Lebedeva, I. V., Sauane, M., Gopalkrishnan, R. V., Dent, P., et al. (2002a). mda-7 (IL-24): Signaling and functional roles. *BioTechniques*, (Suppl), 30–39.

Sarkar, D., Su, Z. Z., Lebedeva, I. V., Sauane, M., Gopalkrishnan, R. V., Valerie, K., et al. (2002b). mda-7 (IL-24) mediates selective apoptosis in human melanoma cells by inducing the coordinated overexpression of the GADD family of genes by means of p38 MAPK. *Proceedings of the National Academy of Sciences of the United States of America*, *99*, 10054–10059.

Sarkar, D., Su, Z. Z., Park, E. S., Vozhilla, N., Dent, P., Curiel, D. T., et al. (2008). A cancer terminator virus eradicates both primary and distant human melanomas. *Cancer Gene Therapy, 15,* 293–302.

Sarkar, D., Su, Z. Z., Vozhilla, N., Park, E. S., Gupta, P., & Fisher, P. B. (2005). Dual cancer-specific targeting strategy cures primary and distant breast carcinomas in nude mice. *Proceedings of the National Academy of Sciences of the United States of America, 102,* 14034–14039.

Sauane, M., Gopalkrishnan, R. V., Choo, H. T., Gupta, P., Lebedeva, I. V., Yacoub, A., et al. (2004). Mechanistic aspects of mda-7/IL-24 cancer cell selectivity analyzed via a bacterial fusion protein. *Oncogene, 23,* 7679–7690.

Sauane, M., Gopalkrishnan, R. V., Lebedeva, I., Mei, M. X., Sarkar, D., Su, Z. Z., et al. (2003). Mda-7/IL-24 induces apoptosis of diverse cancer cell lines through JAK/STAT-independent pathways. *Journal of Cellular Physiology, 196,* 334–345.

Sauane, M., Gopalkrishnan, R. V., Sarkar, D., Su, Z. Z., Lebedeva, I. V., Dent, P., et al. (2003). MDA-7/IL-24: Novel cancer growth suppressing and apoptosis inducing cytokine. *Cytokine & Growth Factor Reviews, 14,* 35–51.

Sauane, M., Gupta, P., Lebedeva, I. V., Su, Z. Z., Sarkar, D., Randolph, A., et al. (2006). N-glycosylation of MDA-7/IL-24 is dispensable for tumor cell-specific apoptosis and "bystander" antitumor activity. *Cancer Research, 66,* 11869–11877.

Sauane, M., Su, Z. Z., Gupta, P., Lebedeva, I. V., Dent, P., Sarkar, D., et al. (2008). Autocrine regulation of mda-7/IL-24 mediates cancer-specific apoptosis. *Proceedings of the National Academy of Sciences of the United States of America, 105,* 9763–9768.

Schaefer, G., Venkataraman, C., & Schindler, U. (2001). Cutting edge: FISP (IL-4-induced secreted protein), a novel cytokine-like molecule secreted by Th2 cells. *Journal of Immunology, 166,* 5859–5863.

Seong, R. K., Choi, Y. K., & Shin, O. S. (2016). MDA7/IL-24 is an anti-viral factor that inhibits influenza virus replication. *Journal of Microbiology, 54,* 695–700.

Shanker, M., Gopalan, B., Patel, S., Bocangel, D., Chada, S., & Ramesh, R. (2007). Vitamin E succinate in combination with mda-7 results in enhanced human ovarian tumor cell killing through modulation of extrinsic and intrinsic apoptotic pathways. *Cancer Letters, 254,* 217–226.

Shapiro, B. A., Vu, N. T., Shultz, M. D., Shultz, J. C., Mietla, J. A., Gouda, M. M., et al. (2016). Melanoma differentiation-associated gene 7/IL-24 exerts cytotoxic effects by altering the alternative splicing of Bcl-x pre-mRNA via the SRC/PKCδ signaling axis. *Journal of Biological Chemistry, 291,* 21669–21681.

Sieger, K. A., Mhashilkar, A. M., Stewart, A., Sutton, R. B., Strube, R. W., Chen, S. Y., et al. (2004). The tumor suppressor activity of MDA-7/IL-24 is mediated by intracellular protein expression in NSCLC cells. *Molecular Therapy, 9,* 355–367.

Su, Z. Z., Emdad, L., Sauane, M., Lebedeva, I. V., Sarkar, D., Gupta, P., et al. (2005). Unique aspects of mda-7/IL-24 antitumor bystander activity: Establishing a role for secretion of MDA-7/IL-24 protein by normal cells. *Oncogene, 24,* 7552–7566.

Su, Z. Z., Lebedeva, I. V., Sarkar, D., Emdad, L., Gupta, P., Kitada, S., et al. (2006). Ionizing radiation enhances therapeutic activity of mda-7/IL-24: Overcoming radiation- and mda-7/IL-24-resistance in prostate cancer cells overexpressing the antiapoptotic proteins bcl-xL or bcl-2. *Oncogene, 25,* 2339–2348.

Su, Z. Z., Lebedeva, I. V., Sarkar, D., Gopalkrishnan, R. V., Sauane, M., Sigmon, C., et al. (2003). Melanoma differentiation associated gene-7, mda-7/IL-24, selectively induces growth suppression, apoptosis and radiosensitization in malignant gliomas in a p53-independent manner. *Oncogene, 22,* 1164–1180.

Su, Z. Z., Madireddi, M. T., Lin, J. J., Young, C. S., Kitada, S., Reed, J. C., et al. (1998). The cancer growth suppressor gene mda-7 selectively induces apoptosis in human breast cancer cells and inhibits tumor growth in nude mice. *Proceedings of the National Academy of Sciences of the United States of America, 95,* 14400–14405.

Suh, Y. J., Chada, S., McKenzie, T., Liu, Y., Swisher, S. G., Lucci, A., et al. (2005). Synergistic tumoricidal effect between celecoxib and adenoviral-mediated delivery of mda-7 in human breast cancer cells. *Surgery, 138,* 422–430.

Talukdar, S., Emdad, L., Das, S. K., Sarkar, D., & Fisher, P. B. (2016). Evolving strategies for therapeutically targeting cancer stem cells. *Advances in Cancer Research, 131,* 159–191.

Tamai, H., Miyake, K., Yamaguchi, H., Takatori, M., Dan, K., Inokuchi, K., et al. (2012). AAV8 vector expressing IL24 efficiently suppresses tumor growth mediated by specific mechanisms in MLL/AF4-positive ALL model mice. *Blood, 119,* 64–71.

Tian, H., Li, L., Zhang, B., Di, J., Chen, F., Li, H., et al. (2012). Critical role of lysine 123 in the ubiquitin-mediated degradation of MDA-7/IL-24. *Journal of Interferon & Cytokine Research, 32,* 575–582.

Tong, A. W., Nemunaitis, J., Su, D., Zhang, Y., Cunningham, C., Senzer, N., et al. (2005). Intratumoral injection of INGN 241, a nonreplicating adenovector expressing the melanoma-differentiation associated gene-7 (mda-7/IL24): Biologic outcome in advanced cancer patients. *Molecular Therapy, 11,* 160–172.

Vargas-Alarcon, G., Posadas-Romero, C., Villarreal-Molina, T., Alvarez-Leon, E., Angeles-Martinez, J., Posadas-Sanchez, R., et al. (2014). IL-24 gene polymorphisms are associated with cardiometabolic parameters and cardiovascular risk factors but not with premature coronary artery disease: The genetics of atherosclerotic disease Mexican study. *Journal of Interferon & Cytokine Research, 34,* 659–666.

Wang, M., & Liang, P. (2005). Interleukin-24 and its receptors. *Immunology, 114,* 166–170.

Wang, Y. H., & Scadden, D. T. (2015). Harnessing the apoptotic programs in cancer stem-like cells. *EMBO Reports, 16,* 1084–1098.

Wang, M., Tan, Z., Zhang, R., Kotenko, S. V., & Liang, P. (2002). Interleukin 24 (MDA-7/MOB-5) signals through two heterodimeric receptors, IL-22R1/IL-20R2 and IL-20R1/IL-20R2. *Journal of Biological Chemistry, 277,* 7341–7347.

Wang, Z., Wang, Y., Chen, Y., & Lv, J. (2016). The IL-24 gene protects human umbilical vein endothelial cells against H(2)O(2)-induced injury and may be useful as a treatment for cardiovascular disease. *International Journal of Molecular Medicine, 37,* 581–592.

Wang, X., Ye, Z., Zhong, J., Xiang, J., & Yang, J. (2007). Adenovirus-mediated Il-24 expression suppresses hepatocellular carcinoma growth via induction of cell apoptosis and cycling arrest and reduction of angiogenesis. *Cancer Biotherapy & Radiopharmaceuticals, 22,* 56–63.

Wang, C. J., Zhang, H., Chen, K., Zheng, J. W., Xiao, C. W., Ji, W. W., et al. (2010). Ad. mda-7 (IL-24) selectively induces apoptosis in hepatocellular carcinoma cell lines, suppresses metastasis, and enhances the effect of doxorubicin on xenograft tumors. *Oncology Research, 18,* 561–574.

Weiss, R., Laengle, J., Sachet, M., Shurygina, A. P., Kiselev, O., Egorov, A., et al. (2015). Interleukin-24 inhibits influenza A virus replication in vitro through induction of toll-like receptor 3 dependent apoptosis. *Antiviral Research, 123,* 93–104.

Weiss, B., Wolk, K., Grunberg, B. H., Volk, H. D., Sterry, W., Asadullah, K., et al. (2004). Cloning of murine IL-22 receptor alpha 2 and comparison with its human counterpart. *Genes Immunology, 5,* 330–336.

Welch, D. R., & Fisher, P. B. (2016). Preface. *Advances in Cancer Research, 132,* xi–xiv.

Whitaker, E. L., Filippov, V. A., & Duerksen-Hughes, P. J. (2012). Interleukin 24: Mechanisms and therapeutic potential of an anti-cancer gene. *Cytokine and Growth Factor Reviews, 23,* 323–331.

Whitaker, E. L., Filippov, V., Filippova, M., Guerrero-Juarez, C. F., & Duerksen-Hughes, P. J. (2011). Splice variants of mda-7/IL-24 differentially affect survival and induce apoptosis in U2OS cells. *Cytokine, 56,* 272–281.

Wolk, K., Haugen, H. S., Xu, W., Witte, E., Waggie, K., Anderson, M., et al. (2009). IL-22 and IL-20 are key mediators of the epidermal alterations in psoriasis while IL-17 and IFN-gamma are not. *Journal of Molecular Medicine (Berlin), 87*, 523–536.

Wolk, K., Kunz, S., Witte, E., Friedrich, M., Asadullah, K., & Sabat, R. (2004). IL-22 increases the innate immunity of tissues. *Immunity, 21*, 241–254.

Wu, Y. M., Zhang, K. J., Yue, X. T., Wang, Y. Q., Yang, Y., Li, G. C., et al. (2009). Enhancement of tumor cell death by combining cisplatin with an oncolytic adenovirus carrying MDA-7/IL-24. *Acta Pharmacologica Sinica, 30*, 467–477.

Xiao, L., Li, X., Niu, N., Qian, J., Xie, G., & Wang, Y. (2010). Dichloroacetate (DCA) enhances tumor cell death in combination with oncolytic adenovirus armed with MDA-7/IL-24. *Molecular and Cellular Biochemistry, 340*, 31–40.

Xu, J., Mo, Y., Wang, X., Liu, J., Zhang, X., Wang, J., et al. (2013). Conditionally replicative adenovirus-based mda-7/IL-24 expression enhances sensitivity of colon cancer cells to 5-fluorouracil and doxorubicin. *Journal of Gastroenterology, 48*, 203–213.

Xu, Z., Robitaille, A. M., Berndt, J. D., Davidson, K. C., Fischer, K. A., Mathieu, J., et al. (2016). Wnt/beta-catenin signaling promotes self-renewal and inhibits the primed state transition in naive human embryonic stem cells. *Proceedings of the National Academy of Sciences of the United States of America, 113*, E6382–E6390.

Yacoub, A., Mitchell, C., Brannon, J., Rosenberg, E., Qiao, L., McKinstry, R., et al. (2003). MDA-7 (interleukin-24) inhibits the proliferation of renal carcinoma cells and interacts with free radicals to promote cell death and loss of reproductive capacity. *Molecular Cancer Therapeutics, 2*, 623–632.

Zhang, Z., Kawamura, K., Jiang, Y., Shingyoji, M., Ma, G., Li, Q., et al. (2013). Heat-shock protein 90 inhibitors synergistically enhance melanoma differentiation-associated gene-7-mediated cell killing of human pancreatic carcinoma. *Cancer Gene Therapy, 20*, 663–670.

Zheng, M., Bocangel, D., Doneske, B., Mhashilkar, A., Ramesh, R., Hunt, K. K., et al. (2007). Human interleukin 24 (MDA-7/IL-24) protein kills breast cancer cells via the IL-20 receptor and is antagonized by IL-10. *Cancer Immunology, Immunotherapy, 56*, 205–215.

Zheng, Y., Danilenko, D. M., Valdez, P., Kasman, I., Eastham-Anderson, J., Wu, J., et al. (2007). Interleukin-22, a T(H)17 cytokine, mediates IL-23-induced dermal inflammation and acanthosis. *Nature, 445*, 648–651.

Zhuo, B., Shi, Y., Qin, H., Sun, Q., Li, Z., Zhang, F., et al. (2017). Interleukin-24 inhibits osteosarcoma cell migration and invasion via the JNK/c-Jun signaling pathways. *Oncology Letters, 13*, 4505–4511.

CHAPTER SIX

Advances and Challenges of HDAC Inhibitors in Cancer Therapeutics

Jesse J. McClure, Xiaoyang Li, C. James Chou[1]

Medical University of South Carolina, College of Pharmacy, Charleston, SC, United States
[1]Corresponding author: e-mail address: chouc@musc.edu

Contents

1. Histone Deacetylase (HDAC) Inhibitors and Their Targets — 184
2. Classes of HDACs — 185
3. Lysine Deacylases — 186
4. Major Classes of HDAC Inhibitors — 187
5. HDAC Inhibitors Have Unique NCI 60 Screening Profiles — 190
6. HDAC Inhibitors in the Clinic — 192
7. Challenges in Solid Tumor — 193
8. Pharmacokinetic Challenges of HDAC Inhibitors — 195
9. The History of Hydrazide-Containing Compounds in Clinic and the Future of Next-Generation HDAC Inhibitors — 200
10. Conclusion — 203
References — 203

Abstract

Since the identification and cloning of human histone deacetylases (HDACs) and the rapid approval of vorinostat (Zolinza®) for the treatment of cutaneous T-cell lymphoma, the field of HDAC biology has met many initial successes. However, many challenges remain due to the complexity involved in the lysine posttranslational modifications, epigenetic transcription regulation, and nonepigenetic cellular signaling cascades. In this chapter, we will: review the discovery of the first HDAC inhibitor and present discussion regarding the future of next-generation HDAC inhibitors, give an overview of different classes of HDACs and their differences in lysine deacylation activity, discuss different classes of HDAC inhibitors and their HDAC isozyme preferences, and review HDAC inhibitors' preclinical studies, their clinical trials, their pharmacokinetic challenges, and future direction. We will also discuss the likely reason for the failure of multiple HDAC inhibitor clinical trials in malignancies other than lymphoma and multiple myeloma. In addition, the potential molecular mechanism(s) that may play a key role in the efficacy and therapeutic response rate in the clinic and the likely patient population for HDAC therapy will be discussed.

1. HISTONE DEACETYLASE (HDAC) INHIBITORS AND THEIR TARGETS

Early studies in murine erythroleukemia cells showed that the polar solvent dimethyl sulfoxide (DMSO) promoted cellular differentiation and growth (Friend, Scher, Holland, & Sato, 1971). Noting this, synthesis of several small molecules mimicking DMSO's structure took place; the most prominent was hexamethylene bisacetamide (HMBA; Fig. 1) (Reuben, Wife, Breslow, Rifkind, & Marks, 1976). The potency of HMBA was greater than DMSO, but its EC_{50} was still in the millimolar range. HMBA changed the gene expression profile in cells it was tested in. This change in profile was determined to be the cause of the observed changes in cellular differentiation and growth; however, the exact molecular target and the associated molecular mechanism were still unclear (Marks & Rifkind, 1978; Marks, Sheffery, & Rifkind, 1987; Richon, Ramsay, Rifkind, & Marks, 1989). HMBA was approved to be tested in clinical trials for myelodysplastic syndrome and acute myelogenous leukemia, but the trials found only marginal and transient responses (Andreeff et al., 1992). Further development of HMBA took place as a result.

The replacement of bisacetamide with a more potent metal-chelating moiety, hydroxamic acid, significantly improved the ligand binding to the unknown molecular target(s). The resulting inhibitor, suberic bishydroxamic acid (SBHA, Fig. 1) (Richon et al., 1996), demonstrated an increase in efficacy using cell culture models of nearly two orders of magnitude compared to HMBA. Eventually, one of the two bishydroxamic acids was replaced with various hydrophobic groups based on the hypothesis that only a single metal-chelating group is required for full potency of the unknown molecular target(s). The lead molecule from this rationale was suberoylanilide hydroxamic acid (SAHA; vorinostat, Zolinza®; Fig. 1) (Richon et al., 1996).

While improvements to vorinostat were occurring, the molecular target of trapoxin was determined, using trapoxin's covalent binding properties, to be HDAC1 (Taunton, Collins, & Schreiber, 1996). Hypotheses formed

Fig. 1 Discovery and evolutions of HDAC inhibitors.

suggesting that vorinostat may also inhibit HDAC1. Further testing of these hypotheses led to its validation (Richon et al., 1998). It is interesting to note that despite being the template from which vorinostat was developed, HMBA does not inhibit HDACs (Richon et al., 1996). This suggests there may be additional molecular target(s) that are critical for leukemia differentiation that also fit into the structure–activity relationships of HDAC inhibitors. After two decades of research, the potential HMBA targets remain undetermined.

2. CLASSES OF HDACs

After the identification of HDAC1, researchers have subsequently identified other HDAC isozymes. There are a total of four major classes of HDACs: three metal-containing classes and one NAD^+-dependent class (Haberland, Montgomery, & Olson, 2009; Parra & Verdin, 2010). Based on their homology to yeast proteins, HDACs 1, 2, 3, and 8 are class I HDACs. HDACs 4, 5, 7, and 9 are class IIa HDACs. HDACs 6 and 10 are class IIb, and HDAC11 is the only member of class IV. Class III deacetylases are sirtuins (SIRTs) that depend on NAD^+ as a cofactor for their function; there are seven SIRTs, SIRT1–7. All tissue types ubiquitously express class I HDACs whereas only select tissues express class IIa HDACs (Barneda-Zahonero & Parra, 2012; Yang & Seto, 2008).

Initially, researchers believed all HDACs and SIRTs could deacetylate lysine substrates. However, two studies showed that only HDACs 1, 2, 3, and 6 had sufficient catalytic activity toward acetylated substrates (Bradner et al., 2010; Lahm et al., 2007). Thus, based on the chemophylogenetic studies of HDAC isozymes, the reactivity of class I HDACs 1, 2, 3, and 6 is significantly different from HDACs 8, 10, 11, and class IIa HDACs. HDACs 1, 2, 3, and 6 are the only metal containing deacetylases that possess observable deacetylase activity. Class IIa HDACs, HDAC8, and HDAC11 only deacetylate an artificial, nonphysiological trifluoroacetyl-lysine substrate. However, they do bind to acetylated substrates, which suggest they serve as either an acetylated lysine reader domain similar to the bromodomain, or they deacylate substrates other than the acetylated lysine (Bradner et al., 2010). Similar to the HDACs, SIRTs 1, 2, and 3 possess the most robust deacetylase activity of the sirtuin family; SIRTs 4–7 all have very weak deacetylase activity, comparatively. SIRT5 can deacylate succinylated and malonylated lysine at far greater rates than acetylated substrate (Du et al., 2011), and SIRT6 has greater catalytic activity for long-chain fatty-acids as such as myristoylated and palmitoylated lysine (Jiang et al., 2013).

3. LYSINE DEACYLASES

Proteomic studies revealed that acetylated histone is not the only substrate for HDACs. Lysine posttranslational modifications (PTMs) are involved in both epigenetic transcription regulation and nonepigenetic cell signaling. Acetylated lysine PTMs are as broad as protein phosphorylation with over 3600 lysine acetylation sites on over 1750 proteins (Choudhary & Grant, 2004), and, like protein phosphorylation, potentially play key roles in cellular signaling pathways. In addition, acetylation is not the only lysine-based PTM that HDACs and SIRTs target. Alternative lysine PTM substrates were first identified for SIRT4 and SIRT5 (Du et al., 2011; Jiang et al., 2013). For the HDACs, HDACs 1 and 2 deacylate propionylated and butyrylated lysines (Chen et al., 2007; McClure et al., 2017; Wei et al., 2017). HDAC3 deacylates lysines modified by formylation and acylation of short-chain fatty acid up to eight carbons in length (McClure et al., 2017). In particular, HDAC3 has specific activity toward crotonylated and valerylated lysines (Madsen & Olsen, 2012; McClure et al., 2017). HDAC6 can also catalyze the deformylation of lysines. HDAC8, belonging to class I HDACs, shows weak catalytic activity toward acetylated lysine substrates and has catalytic preference similar to class IIa HDACs. Recently, Aramsangtienchai et al. have shown that HDAC8 catalyzes the hydrolysis of lysines from long-chain fatty acid modifications such as octanoyl, dodecanoyl, and myristoyl groups (Aramsangtienchai et al., 2016). They also demonstrate this catalytic ability is several fold more efficient than its catalytic ability toward acetylated lysines. Taken together, these findings make lysine-based PTMs biology significantly more complex than previously imagined. One could argue that calling these enzymes HDACs is a misnomer that does not fully reflect the capacity of this group. Instead, it may be more relevant to call them lysine deacylases, as different HDACs and SIRTs have specific acylation substrates covering a broad range of biological processes, not just limited to histone lysine tails. The multitude of lysine PTMs and pathways involved further adds to the complexity of the roles of HDACs and SIRTs in various physiological conditions. Ongoing studies continue to identify the key biological roles performed by HDACs and the relevance of these roles in different pathological conditions.

4. MAJOR CLASSES OF HDAC INHIBITORS

There are several major classes of HDAC inhibitors including hydroxamic acid-based, cyclic tetra/depsipeptides, amino-benzamide-based, and short-chain fatty acid-derived inhibitors (Falkenberg & Johnstone, 2014). A recently discovered hydrazide-based HDAC inhibitor further increased the diversity of the HDAC inhibitors (McClure et al., 2016; Wang et al., 2015) (Fig. 2).

Hydroxamic acid-based HDAC inhibitors were the first class of HDAC inhibitor to be developed. Vorinostat (Figs. 1 and 2), the first marketed HDAC inhibitor, has nanomolar affinity toward HDACs. Initially, researchers believed this compound was capable of inhibiting all HDACs. However, further testing demonstrated it could only inhibit HDACs 1, 2, 3, and 6 at reasonable concentrations (Bradner et al., 2010). Owing to its potency in an orphaned disease, vorinostat rapidly moved through preclinical and clinical studies. After completion of these studies, vorinostat earned FDA approval for the treatment of cutaneous T-cell lymphoma (CTCL). Soon after this

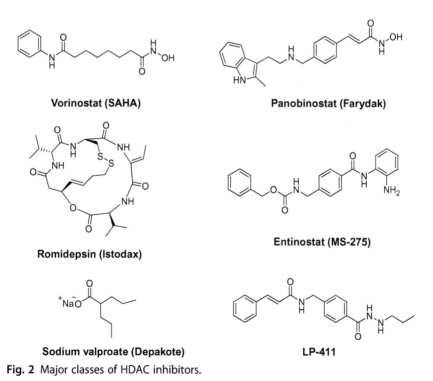

Fig. 2 Major classes of HDAC inhibitors.

approval, tetra/depsipeptides were also found to be capable of inhibiting HDACs. Romidepsin, a natural product isolated from the bacterium *Chromobacterium violaceum* (Falkenberg & Johnstone, 2014) (Fig. 2), gained approval to treat both CTCL and peripheral T-cell lymphomas (PTCL). However, the exact molecular mechanism affecting efficacy of both vorinostat and romidepsin on CTCL and PTCL currently remains unclear. Recent clinical analysis on CTCL patients demonstrates that chromatin accessibility landscape closely correlates to the treatment responses of HDAC inhibitors (Qu et al., 2017). Despite the yet unknown mechanism of efficacy, pharmaceutical companies studied vorinostat and many other hydroxamic acid-based analogs in clinical settings for various cancers, but they have only shown limited efficacy. We will discuss hypotheses for the uneven treatment responses by HDAC inhibitors in further detail in later sections of this chapter.

Nearly a decade after vorinostat's initial approval, panobinostat (Farydak®, Fig. 2) became the first HDAC inhibitor to gain approval for a nonlymphoma cancer. The pan-HDAC inhibitor was approved as an adjunctive therapy for the treatment of refractory/relapsed multiple myeloma. This signified the first step toward breaking out of the lymphoma "bubble" that housed HDAC inhibitors, while simultaneously shedding light on the ability to design inhibitors capable of inhibiting all HDACs yet remaining tolerable enough for use in humans.

A lesser utilized class of HDAC inhibitors, short-chain fatty acids, also have HDAC inhibition at high millimolar concentrations. Valproic acid (valproate, Depakote®; Fig. 2) is used to treat epilepsy and bipolar disorder in clinic (Davie, 2003; Ghodke-Puranik et al., 2013). Hypotheses suggest that inhibition of HDACs plays a mechanistic role in valproic acid's antiepileptic effect. This is mainly because valproic acid's antiepileptic effect does not take effect immediately, suggesting a longer duration cellular molecular mechanism is involved, i.e., transcription regulation (Ghodke-Puranik et al., 2013). Unfortunately, due to short-chain fatty acids' weak inhibitory activity for HDACs, valproic acid and other derivatives have limited applications in the clinic.

Amino-benzamide-based HDAC inhibitors were the first inhibitors to target class I HDACs selectively (Hu et al., 2003; Suzuki et al., 1999). Enzyme kinetic studies of amino-benzamide-based HDAC inhibitors demonstrated that the amino-benzamide motif possesses a tight-binding mechanism (slow-on/slow-off), unlike the classic fast-on/fast-off kinetics associated with hydroxamic acid-based HDAC inhibitors (Chou, Herman, & Gottesfeld, 2008). Entinostat (MS-275, Fig. 2) is the first amino-benzamide-based HDAC inhibitor to reach clinical trials (Ryan et al., 2005). Interestingly, entinostat has a lower therapeutic index than vorinostat in the clinic (Ryan et al., 2005). This is rather surprising, as entinostat possesses much more selectivity toward HDACs compared to vorinostat. Entinostat's low therapeutic index is likely due to lower affinity for targeted HDACs, off-target toxicity, and poor erratic pharmacokinetic properties.

Recently, a new HDAC inhibitor family with a previously unutilized motif in HDAC inhibition has been reported (McClure et al., 2016; Wang et al., 2015). Hydrazide-based HDAC inhibitors are potent HDACs 1, 2, and 3 inhibitors with low nanomolar to picomolar affinity (Fig. 2). Hydrazides are well-known chemical groups for several FDA-approved drugs such as first-generation monoamine oxidase inhibitors including phenelzine or isocarboxazid and the well-known antituberculosis agent, isoniazid. Hydrazide toxicity and safety profiles are well known; however, the mechanism behind their toxicity remains unknown. Hypotheses surrounding this unknown mechanism will be discussed later in this review. Hydrazide-based HDAC inhibitors also have a unique inhibition mechanism. Based on in vitro kinetic studies and Lineweaver–Burk plots derived from them, they display a mixed enzyme inhibition mechanism. This suggests there may be additional allosteric inhibition site(s) within the HDAC isozyme and its coenzymes (McClure et al., 2016). In addition, preliminary studies demonstrate that hydrazides are likely to be nonmutagenic via Ames tests (Fig. 3). Positive Ames tests, having a propensity to cause

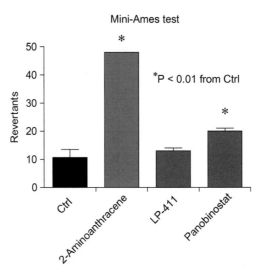

Fig. 3 Ames test of prominent HDAC inhibitors and LP-411.

mutagenicity, is a major disadvantage of many hydroxamic acid-based HDAC inhibitors that limits their uses solely to oncology. Use of agents capable of inducing mutagenicity may also cause treatment-induced leukemia (Shen & Kozikowski, 2016).

5. HDAC INHIBITORS HAVE UNIQUE NCI 60 SCREENING PROFILES

From the early studies of vorinostat, researchers observed that HDAC inhibitors only induce lethality against selective cancer cell types. This is particularly apparent when comparing the results of FDA-approved HDAC inhibitors and HDAC inhibitors in clinic from the NCI 60 screening profiles (Fig. 4). The results show that HDAC inhibitors have surprisingly similar profiles of inducing cell death in a select subpopulation of cells. Vorinostat, entinostat, and the hydrazide-based inhibitor LP-411 share almost identical profiles; comparing the profile of LP-411 to the profiles of other well-studied HDAC inhibitors with COMPARE analysis yields a Pearson correlation coefficient of >0.8. Vorinostat, entinostat, and LP-411 all have growth inhibition (GI_{50}) in the nanomolar range. However, these inhibitors do not induce cell death at these concentrations. These three inhibitors only induce lethality in 8 out of 60 cell lines. Vorinostat and entinostat have an average lethal concentration (LC_{50}) of 17.1–38.6 µM, respectively, for cancer cell lines where they show efficacy. This is far greater than LP-411's LC_{50} values in comparable cell lines of 2.8 µM (Fig. 4). Unlike other HDAC inhibitors, the NCI 60 screening profile for panobinostat appears atypical compared to the other HDAC inhibitors. It induces lethality in most cell lines, indicating its nonselective nature. It is interesting to note that although LP-411 has similar HDAC inhibition potency to panobinostat and greater than 10-fold potency compared to vorinostat, it still maintains its selective profile in the NCI 60 cell screen. Researchers initially thought maintenance of selectivity with greatly increased potency was impossible (Marks & Breslow, 2007).

Further study of the COMPARE analysis results from the NCI 60 cell line screen demonstrates high correlation (Pearson $R > 0.6$) between HDAC inhibitors and the hypomethylating agent azacitidine. This suggests that histone acetylation is likely to be closely intertwined with molecular mechanism(s) involving DNA methylation, ultimately allowing the survival of malignant cells. Not surprisingly, HDAC inhibitors are synergistic with

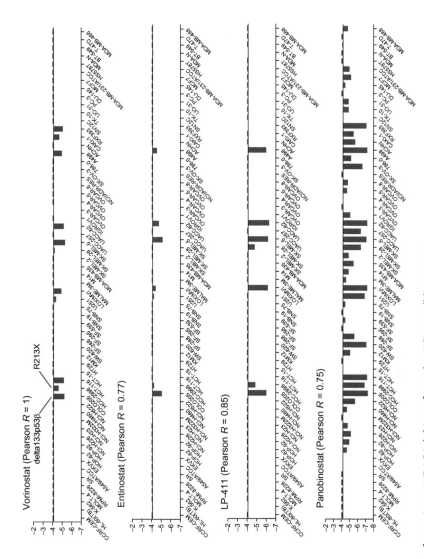

Fig. 4 Profiles of prominent HDAC inhibitors from the NCI 60 cell line screen.

hypomethylating agents (Marchi et al., 2015). Determining which pathways are critically relevant for the efficacy of HDAC inhibitors remains an unfinished task.

Deeper analysis of the NCI 60 cell line screen revealed that almost all cell lines that are sensitive to HDAC inhibitor-induced lethality are p53 wild type (wt-p53, Fig. 4). The two exceptions to this are mutations that generate p53Δ133p53β and p53 R213X mutants. The Δ133p53β mutation, found in the Colo-205 cell line, is a gain-of-function mutation that inhibits proapoptotic signaling (Arsic et al., 2017). The R213X mutation is the most common p53 mutation, generating a transcriptionally altered p53 that augments transcription of MDM2, a p53 inhibitor (Pan & Haines, 2000). However, not all wt-p53 cell lines are sensitive to the HDAC inhibitor treatment, indicating additional triggering molecular mechanism(s) are also required for HDAC inhibitors to be effective. It is still unclear what other molecular mechanistic requirements will be needed to identify critical pathways relevant to developing effective HDAC inhibitor treatments. Furthermore, this research may lead to the identification of the patient population who will benefit most from HDAC inhibitor-based therapies.

6. HDAC INHIBITORS IN THE CLINIC

The majority of the currently FDA-approved HDAC inhibitors are for the treatment of lymphomas, CTCL and PTCL. Vorinostat has an effective concentration against CTCL cell lines in the low micromolar range in vitro. Reaching this effective concentration is possible in vivo despite its observed poor pharmacokinetic profile (Al-Yacoub et al., 2012). A currently lead-in-class hydroxamic acid-based HDAC inhibitor, pracinostat, demonstrates significantly improved in vivo pharmacokinetic profiles. Early clinical trials report a median overall survival of 19.1 months and a complete response (CR) rate of 42% (21 of 50 patients) in newly diagnosed acute myeloid leukemia patients who are not candidates for induction chemotherapy, when coadministered with the hypomethylating agent azacitidine. This is a significant improvement over the recent phase III study of monoazacitidine therapy (Dombret et al., 2015) which showed a median overall survival of 10.4 months with azacitidine alone and a CR rate of 19.5%. This is also a significant improvement over vorinostat clinical trials which had 0%–4.5% CR rate (Schaefer et al., 2009). These studies suggest that HDAC inhibitors play a key part in inducing clinical response rates, especially CR, in the clinical setting. Hence, to develop an HDAC inhibitor that produces clinical responses, both its potency and in vivo pharmacokinetic profiles need further consideration.

7. CHALLENGES IN SOLID TUMOR

Despite the demonstrated effect in hematological tumors, HDAC inhibitor monotherapy has unfortunately not demonstrated such success in solid tumors. A phase I study showed vorinostat was well tolerated and had encouraging anticancer activity in solid tumors with both intravenous and oral administration. Other clinically used HDAC inhibitors also showed efficacy as seen with romidepsin (Sandor et al., 2002), belinostat (Steele et al., 2008), panobinostat (Jones et al., 2011), and entinostat (Ryan et al., 2005). These HDAC inhibitors were subsequently evaluated as monotherapies in phase II trials against solid tumors. However, a negligible amount of patients in these clinical trials reached CR or partial response. Furthermore, the inhibitors also induced serious side effects (Table 1). It is notable that there are successful examples of HDAC inhibitors in the treatment of solid tumors when combined with other chemotherapeutic agents. The tripartite therapy of belinostat, carboplatin, and paclitaxel in heavily pretreated patients with epithelial ovarian cancer was well tolerated and demonstrated clinical benefit with an ORR of 43% (Dizon et al., 2012). Objective responses were also observed in a phase I/II trial of combined epigenetic therapies with azacitidine and entinostat in extensively pretreated patients with recurrent, metastatic nonsmall cell lung cancer (NSCLC) (Rodríguez-Paredes & Esteller, 2011).

Up to now, why HDAC inhibitors display greater efficacy in hematological malignancies compared to solid tumors remains poorly understood. One possibility could be due to the poor pharmacokinetics many HDAC inhibitors possess that ultimately leads to insufficient ability to reach therapeutic concentrations. Most of the hydroxamic acid-based HDAC inhibitors have a short half-life in vivo; for example, vorinostat and belinostat have a $t_{1/2}$ of 0.8–3.9 h (Rubin et al., 2006) and 0.9 h (Steele et al., 2011), respectively (Table 2). In hematological malignancies such as CTCL, PTCL, and multiple myeloma, it is easier for HDAC inhibitors to reach their therapeutic concentrations, so the short half-life may not obstruct their activity. Furthermore, the inhibitors display low permeability, which likely slows their accumulation in solid tumors. Coupling this with vorinostat and belinostat's short half-life would signify that the agents simply cannot get

Table 1 Clinical Trials of HDAC Inhibitors in Solid Tumors

	Cancer Type	Phase	Grade ≥ 3 Adverse Events
Vorinostat	Nonsmall cell lung cancer (NSCLC). Breast cancer, colorectal cancer, thyroid carcinoma, head and neck cancer, ovarian or primary peritoneal carcinoma, glioblastoma multiforme (Blumenschein et al., 2008; Galanis et al., 2009; Luu et al., 2008; Modesitt, Sill, Hoffman, & Bender, 2008; Vansteenkiste et al., 2008; Woyach et al., 2009)	II	Thrombocytopenia, anemia, asthenia, nausea, fatigue, dehydration, ataxia, pneumonia, bruises, and deep vein thrombosis
Romidepsin	Renal cell carcinoma (RCC) lung cancer, colorectal cancer, prostate cancer, small cell lung cancer, malignant glioma, thyroid carcinoma, head and neck cancer (Haigentz et al., 2012; Iwamoto et al., 2011; Jones et al., 2012; Molife et al., 2009; Otterson, Hodgson, Pang, Vokes, & Cancer and Leukemia Group B, 2010; Schrump et al., 2008; Sherman et al., 2013; Stadler, Margolin, Ferber, McCulloch, & Thompson, 2006; Whitehead et al., 2009)	II	Atrial fibrillation and tachycardia, anemia, neutropenia, thrombocytopenia, hypoxia
Belinostat	Malignant pleural mesothelioma, ovarian cancer, thymic epithelial tumors, hepatocellular carcinoma (Giaccone et al., 2011; Lassen et al., 2010; Mackay et al., 2010; Ramalingam et al., 2009; Yeo et al., 2012)	I + II	Leucopenia, neutropenia, thrombocytopenia anemia
Panobinostat	High-grade glioma, RCC, pancreatic cancer (Drappatz et al., 2012; Hainsworth et al., 2011; Strickler et al., 2012; Wang, Cao, & Dudek, 2012)	I + II	Hyperglycemia, thrombocytopenia nausea, vomiting, diarrhea, anorexia, hypertension
Entinostat	Pretreated metastatic melanoma; NSCLC (Hauschild et al., 2008; Witta et al., 2012)	II	Hyponatremia, neutropenia, anemia, hypophosphatemia, and hypoalbuminemia

Table 2 Oral Clinical Pharmacokinetics of HDAC Inhibitors

Parameter	Vorinostat (Rubin et al., 2006)	Romidepsin (Byrd et al., 2005)	Belinostat (Steele et al., 2011)	Panobinostat (Giles et al., 2006)	Entinostat (Ryan et al., 2005)
Dose (mg/m^2)	400	13	900	14	10
$t_{1/2}$ (h)	0.8–3.9	3.67	0.9	12.0	51.58
AUC$_{0\text{-inf}}$ (ng/h/mL)	1716	2107	6266	459.6	528.87
C_{max} (ng/mL)	269	485	3299	565.6	45.07

into solid tumors at reasonable concentrations before they are degraded. Unlike vorinostat and belinostat, the benzamide-based HDAC inhibitor entinostat possesses a much longer half-life of 51.58 h (Ryan et al., 2005); however, its AUC$_{0\text{-inf}}$ and C_{max} are much lower than vorinostat and belinostat at its recommended dose (Table 2), which may be the reason for its poor therapeutic effect.

8. PHARMACOKINETIC CHALLENGES OF HDAC INHIBITORS

The field of HDAC inhibitors centers on the use of a few major classes of compounds. These classes are the hydroxamic acids, the benzamides, and the cyclic tetra/depsipeptides. There are other, less commonly used or less promising classes, such as the short-chain fatty acids; however, the only class of HDAC inhibitors that have shown promise past phase II clinical trials are the aforementioned three. The hydroxamic acids in particular have shown the most widespread use, comprising three of the four FDA approved HDAC inhibitors. These agents, however, suffer from a myriad of metabolic and kinetic issues that possibly lead to their diminished efficacy in clinic despite all having promising preclinical efficacy.

Arguably, the best-known HDAC inhibitor, vorinostat, was the first to gain FDA approval for refractory CTCL (Mann, Johnson, Cohen, Justice, & Pazdur, 2007). Despite this, it displays less than desirable metabolic and kinetic profiles in vivo. Specifically, it has an aqueous solubility of approximately 190 μg/mL (Cai et al., 2010) and a cell permeability of approximately 2×10^{-6} cm/s in Caco-2 cells (Kantharaj & Jayaraman, 2011). Taken together, this places the compound in the Biopharmaceutical Classification System as a class IV drug having low water solubility and low cell permeability.

In animals, vorinostat displayed high clearance in dogs of 7.8 L/kg/h, higher than the dogs' basal liver flow rate of approximately 1.9 L/kg/h (Sandhu et al., 2007). Vorinostat also displayed high clearance in rat of 3.3 L/kg/h, equal to the rats' basal liver flow rate of approximately 3.3 L/kg/h. Furthermore, the compound has relatively short half-life of 12 min in dog and rat alike (Sandhu et al., 2007). Last, vorinostat has poor oral bioavailability in both dog and rat (11%–2%, respectively). Further inquiry showed this poor profile was likely due to extensive metabolic inactivation and high first-pass effect with over 80% of the drug being absorbed, and likely sent to the liver, from the intestines (Sandhu et al., 2007).

The major metabolites found with radiolabeled vorinostat suggest its inactivation rapidly begins with o-glucuronidation. This bulky inactivation, performed by primarily by UDP-glucuronosyltransferase 2B17, not only prevents the compound from fitting into the binding tunnel of HDACs, but also prevents the resonance required for the hydroxamic acid moiety to form a dianionic structure capable of neutralizing the dicationic Zn^{2+} in the catalytic pocket of the enzymes (Balliet et al., 2009). Furthermore, this inactivation leads to two further metabolites, a β-oxidation product, 4-anilino-4-oxobutanoic acid and a hydrolysis product, 8-anilino-8-oxobutanoic acid (Kantharaj & Jayaraman, 2011). Seemingly its best kinetic attribute, vorinostat did not inhibit nor induce any of the major CYP families (Kantharaj & Jayaraman, 2011).

Approved in 2014, belinostat was the second hydroxamic acid-based HDAC inhibitor to receive FDA approval. Its indicated use is for relapsed/refractory PTCL. Compared to vorinostat, belinostat is slightly less soluble in water with an aqueous solubility of 140 μg/mL according to its package insert. Cell permeability data were unavailable. The poor water solubility of this compound translates to belinostat either being a Biopharmaceutical Classification System class II or IV depending on its cell permeability.

Preclinical ADME data for belinostat are lacking compared to vorinostat. It seems that completed studies largely rely on IV or IP dosing (Marquard et al., 2008; Plumb et al., 2003; Warren et al., 2008). This use of IV rather than oral administration could be due to its poor aqueous solubility, or a poor bioavailability issue. It is worth noting that at least one study with oral administration in dogs was performed, which reported an oral bioavailability of 30%–35% (Steele et al., 2011). Clearance in rhesus monkeys was 25.5 $L/m^2/h$ (Warren et al., 2008) after 30 min IV infusion. The reported half-life in these animals was approximately 60 min. In mice, the half-life was 24 min (Marquard et al., 2008).

Like vorinostat, glucuronidation is the most significant metabolic pathway for belinostat. This inactivating metabolism is largely due to UDP-glucuronosyltransferase 1A1 (Wang et al., 2013). Indeed, this inactivation is great enough, that when coupled with genomic differences from patient to patient, the FDA recommends a reduced dose for individuals with mutant alleles of this enzyme reflected in the package insert for belinostat.

The most recently FDA approved hydroxamic acid-based HDAC inhibitor is panobinostat. Panobinostat was the first to break the T-cell lymphoma approvals by gaining acceptance as adjunctive therapy in relapsed/refractory multiple myeloma. Like belinostat, the amount of publicly available preclinical ADME data for panobinostat is relatively small compared to that of vorinostat. However, the Investigative New Drug packet submitted to the European Medicines Agency gave a plethora of preclinical data surrounding panobinostat's pharmacokinetics.

First, the lactate salt of panobinostat is only slightly soluble in water and is heavily pH dependent. Interestingly, this compound is considered as a Biopharmaceutics Classification System class I like compound (Shapiro et al., 2012). Clearance in rats was determined to be approximately 22.1 and 3.3 L/kg/h in dogs; however, it has poor bioavailability in rats of approximately 6% and moderate bioavailability in dogs at 35% (Konsoula, Cao, Velena, & Jung, 2009). The half-life in mice is longer than that of vorinostat or belinostat at 174 min (Kantharaj & Jayaraman, 2011). In rats, this half-life is approximately 30 h when IV panobinostat is administered. Panobinostat remained highly bound to plasma proteins across all species tested. In mice, rats, and dogs, the percent protein bound was approximately 88.9%, 89.6%, and 78.7%. Furthermore, panobinostat is also extensively metabolized in vivo. The parent compound comprises less than 3% of the excreted dose (Clive et al., 2009).

Similar to the hydroxamic acid-based vorinostat and belinostat, panobinostat is extensively glucuronidated in vivo. The major metabolites found are either the *o*-glucuronidated compound, or the β-oxidation or hydrolysis products that are a result of this glucuronidation. A wide array of UDP-glucuronosyltransferases acts on panobinostat as compared to vorinostat or belinostat. Active transferases include 1A1, 1A3, 1A7, 1A8, 1A9, and 2B4 according to panobinostat's package insert. The cytochrome p450 3A family is responsible for approximately 40% of panobinostat's elimination. As such, only 4.1% of parent panobinostat is responsible for the AUC of circulating metabolites after oral radiolabeled administration in rats and ~10% in dogs.

The second HDAC inhibitor to receive FDA approval, and the only cyclic tetra/depsipeptide, romidepsin is used in refractory CTCL (FDA data access). Romidepsin's insolubility in water necessitates its injection only dosage form. Romidepsin showed a high clearance of 12.2 L/m^2/h in mice (Graham et al., 2006) coupled with a relatively longer half-life of 348 min. Owing to its poor solubility, no studies have been published, to the best of our knowledge, which utilized oral administration of this compound in animals. As such, we can assume the bioavailability to be negligible.

Upon administration, romidepsin is highly bound by proteins, 92%–94% in humans from its package insert. Furthermore, it is a substrate of P-glycoprotein (Xiao et al., 2005). Last, extensive metabolism of romidepsin occurs by the liver, in particular cytochrome p450 3A4 (Shiraga, Tozuka, Ishimura, Kawamura, & Kagayama, 2005). However, it does not appear to affect CYP activity in humans. Taken together, romidepsin's poor kinetics, high levels of protein binding, and efflux via P-glycoprotein prevent this compound from working as efficaciously as it could and leaves much to be desired in clinic.

Entinostat is likely the best-known and well-studied benzamide-based HDAC inhibitor to gain traction as a potential new therapy. It utilizes a similar strategy mechanistically to the other HDAC inhibitors; however, unlike the hydroxamic acid-based inhibitors, its binding kinetics in vitro are more complicated. Entinostat, among other benzamide inhibitors, exhibits a slow-on binding mode (Chou et al., 2008). Unfortunately, not much publicly accessible data exist in regard to preclinical kinetic studies. We were, however, able to find that entinostat is only moderately plasma protein bound (~40%) in nonhuman animals. Interestingly, entinostat had a relatively short half-life of approximately 1 h in mice, rats, and dogs. Its half-life in humans, however, is approximately 52 h (Ryan et al., 2005). The oral bioavailability of entinostat is quite high at 85%; however, its interpatient variability in pharmacokinetics makes it rather difficult to dose in clinic (Ryan et al., 2005). Patients displayed t_{max} values between 0.5 and 60 h. Last, entinostat, unlike the other FDA-approved HDAC inhibitors, displays much lower susceptibility to metabolism in humans (Acharya et al., 2006). Taken together, entinostat's pharmacokinetic variability will likely hamper its use in clinic unless a predictor of its absorption can be determined prior to dosing.

Given the preanimal efficacy that HDAC inhibitors show, coupled with their lackluster pharmacokinetic, absorption, metabolism, and bioavailability issues surrounding these otherwise efficacious inhibitors, pracinostat was designed with the intention of developing an inhibitor with superior ADME properties. Like the majority of FDA-approved HDAC inhibitors, it is a hydroxamic acid-based inhibitor. It is highly soluble in water and is highly absorbed in the intestines (Wang et al., 2011). As such, it is a Biopharmaceutical Classification System class I compound with high solubility and high absorption. Its oral bioavailability of 34% in mice and 65% in dogs makes it superior to all FDA-approved HDAC inhibitors in this regard (Kantharaj & Jayaraman, 2011). Furthermore, likely due to this increased bioavailability, its C_{max} in dogs and humans alike is over fivefold greater than other FDA-approved and well-studied HDAC inhibitors. Its half-life in mice of 2.4 and 4.1 h in dogs is longer than that of vorinostat and belinostat, and roughly equivalent to panobinostat. Pracinostat showed high systemic clearance of 9.2, 4.5, and 1.5 L/kg/h in mice, rats, and dogs, respectively. It was mainly metabolized through CYP 3A4 and 1A2, but did not inhibit major human CYPs. However, like other hydroxamic acid-based HDAC inhibitors, it is extensively glucuronidated, as well as subsequently β-oxidized and hydrolyzed. Similar to panobinostat, pracinostat displayed stable, yet high plasma protein binding of ∼90% in preclinical studies (Kantharaj & Jayaraman, 2011).

The desirable and predictable kinetics profile of pracinostat suggests it may be a superior candidate in terms of clinical efficacy. One group has argued that tumor growth inhibition correlates to the ratio of AUC/IC_{50} as well as C_{max}/IC_{50} (Kantharaj & Jayaraman, 2011). The authors argue that an increased AUC/IC_{50} ratio, or an increased C_{max}/IC_{50} ratio, or increased time above the IC_{50} in vivo all lead to increased tumor growth inhibition. If this was to be true, an agent with a high C_{max} "spike" relative to its IC_{50} coupled with a long half-life would lead to an ideal HDAC inhibitor. This hypothesis, however, only deals with halting tumor growth. It does not factor in that having a high spike in serum concentration of an agent, followed by a long clearance may lead to pronounced toxicity. As such, any agent with this profile would have to be rather selective in its killing prowess to avoid systemic toxicities and intolerabilities in vivo.

The profile of LP-411 seems to fit exactly this bill. Preliminary in vivo data suggest that its C_{max} and AUC are at least 200 times greater than all FDA-approved and well-tested HDAC inhibitors (Fig. 5, Table 3). The data demonstrate that, at similar concentrations, LP-411 is several magnitudes more available while maintaining similar half-lives. Follow-up studies confirming these results as well as further elucidating the pharmacokinetic profile of this promising new inhibitors are required.

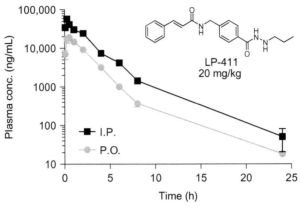

Fig. 5 In vivo pharmacokinetics of next-generation HDAC inhibitor, LP-411.

Table 3 In Vivo Pharmacokinetic Parameters of LP-411 Compared to Similar, Published Data

Compound	LP-411	Panobinostat	Pracinostat
Administered dose (mg/kg)	20	20	50
C_{max} (ng/mL)	19,700	116	2632
t_{max} (h)	0.42	0.17	0.17
$t_{1/2}$ (h)	3.3	2.9	2.4
$AUC_{0\text{-}inf}$ (ng/h/mL)	48,416	126	1840
%F	38%	4.6	34%

9. THE HISTORY OF HYDRAZIDE-CONTAINING COMPOUNDS IN CLINIC AND THE FUTURE OF NEXT-GENERATION HDAC INHIBITORS

The discovery of hydrazide-based HDAC inhibitors as promising therapeutics has opened many doors regarding the use of these agents in clinic (McClure et al., 2016; Wang et al., 2015). These agents' efficacy remains unproven in vivo; however, their in vitro binding dynamics, ex vivo metabolism studies, and rudimentary in vivo pharmacokinetic studies have proved to be quite interesting (McClure et al., 2016) (Fig. 5). Unlike hydroxamic acid-based HDAC inhibitors, the hydrazide-based inhibitors are seemingly impervious to glucuronidation (McClure et al., 2016).

As mentioned earlier, the C_{max}, AUC, and half-life of these agents in preliminary mouse studies suggest they could demonstrate high levels of in vivo efficacy.

One question brought up during the advent of these agents' use was the potential toxicity of the hydrazine/hydrazide motif. Indeed, several agents containing this moiety were used in the early 1960s for depression. These agents included benmoxin, iproclozide, iproniazid, isocarboxazid, mebanazine, nialamide, octamoxin, phenelzine, pheniprazine, phenoxypropazine, pivhydrazine, and safrazine. Removal of many of these agents from the market occurred after reports of fatal cases of fulminant hepatitis and liver failure (Koechlin, Schwartz, & Oberhaensli, 1962; Nelson, Mitchell, Snodgrass, & Timbrell, 1978; Nelson, Mitchell, Timbrell, Snodgrass, & Corcoran, 1976). Despite the majority of these agents displaying this toxicity, two of these antidepressants remain on the market today, isocarboxazid and phenelzine. Further, perhaps the most recognizable hydrazide-containing agent is isoniazid. This medication, still used as an antituberculosis agent either alone or in combination with other agents, has the propensity to cause hepatotoxicity, yet the cases vary in prevalence, and are rarely lethal. Estimates of isoniazid-induced hepatotoxicity range from 0.5%–2% (Nolan, Goldberg, & Buskin, 1999) to 4% (McNeill, Allen, Estrada, & Cook, 2003) while other reports demonstrate it may be as low as 0.2% (Fountain, Tolley, Chrisman, & Self, 2005). The rate of fatality of this fraction of patients who experience hepatotoxicity is between 0.05% and 0.1% (Chalasani et al., 2015; Chan, Or, Cheung, & Woo, 1995; Hayashi et al., 2015; LoBue & Moser, 2003; Stead, To, Harrison, & Abraham 3rd., 1987; Young, Wessolossky, Ellis, Kaminski, & Daly, 2009). As such, at the highest reported incidence and fatality, approximately 1 in 25,000 patients given isoniazid will experience a fatal case of hepatotoxicity.

As such, much research surrounding this potential lethality is available. Early research led to the belief that this hepatotoxicity was due to covalent binding of a metabolite, acetylhydrazine, to liver tissue (Lauterburg, Smith, Todd, & Mitchell, 1985; Timbrell, Mitchell, Snodgrass, & Nelson, 1980). However, these studies examined rat livers and used doses well outside of the normal dosing range used in clinic. Making the question of mechanism more complicated, many of the agents that are included with tuberculosis treatment are themselves hepatotoxic (Tostmann et al., 2008). It remains difficult to ascertain a true image of isoniazid's propensity to induce serious liver damage as a single agent due to the lack of defined single agent hepatotoxic studies in humans and the concomitant clinical use of multiple agents also known to induce liver damage.

As such, a more thorough look at isoniazid's toxicity is required. Competing hypotheses suggest that isoniazid itself is capable of covalently binding liver protein (Metushi, Cai, Zhu, Nakagawa, & Uetrecht, 2011). This mechanism may be unique to isoniazid as it is the only FDA-approved agent with a primary nitrogen. Isocarboxazid contains a benzyl group attached to its ß-nitrogen in its hydrazide group. Phenelzine, while containing a primary nitrogen, is a hydrazine rather than hydrazide. This lack of a carbonyl may substantially change its metabolism pathway in vivo. Another possibility is that isoniazid damages the liver through mitochondrial injury rather than direct hepatic tissue insult. Proponents of this argument posit that a majority of drug-induced hepatotoxicity stems from mitochondrial alterations or harm (Pessayre, Mansouri, Berson, & Fromenty, 2010). Perhaps the most interesting hypothesis surrounding this insult involves immune response to isoniazid in a small subpopulation of patients. Metushi et al. argue that an immune-mediated response raised against metabolites of isoniazid may be at play. They also demonstrate the presence of antibodies reactive toward isoniazid, further bolstering their claim (Metushi, Uetrecht, & Phillips, 2016). Their work has led to the hypothesis that the lower grade liver damage seen with many patients is isoniazid induced; however, it is far less serious than the immune-mediated response, which can lead to fulminant liver failure.

To make an already complicated situation more complicated, a genetic component may be responsible for the observed difference in reactivity. Conflicting evidence regarding the involvement of slow and fast acetylators has demonstrated the potential role of CYP 2E1 in isoniazid-associated hepatotoxicity (Huang et al., 2003; Lee et al., 2010; Sheng et al., 2014; Singla, Gupta, Birbian, & Singh, 2014; Vuilleumier et al., 2006; Yue, Peng, Yang, Kong, & Liu, 2004).

Finally, an interesting hypothesis involving the patient population receiving isoniazid may be worth further pursuit. A recent review demonstrated the stark correlation between patients with tuberculosis and those with hepatitis C (Chan et al., 2017). As such, isoniazid may cause toxicity through a "two hit" model. That is, the patient population may be suffering from an already damaged or injured liver. When a second insult is introduced, some patients may develop liver damage.

Coupling these diverse hypotheses to explain isoniazid-induced hepatotoxicity with the common use of other hydrazide-containing compounds such as nitrofurantoin, dantrolene, streptozocin, and atazanavir, which do not seem to induce or cause liver damage to the extent that isoniazid does, demonstrates the possibility of success for novel agents containing this motif in clinic as long as caution is practiced and liver function is extensively analyzed.

10. CONCLUSION

The future of HDAC inhibitors for cancer therapeutics will heavily rely on two major advancements in the field. One will be to significantly improve the in vivo pharmacokinetic properties of next-generation HDAC inhibitors. The potency and selectivity of the HDAC inhibitors will also need to be improved. Second, because HDAC inhibitors have unique and correlative cellular toxicity profiles, further understanding of the complicated HDAC biology involved in cancer and identifying biomarker(s) relative to therapeutic effects will allow the identification of the patient population most suited for HDAC inhibitor therapy. This will be a key turning point for HDAC therapeutics and will likely significantly improve the outcome of future HDAC inhibitor clinical trials.

REFERENCES

Acharya, M. R., Sparreboom, A., Sausville, E. A., Conley, B. A., Doroshow, J. H., Venitz, J., et al. (2006). Interspecies differences in plasma protein binding of MS-275, a novel histone deacetylase inhibitor. *Cancer Chemotherapy and Pharmacology, 57*(3), 275–281. https://doi.org/10.1007/s00280-005-0058-8.

Al-Yacoub, N., Fecker, L. F., Mobs, M., Plotz, M., Braun, F. K., Sterry, W., et al. (2012). Apoptosis induction by SAHA in cutaneous T-cell lymphoma cells is related to downregulation of c-FLIP and enhanced TRAIL signaling. *The Journal of Investigative Dermatology, 132*(9), 2263–2274. https://doi.org/10.1038/jid.2012.125.

Andreeff, M., Stone, R., Michaeli, J., Young, C. W., Tong, W. P., Sogoloff, H., et al. (1992). Hexamethylene bisacetamide in myelodysplastic syndrome and acute myelogenous leukemia: A phase II clinical trial with a differentiation-inducing agent. *Blood, 80*(10), 2604–2609.

Aramsangtienchai, P., Spiegelman, N. A., He, B., Miller, S. P., Dai, L., Zhao, Y., et al. (2016). HDAC8 catalyzes the hydrolysis of long chain fatty acyl lysine. *ACS Chemical Biology, 11*(10), 2685–2692. https://doi.org/10.1021/acschembio.6b00396.

Arsic, N., Ho-Pun-Cheung, A., Evelyne, C., Assenat, E., Jarlier, M., Anguille, C., et al. (2017). The p53 isoform delta133p53ss regulates cancer cell apoptosis in a RhoB-dependent manner. *PLoS One, 12*(2), e0172125. https://doi.org/10.1371/journal.pone.0172125.

Balliet, R. M., Chen, G., Gallagher, C. J., Dellinger, R. W., Sun, D., & Lazarus, P. (2009). Characterization of UGTs active against SAHA and association between SAHA glucuronidation activity phenotype with UGT genotype. *Cancer Research, 69*(7), 2981–2989. https://doi.org/10.1158/0008-5472.can-08-4143.

Barneda-Zahonero, B., & Parra, M. (2012). Histone deacetylases and cancer. *Molecular Oncology, 6*(6), 579–589. https://doi.org/10.1016/j.molonc.2012.07.003.

Blumenschein, G. R., Kies, M. S., Papadimitrakopoulou, V. A., Lu, C., Kumar, A. J., Ricker, J. L., et al. (2008). Phase II trial of the histone deacetylase inhibitor vorinostat (Zolinza™, suberoylanilide hydroxamic acid, SAHA) in patients with recurrent and/or metastatic head and neck cancer. *Investigational New Drugs*, *26*(1), 81–87.

Bradner, J. E., West, N., Grachan, M. L., Greenberg, E. F., Haggarty, S. J., Warnow, T., et al. (2010). Chemical phylogenetics of histone deacetylases. *Nature Chemical Biology*, *6*(3), 238–243. https://doi.org/10.1038/nchembio.313.

Byrd, J. C., Marcucci, G., Parthun, M. R., Xiao, J. J., Klisovic, R. B., Moran, M., et al. (2005). A phase 1 and pharmacodynamic study of depsipeptide (FK228) in chronic lymphocytic leukemia and acute myeloid leukemia. *Blood*, *105*(3), 959–967.

Cai, Y. Y., Yap, C. W., Wang, Z., Ho, P. C., Chan, S. Y., Ng, K. Y., et al. (2010). Solubilization of vorinostat by cyclodextrins. *Journal of Clinical Pharmacy and Therapeutics*, *35*(5), 521–526. https://doi.org/10.1111/j.1365-2710.2009.01095.x.

Chalasani, N., Bonkovsky, H. L., Fontana, R., Lee, W., Stolz, A., Talwalkar, J., et al. (2015). Features and outcomes of 899 patients with drug-induced liver injury: The DILIN prospective study. *Gastroenterology*, *148*(7) 1340–1352.e1347. https://doi.org/10.1053/j.gastro.2015.03.006.

Chang, T. E., Huang, Y. S., Chang, C. H., Perng, C. L., Huang, Y. H., & Hou, M. C. (2017). The susceptibility of anti-tuberculosis drug-induced liver injury and chronic hepatitis C infection: A systematic review and meta-analysis. *Journal of the Chinese Medical Association*, *81*(2), 111–118. https://doi.org/10.1016/j.jcma.2017.10.002.

Chan, C. H., Or, K. K., Cheung, W., & Woo, J. (1995). Adverse drug reactions and outcome of elderly patients on antituberculosis chemotherapy with and without rifampicin. *Journal of Medicine*, *26*(1–2), 43–52.

Chen, Y., Sprung, R., Tang, Y., Ball, H., Sangras, B., Kim, S. C., et al. (2007). Lysine propionylation and butyrylation are novel post-translational modifications in histones. *Molecular & Cellular Proteomics*, *6*(5), 812–819. https://doi.org/10.1074/mcp.M700021-MCP200.

Chou, C. J., Herman, D., & Gottesfeld, J. M. (2008). Pimelic diphenylamide 106 is a slow, tight-binding inhibitor of class I histone deacetylases. *The Journal of Biological Chemistry*, *283*(51), 35402–35409. https://doi.org/10.1074/jbc.M807045200.

Choudhary, J., & Grant, S. G. (2004). Proteomics in postgenomic neuroscience: The end of the beginning. *Nature Neuroscience*, *7*(5), 440–445. https://doi.org/10.1038/nn1240.

Clive, S., Woo, M. M., Stewart, M., Nydam, T., Hirawat, S., & Kagan, M. (2009). Elucidation of the metabolic and elimination pathways of panobinostat (LBH589) using [C-14]-panobinostat. *Journal of Clinical Oncology*, *27*(15), 2549.

Davie, J. R. (2003). Inhibition of histone deacetylase activity by butyrate. *The Journal of Nutrition*, *133*(7 Suppl), 2485s–2493s.

Dizon, D. S., Blessing, J. A., Penson, R. T., Drake, R. D., Walker, J. L., Johnston, C. M., et al. (2012). A phase II evaluation of belinostat and carboplatin in the treatment of recurrent or persistent platinum-resistant ovarian, fallopian tube, or primary peritoneal carcinoma: A gynecologic oncology group study. *Gynecologic Oncology*, *125*(2), 367–371. https://doi.org/10.1016/j.ygyno.2012.02.019.

Dombret, H., Seymour, J. F., Butrym, A., Wierzbowska, A., Selleslag, D., Jang, J. H., et al. (2015). International phase 3 study of azacitidine vs conventional care regimens in older patients with newly diagnosed AML with >30% blasts. *Blood*, *126*(3), 291–299. https://doi.org/10.1182/blood-2015-01-621664.

Drappatz, J., Lee, E., Hammond, S., Grimm, S., Norden, A., Beroukhim, R., et al. (2012). Phase I study of panobinostat in combination with bevacizumab for recurrent high-grade glioma. *Journal of Neuro-Oncology*, *107*(1), 133–138.

Du, J., Zhou, Y., Su, X., Yu, J. J., Khan, S., Jiang, H., et al. (2011). Sirt5 is a NAD-dependent protein lysine demalonylase and desuccinylase. *Science, 334*(6057), 806–809. 334/6057/806 [pii]. https://doi.org/10.1126/science.1207861.

Falkenberg, K. J., & Johnstone, R. W. (2014). Histone deacetylases and their inhibitors in cancer, neurological diseases and immune disorders. *Nature Reviews. Drug Discovery, 13*(9), 673–691. https://doi.org/10.1038/nrd4360.

Fountain, F. F., Tolley, E., Chrisman, C. R., & Self, T. H. (2005). Isoniazid hepatotoxicity associated with treatment of latent tuberculosis infection: A 7-year evaluation from a public health tuberculosis clinic. *Chest, 128*(1), 116–123. https://doi.org/10.1378/chest.128.1.116.

Friend, C., Scher, W., Holland, J. G., & Sato, T. (1971). Hemoglobin synthesis in murine virus-induced leukemic cells in vitro: Stimulation of erythroid differentiation by dimethyl sulfoxide. *Proceedings of the National Academy of Sciences of the United States of America, 68*(2), 378–382.

Galanis, E., Jaeckle, K. A., Maurer, M. J., Reid, J. M., Ames, M. M., Hardwick, J. S., et al. (2009). Phase II trial of vorinostat in recurrent glioblastoma multiforme: A north central cancer treatment group study. *Journal of Clinical Oncology, 27*(12), 2052–2058.

Ghodke-Puranik, Y., Thorn, C. F., Lamba, J. K., Leeder, J. S., Song, W., Birnbaum, A. K., et al. (2013). Valproic acid pathway: Pharmacokinetics and pharmacodynamics. *Pharmacogenetics and Genomics, 23*(4), 236–241. https://doi.org/10.1097/FPC.0b013e32835ea0b2.

Giaccone, G., Rajan, A., Berman, A., Kelly, R. J., Szabo, E., Lopez-Chavez, A., et al. (2011). Phase II study of belinostat in patients with recurrent or refractory advanced thymic epithelial tumors. *Journal of Clinical Oncology, 29*(15), 2052–2059.

Giles, F., Fischer, T., Cortes, J., Garcia-Manero, G., Beck, J., Ravandi, F., et al. (2006). A phase I study of intravenous LBH589, a novel cinnamic hydroxamic acid analogue histone deacetylase inhibitor, in patients with refractory hematologic malignancies. *Clinical Cancer Research, 12*(15), 4628–4635. https://doi.org/10.1158/1078-0432.CCR-06-0511.

Graham, C., Tucker, C., Creech, J., Favours, E., Billups, C. A., Liu, T., et al. (2006). Evaluation of the antitumor efficacy, pharmacokinetics, and pharmacodynamics of the histone deacetylase inhibitor depsipeptide in childhood cancer models in vivo. *Clinical Cancer Research, 12*(1), 223–234. https://doi.org/10.1158/1078-0432.ccr-05-1225.

Haberland, M., Montgomery, R. L., & Olson, E. N. (2009). The many roles of histone deacetylases in development and physiology: Implications for disease and therapy. *Nature Reviews. Genetics, 10*(1), 32–42. https://doi.org/10.1038/nrg2485.

Haigentz, M., Kim, M., Sarta, C., Lin, J., Keresztes, R. S., Culliney, B., et al. (2012). Phase II trial of the histone deacetylase inhibitor romidepsin in patients with recurrent/metastatic head and neck cancer. *Oral Oncology, 48*(12), 1281–1288.

Hainsworth, J. D., Infante, J. R., Spigel, D. R., Arrowsmith, E. R., Boccia, R. V., & Burris, H. A. (2011). A phase II trial of panobinostat, a histone deacetylase inhibitor, in the treatment of patients with refractory metastatic renal cell carcinoma. *Cancer Investigation, 29*(7), 451–455.

Hauschild, A., Trefzer, U., Garbe, C., Kaehler, K. C., Ugurel, S., Kiecker, F., et al. (2008). Multicenter phase II trial of the histone deacetylase inhibitor pyridylmethyl-N-{4-[(2-aminophenyl)-carbamoyl]-benzyl}-carbamate in pretreated metastatic melanoma. *Melanoma Research, 18*(4), 274–278.

Hayashi, P. H., Fontana, R. J., Chalasani, N. P., Stolz, A. A., Talwalkar, J. A., Navarro, V. J., et al. (2015). Under-reporting and poor adherence to monitoring guidelines for severe cases of isoniazid hepatotoxicity. *Clinical Gastroenterology and Hepatology, 13*(9) 1676–1682.e1671. https://doi.org/10.1016/j.cgh.2015.02.024.

Hu, E., Dul, E., Sung, C. M., Chen, Z., Kirkpatrick, R., Zhang, G. F., et al. (2003). Identification of novel isoform-selective inhibitors within class I histone deacetylases. *The Journal of Pharmacology and Experimental Therapeutics, 307*(2), 720–728. https://doi.org/10.1124/jpet.103.055541.

Huang, Y. S., Chern, H. D., Su, W. J., Wu, J. C., Chang, S. C., Chiang, C. H., et al. (2003). Cytochrome P450 2E1 genotype and the susceptibility to antituberculosis drug-induced hepatitis. *Hepatology, 37*(4), 924–930. https://doi.org/10.1053/jhep.2003.50144.

Iwamoto, F. M., Lamborn, K. R., Kuhn, J. G., Wen, P. Y., Alfred Yung, W., Gilbert, M. R., et al. (2011). A phase I/II trial of the histone deacetylase inhibitor romidepsin for adults with recurrent malignant glioma: North American brain tumor consortium study 03-03. *Neuro-Oncology, 13*(5), 509–516.

Jiang, H., Khan, S., Wang, Y., Charron, G., He, B., Sebastian, C., et al. (2013). SIRT6 regulates TNF-alpha secretion through hydrolysis of long-chain fatty acyl lysine. *Nature, 496*(7443), 110–113. https://doi.org/10.1038/nature12038.

Jones, S. F., Bendell, J., Infante, J., Spigel, D., Thompson, D., Yardley, D., et al. (2011). A phase I study of panobinostat in combination with gemcitabine in the treatment of solid tumors. *Clinical Advances in Hematology & Oncology, 9*(3), 225–230.

Jones, S. F., Infante, J. R., Spigel, D. R., Peacock, N. W., Thompson, D. S., Greco, F. A., et al. (2012). Phase 1 results from a study of romidepsin in combination with gemcitabine in patients with advanced solid tumors. *Cancer Investigation, 30*(6), 481–486.

Kantharaj, E., & Jayaraman, R. (2011). Histone deacetylase inhibitors as therapeutic agents for cancer therapy: Drug metabolism and pharmacokinetic properties. In *Drug development—A case study based insight into modern strategies* (pp. 101–120). InTechOpen. https://doi.org/10.5772/27799.

Koechlin, B. A., Schwartz, M. A., & Oberhaensli, W. E. (1962). Metabolism of C-14-iproniazid and C-14-isocarboxazid in man. *The Journal of Pharmacology and Experimental Therapeutics, 138*, 11–20.

Konsoula, Z., Cao, H., Velena, A., & Jung, M. (2009). Pharmacokinetics–pharmacodynamics and antitumor activity of mercaptoacetamide-based histone deacetylase inhibitors. *Molecular Cancer Therapeutics, 8*(10), 2844–2851. https://doi.org/10.1158/1535-7163.mct-09-0629.

Lahm, A., Paolini, C., Pallaoro, M., Nardi, M. C., Jones, P., Neddermann, P., et al. (2007). Unraveling the hidden catalytic activity of vertebrate class IIa histone deacetylases. *Proceedings of the National Academy of Sciences of the United States of America, 104*(44), 17335–17340. https://doi.org/10.1073/pnas.0706487104.

Lassen, U., Molife, L., Sorensen, M., Engelholm, S., Vidal, L., Sinha, R., et al. (2010). A phase I study of the safety and pharmacokinetics of the histone deacetylase inhibitor belinostat administered in combination with carboplatin and/or paclitaxel in patients with solid tumours. *British Journal of Cancer, 103*(1), 12–17.

Lauterburg, B. H., Smith, C. V., Todd, E. L., & Mitchell, J. R. (1985). Oxidation of hydrazine metabolites formed from isoniazid. *Clinical Pharmacology and Therapeutics, 38*(5), 566–571.

Lee, S. W., Chung, L. S., Huang, H. H., Chuang, T. Y., Liou, Y. H., & Wu, L. S. (2010). NAT2 and CYP2E1 polymorphisms and susceptibility to first-line anti-tuberculosis drug-induced hepatitis. *The International Journal of Tuberculosis and Lung Disease, 14*(5), 622–626.

LoBue, P. A., & Moser, K. S. (2003). Use of isoniazid for latent tuberculosis infection in a public health clinic. *American Journal of Respiratory and Critical Care Medicine, 168*(4), 443–447. https://doi.org/10.1164/rccm.200303-390OC.

Luu, T. H., Morgan, R. J., Leong, L., Lim, D., McNamara, M., Portnow, J., et al. (2008). A phase II trial of vorinostat (suberoylanilide hydroxamic acid) in metastatic breast cancer: A California cancer consortium study. *Clinical Cancer Research, 14*(21), 7138–7142.

Mackay, H. J., Hirte, H., Colgan, T., Covens, A., MacAlpine, K., Grenci, P., et al. (2010). Phase II trial of the histone deacetylase inhibitor belinostat in women with platinum resistant epithelial ovarian cancer and micropapillary (LMP) ovarian tumours. *European Journal of Cancer, 46*(9), 1573–1579.

Madsen, A. S., & Olsen, C. A. (2012). Profiling of substrates for zinc-dependent lysine deacylase enzymes: HDAC3 exhibits decrotonylase activity in vitro. *Angewandte Chemie (International Ed. in English), 51*(36), 9083–9087. https://doi.org/10.1002/anie.201203754.

Mann, B. S., Johnson, J. R., Cohen, M. H., Justice, R., & Pazdur, R. (2007). FDA approval summary: Vorinostat for treatment of advanced primary cutaneous T-cell lymphoma. *The Oncologist, 12*(10), 1247–1252. https://doi.org/10.1634/theoncologist.12-10-1247.

Marchi, E., Zullo, K. M., Amengual, J. E., Kalac, M., Bongero, D., McIntosh, C. M., et al. (2015). The combination of hypomethylating agents and histone deacetylase inhibitors produce marked synergy in preclinical models of T-cell lymphoma. *British Journal of Haematology*. https://doi.org/10.1111/bjh.13566.

Marks, P. A., & Breslow, R. (2007). Dimethyl sulfoxide to vorinostat: Development of this histone deacetylase inhibitor as an anticancer drug. *Nature Biotechnology, 25*(1), 84–90. https://doi.org/10.1038/nbt1272.

Marks, P. A., & Rifkind, R. A. (1978). Erythroleukemic differentiation. *Annual Review of Biochemistry, 47*, 419–448. https://doi.org/10.1146/annurev.bi.47.070178.002223.

Marks, P. A., Sheffery, M., & Rifkind, R. A. (1987). Induction of transformed cells to terminal differentiation and the modulation of gene expression. *Cancer Research, 47*(3), 659–666.

Marquard, L., Petersen, K. D., Persson, M., Hoff, K. D., Jensen, P. B., & Sehested, M. (2008). Monitoring the effect of belinostat in solid tumors by H4 acetylation. *APMIS, 116*(5), 382–392. https://doi.org/10.1111/j.1600-0463.2008.00957.x.

McClure, J. J., Inks, E. S., Zhang, C., Peterson, Y. K., Li, J., Chundru, K., et al. (2017). Comparison of the deacylase and deacetylase activity of zinc-dependent HDACs. *ACS Chemical Biology, 12*(6), 1644–1655. https://doi.org/10.1021/acschembio.7b00321.

McClure, J. J., Zhang, C., Inks, E. S., Peterson, Y. K., Li, J., & Chou, C. J. (2016). Development of allosteric hydrazide-containing class I histone deacetylase inhibitors for use in acute myeloid leukemia. *Journal of Medicinal Chemistry, 59*(21), 9942–9959. https://doi.org/10.1021/acs.jmedchem.6b01385.

McNeill, L., Allen, M., Estrada, C., & Cook, P. (2003). Pyrazinamide and rifampin vs isoniazid for the treatment of latent tuberculosis: Improved completion rates but more hepatotoxicity. *Chest, 123*(1), 102–106.

Metushi, I. G., Cai, P., Zhu, X., Nakagawa, T., & Uetrecht, J. P. (2011). A fresh look at the mechanism of isoniazid-induced hepatotoxicity. *Clinical Pharmacology and Therapeutics, 89*(6), 911–914. https://doi.org/10.1038/clpt.2010.355.

Metushi, I., Uetrecht, J., & Phillips, E. (2016). Mechanism of isoniazid-induced hepatotoxicity: Then and now. *British Journal of Clinical Pharmacology, 81*(6), 1030–1036. https://doi.org/10.1111/bcp.12885.

Modesitt, S. C., Sill, M., Hoffman, J. S., & Bender, D. P. (2008). A phase II study of vorinostat in the treatment of persistent or recurrent epithelial ovarian or primary peritoneal carcinoma: A gynecologic oncology group study. *Gynecologic Oncology, 109*(2), 182–186.

Molife, L., Attard, G., Fong, P., Karavasilis, V., Reid, A., Patterson, S., et al. (2009). Phase II, two-stage, single-arm trial of the histone deacetylase inhibitor (HDACi) romidepsin in metastatic castration-resistant prostate cancer (CRPC). *Annals of Oncology, 21*(1), 109–113.

Nelson, S. D., Mitchell, J. R., Snodgrass, W. R., & Timbrell, J. A. (1978). Hepatotoxicity and metabolism of iproniazid and isopropylhydrazine. *The Journal of Pharmacology and Experimental Therapeutics, 206*(3), 574–585.

Nelson, S. D., Mitchell, J. R., Timbrell, J. A., Snodgrass, W. R., & Corcoran, G. B., 3rd. (1976). Isoniazid and iproniazid: Activation of metabolites to toxic intermediates in man and rat. *Science, 193*(4256), 901–903.

Nolan, C. M., Goldberg, S. V., & Buskin, S. E. (1999). Hepatotoxicity associated with isoniazid preventive therapy: A 7-year survey from a public health tuberculosis clinic. *JAMA, 281*(11), 1014–1018.

Otterson, G. A., Hodgson, L., Pang, H., Vokes, E. E., & Cancer and Leukemia Group B. (2010). Phase II study of the histone deacetylase inhibitor romidepsin in relapsed small cell lung cancer (cancer and leukemia group B 30304). *Journal of Thoracic Oncology, 5*(10), 1644–1648.

Pan, Y., & Haines, D. S. (2000). Identification of a tumor-derived p53 mutant with novel transactivating selectivity. *Oncogene, 19*(27), 3095–3100. https://doi.org/10.1038/sj.onc.1203663.

Parra, M., & Verdin, E. (2010). Regulatory signal transduction pathways for class IIa histone deacetylases. *Current Opinion in Pharmacology, 10*(4), 454–460. https://doi.org/10.1016/j.coph.2010.04.004.

Pessayre, D., Mansouri, A., Berson, A., & Fromenty, B. (2010). Mitochondrial involvement in drug-induced liver injury. *Handbook of Experimental Pharmacology, 196,* 311–365. https://doi.org/10.1007/978-3-642-00663-0_11.

Plumb, J. A., Finn, P. W., Williams, R. J., Bandara, M. J., Romero, M. R., Watkins, C. J., et al. (2003). Pharmacodynamic response and inhibition of growth of human tumor xenografts by the novel histone deacetylase inhibitor PXD101. *Molecular Cancer Therapeutics, 2*(8), 721–728.

Qu, K., Zaba, L. C., Satpathy, A. T., Giresi, P. G., Li, R., Jin, Y., et al. (2017). Chromatin accessibility landscape of cutaneous T cell lymphoma and dynamic response to HDAC Inhibitors. *Cancer Cell, 32*(1) 27–41.e24. https://doi.org/10.1016/j.ccell.2017.05.008.

Ramalingam, S. S., Belani, C. P., Ruel, C., Frankel, P., Gitlitz, B., Koczywas, M., et al. (2009). Phase II study of belinostat (PXD101), a histone deacetylase inhibitor, for second line therapy of advanced malignant pleural mesothelioma. *Journal of Thoracic Oncology, 4*(1), 97–101.

Reuben, R. C., Wife, R. L., Breslow, R., Rifkind, R. A., & Marks, P. A. (1976). A new group of potent inducers of differentiation in murine erythroleukemia cells. *Proceedings of the National Academy of Sciences of the United States of America, 73*(3), 862–866.

Richon, V. M., Emiliani, S., Verdin, E., Webb, Y., Breslow, R., Rifkind, R. A., et al. (1998). A class of hybrid polar inducers of transformed cell differentiation inhibits histone deacetylases. *Proceedings of the National Academy of Sciences of the United States of America, 95*(6), 3003–3007.

Richon, V. M., Ramsay, R. G., Rifkind, R. A., & Marks, P. A. (1989). Modulation of the c-myb, c-myc and p53 mRNA and protein levels during induced murine erythroleukemia cell differentiation. *Oncogene, 4*(2), 165–173.

Richon, V. M., Webb, Y., Merger, R., Sheppard, T., Jursic, B., Ngo, L., et al. (1996). Second generation hybrid polar compounds are potent inducers of transformed cell differentiation. *Proceedings of the National Academy of Sciences of the United States of America, 93*(12), 5705–5708.

Rodríguez-Paredes, M., & Esteller, M. (2011). A combined epigenetic therapy equals the efficacy of conventional chemotherapy in refractory advanced non-small cell lung cancer. *Cancer Discovery, 1*(7), 557–559.

Rubin, E. H., Agrawal, N. G., Friedman, E. J., Scott, P., Mazina, K. E., Sun, L., et al. (2006). A study to determine the effects of food and multiple dosing on the pharmacokinetics of vorinostat given orally to patients with advanced cancer. *Clinical Cancer Research, 12*(23), 7039–7045.

Ryan, Q. C., Headlee, D., Acharya, M., Sparreboom, A., Trepel, J. B., Ye, J., et al. (2005). Phase I and pharmacokinetic study of MS-275, a histone deacetylase inhibitor, in patients with advanced and refractory solid tumors or lymphoma. *Journal of Clinical Oncology, 23*(17), 3912–3922. https://doi.org/10.1200/JCO.2005.02.188.

Sandhu, P., Andrews, P. A., Baker, M. P., Koeplinger, K. A., Soli, E. D., Miller, T., et al. (2007). Disposition of vorinostat, a novel histone deacetylase inhibitor and anticancer agent, in preclinical species. *Drug Metabolism Letters, 1*(2), 153–161.

Sandor, V., Bakke, S., Robey, R. W., Kang, M. H., Blagosklonny, M. V., Bender, J., et al. (2002). Phase I trial of the histone deacetylase inhibitor, depsipeptide (FR901228, NSC 630176), in patients with refractory neoplasms. *Clinical Cancer Research, 8*(3), 718–728.

Schaefer, E. W., Loaiza-Bonilla, A., Juckett, M., DiPersio, J. F., Roy, V., Slack, J., et al. (2009). A phase 2 study of vorinostat in acute myeloid leukemia. *Haematologica, 94*(10), 1375–1382. https://doi.org/10.3324/haematol.2009.009217.

Schrump, D. S., Fischette, M. R., Nguyen, D. M., Zhao, M., Li, X., Kunst, T. F., et al. (2008). Clinical and molecular responses in lung cancer patients receiving romidepsin. *Clinical Cancer Research, 14*(1), 188–198.

Shapiro, G. I., Frank, R., Dandamudi, U. B., Hengelage, T., Zhao, L., Gazi, L., et al. (2012). The effect of food on the bioavailability of panobinostat, an orally active pan-histone deacetylase inhibitor, in patients with advanced cancer. *Cancer Chemotherapy and Pharmacology, 69*(2), 555–562. https://doi.org/10.1007/s00280-011-1758-x.

Shen, S., & Kozikowski, A. P. (2016). Why hydroxamates may not be the best histone deacetylase inhibitors—What some may have forgotten or would rather forget? *ChemMedChem, 11*(1), 15–21. https://doi.org/10.1002/cmdc.201500486.

Sheng, Y. J., Wu, G., He, H. Y., Chen, W., Zou, Y. S., Li, Q., et al. (2014). The association between CYP2E1 polymorphisms and hepatotoxicity due to anti-tuberculosis drugs: A meta-analysis. *Infection, Genetics and Evolution, 24*, 34–40. https://doi.org/10.1016/j.meegid.2014.01.034.

Sherman, E. J., Su, Y. B., Lyall, A., Schöder, H., Fury, M. G., Ghossein, R. A., et al. (2013). Evaluation of romidepsin for clinical activity and radioactive iodine reuptake in radioactive iodine–refractory thyroid carcinoma. *Thyroid, 23*(5), 593–599.

Shiraga, T., Tozuka, Z., Ishimura, R., Kawamura, A., & Kagayama, A. (2005). Identification of cytochrome P450 enzymes involved in the metabolism of FK228, a potent histone deacetylase inhibitor, in human liver microsomes. *Biological & Pharmaceutical Bulletin, 28*(1), 124–129.

Singla, N., Gupta, D., Birbian, N., & Singh, J. (2014). Association of NAT2, GST and CYP2E1 polymorphisms and anti-tuberculosis drug-induced hepatotoxicity. *Tuberculosis (Edinburgh, Scotland), 94*(3), 293–298. https://doi.org/10.1016/j.tube.2014.02.003.

Stadler, W. M., Margolin, K., Ferber, S., McCulloch, W., & Thompson, J. A. (2006). A phase II study of depsipeptide in refractory metastatic renal cell cancer. *Clinical Genitourinary Cancer, 5*(1), 57–60.

Stead, W. W., To, T., Harrison, R. W., & Abraham, J. H., 3rd. (1987). Benefit-risk considerations in preventive treatment for tuberculosis in elderly persons. *Annals of Internal Medicine, 107*(6), 843–845.

Steele, N., Plumb, J., Vidal, L., Tjørnelund, J., Knoblauch, P., Buhl-Jensen, P., et al. (2011). Pharmacokinetic and pharmacodynamic properties of an oral formulation of the histone deacetylase inhibitor belinostat (PXD101). *Cancer Chemotherapy and Pharmacology, 67*(6), 1273–1279. https://doi.org/10.1007/s00280-010-1419-5.

Steele, N. L., Plumb, J. A., Vidal, L., Tjørnelund, J., Knoblauch, P., Rasmussen, A., et al. (2008). A phase 1 pharmacokinetic and pharmacodynamic study of the histone deacetylase inhibitor belinostat in patients with advanced solid tumors. *Clinical Cancer Research, 14*(3), 804–810.

Strickler, J. H., Starodub, A. N., Jia, J., Meadows, K. L., Nixon, A. B., Dellinger, A., et al. (2012). Phase I study of bevacizumab, everolimus, and panobinostat (LBH-589) in advanced solid tumors. *Cancer Chemotherapy and Pharmacology, 70*(2), 251–258.

Suzuki, T., Ando, T., Tsuchiya, K., Fukazawa, N., Saito, A., Mariko, Y., et al. (1999). Synthesis and histone deacetylase inhibitory activity of new benzamide derivatives. *Journal of Medicinal Chemistry, 42*(15), 3001–3003. https://doi.org/10.1021/jm980565u.

Taunton, J., Collins, J. L., & Schreiber, S. L. (1996). Synthesis of natural and modified trapoxins, useful reagents for exploring histone deacetylase function. *Journal of the American Chemical Society, 118*(43), 10412–10422. https://doi.org/10.1021/ja9615841.

Timbrell, J. A., Mitchell, J. R., Snodgrass, W. R., & Nelson, S. D. (1980). Isoniazid hepatoxicity: The relationship between covalent binding and metabolism in vivo. *The Journal of Pharmacology and Experimental Therapeutics, 213*(2), 364–369.

Tostmann, A., Boeree, M. J., Aarnoutse, R. E., de Lange, W. C., van der Ven, A. J., & Dekhuijzen, R. (2008). Antituberculosis drug-induced hepatotoxicity: Concise up-to-date review. *Journal of Gastroenterology and Hepatology, 23*(2), 192–202. https://doi.org/10.1111/j.1440-1746.2007.05207.x.

Vansteenkiste, J., Van Cutsem, E., Dumez, H., Chen, C., Ricker, J. L., Randolph, S. S., et al. (2008). Early phase II trial of oral vorinostat in relapsed or refractory breast, colorectal, or non-small cell lung cancer. *Investigational New Drugs, 26*(5), 483–488.

Vuilleumier, N., Rossier, M. F., Chiappe, A., Degoumois, F., Dayer, P., Mermillod, B., et al. (2006). CYP2E1 genotype and isoniazid-induced hepatotoxicity in patients treated for latent tuberculosis. *European Journal of Clinical Pharmacology, 62*(6), 423–429. https://doi.org/10.1007/s00228-006-0111-5.

Wang, H., Cao, Q., & Dudek, A. Z. (2012). Phase II study of panobinostat and bortezomib in patients with pancreatic cancer progressing on gemcitabine-based therapy. *Anticancer Research, 32*(3), 1027–1031.

Wang, L. Z., Ramirez, J., Yeo, W., Chan, M. Y., Thuya, W. L., Lau, J. Y., et al. (2013). Glucuronidation by UGT1A1 is the dominant pathway of the metabolic disposition of belinostat in liver cancer patients. *PLoS One, 8*(1), e54522. https://doi.org/10.1371/journal.pone.0054522.

Wang, Y., Stowe, R. L., Pinello, C. E., Tian, G., Madoux, F., Li, D., et al. (2015). Identification of histone deacetylase inhibitors with benzoylhydrazide scaffold that selectively inhibit class I histone deacetylases. *Chemistry & Biology, 22*(2), 273–284. https://doi.org/10.1016/j.chembiol.2014.12.015.

Wang, H., Yu, N., Chen, D., Lee, K. C., Lye, P. L., Chang, J. W., et al. (2011). Discovery of (2E)-3-{2-butyl-1-[2-(diethylamino)ethyl]-1H-benzimidazol-5-yl}-N-hydroxyacrylami de (SB939), an orally active histone deacetylase inhibitor with a superior preclinical profile. *Journal of Medicinal Chemistry, 54*(13), 4694–4720. https://doi.org/10.1021/jm2003552.

Warren, K. E., McCully, C., Dvinge, H., Tjornelund, J., Sehested, M., Lichenstein, H. S., et al. (2008). Plasma and cerebrospinal fluid pharmacokinetics of the histone deacetylase inhibitor, belinostat (PXD101), in non-human primates. *Cancer Chemotherapy and Pharmacology, 62*(3), 433–437. https://doi.org/10.1007/s00280-007-0622-5.

Wei, W., Liu, X., Chen, J., Gao, S., Lu, L., Zhang, H., et al. (2017). Class I histone deacetylases are major histone decrotonylases: Evidence for critical and broad function of histone crotonylation in transcription. *Cell Research, 27*(7), 898–915. https://doi.org/10.1038/cr.2017.68.

Whitehead, R. P., Rankin, C., Hoff, P. M., Gold, P. J., Billingsley, K. G., Chapman, R. A., et al. (2009). Phase II trial of romidepsin (NSC-630176) in previously treated colorectal cancer patients with advanced disease: A southwest oncology group study (S0336). *Investigational New Drugs, 27*(5), 469.

Witta, S. E., Jotte, R. M., Konduri, K., Neubauer, M. A., Spira, A. I., Ruxer, R. L., et al. (2012). Randomized phase II trial of erlotinib with and without entinostat in patients with advanced non-small-cell lung cancer who progressed on prior chemotherapy. *Journal of Clinical Oncology, 30*(18), 2248–2255.

Woyach, J. A., Kloos, R. T., Ringel, M. D., Arbogast, D., Collamore, M., Zwiebel, J. A., et al. (2009). Lack of therapeutic effect of the histone deacetylase inhibitor vorinostat in patients with metastatic radioiodine-refractory thyroid carcinoma. *The Journal of Clinical Endocrinology & Metabolism, 94*(1), 164–170.

Xiao, J. J., Foraker, A. B., Swaan, P. W., Liu, S., Huang, Y., Dai, Z., et al. (2005). Efflux of depsipeptide FK228 (FR901228, NSC-630176) is mediated by P-glycoprotein and multidrug resistance-associated protein 1. *The Journal of Pharmacology and Experimental Therapeutics, 313*(1), 268–276. https://doi.org/10.1124/jpet.104.072033.

Yang, X. J., & Seto, E. (2008). The Rpd3/Hda1 family of lysine deacetylases: From bacteria and yeast to mice and men. *Nature Reviews. Molecular Cell Biology, 9*(3), 206–218. https://doi.org/10.1038/nrm2346.

Yeo, W., Chung, H. C., Chan, S. L., Wang, L. Z., Lim, R., Picus, J., et al. (2012). Epigenetic therapy using belinostat for patients with unresectable hepatocellular carcinoma: A multicenter phase I/II study with biomarker and pharmacokinetic analysis of tumors from patients in the Mayo phase II consortium and the cancer therapeutics research group. *Journal of Clinical Oncology, 30*(27), 3361–3367.

Young, H., Wessolossky, M., Ellis, J., Kaminski, M., & Daly, J. S. (2009). A retrospective evaluation of completion rates, total cost, and adverse effects for treatment of latent tuberculosis infection in a public health clinic in central Massachusetts. *Clinical Infectious Diseases, 49*(3), 424–427. https://doi.org/10.1086/600394.

Yue, J., Peng, R. X., Yang, J., Kong, R., & Liu, J. (2004). CYP2E1 mediated isoniazid-induced hepatotoxicity in rats. *Acta Pharmacologica Sinica, 25*(5), 699–704.

CHAPTER SEVEN

Prospects of Gene Therapy to Treat Melanoma

Mitchell E. Menezes*,[2], Sarmistha Talukdar*,[2], Stephen L. Wechman*,[2], Swadesh K. Das*,[†,‡], Luni Emdad*,[†,‡], Devanand Sarkar*,[†,‡], Paul B. Fisher*,[†,‡,1]

*Virginia Commonwealth University, School of Medicine, Richmond, VA, United States
[†]VCU Institute of Molecular Medicine, Virginia Commonwealth University, School of Medicine, Richmond, VA, United States
[‡]VCU Massey Cancer Center, Virginia Commonwealth University, School of Medicine, Richmond, VA, United States
[1]Corresponding author: e-mail address: paul.fisher@vcuhealth.org

Contents

1. Introduction	214
2. Targets for Gene Therapy	215
3. Gene Therapy in Melanoma	219
4. Challenges of Gene Therapy	222
5. Immunotherapy and Combination Therapy	227
6. Conclusions and Future Directions	231
Acknowledgments	232
Conflict of Interest	232
References	232

Abstract

The incidence of melanoma has continued to increase over the past 30 years. Hence, developing effective therapies to treat both primary and metastatic melanoma are essential. While advances in targeted therapy and immunotherapy have provided novel therapeutic options to treat melanoma, gene therapy may provide additional strategies for the treatment of metastatic melanoma clinically. This review focuses upon the challenges and opportunities that gene therapy provides for targeting melanoma. We begin with a discussion of the various gene therapy targets which are relevant to melanoma. Next, we explore the gene therapy clinical trials that have been conducted for treating melanoma. Finally, challenges faced in gene therapy as well as combination therapies for targeting melanoma, which may circumvent these obstacles, will be discussed. Targeted combination gene therapy strategies hold significant promise for developing the most effective therapeutic outcomes, while reducing the toxicity to noncancerous

[2] Contributed equally to this review: M.E.M and S.T.

cells, and would integrate the patient's immune system to diminish melanoma progression. Next-generation vectors designed to embody required safety profiles and "theranostic" attributes, combined with immunotherapeutic strategies would be critical in achieving beneficial management and therapeutic outcomes in melanoma patients.

1. INTRODUCTION

It is estimated that 87,110 people will be diagnosed with melanoma and 9730 people are expected to die from the disease in the United States alone in 2017 (American Cancer Society, Cancer Facts and Figures 2017). The stage of melanoma at diagnosis is a prominent mitigating factor in overall patient survival. Patients identified with lower stage tumors (stage I or II) have significantly higher 5- and 10-year overall survival rates, respectively. On the other hand, patients with later stage tumors (stage III or IV) at the time of diagnosis, who also frequently have distant metastases, have a significantly lower overall survival rate (Tas, 2012). The primary therapeutic option for melanoma is surgical resection of the tumor, however, for patients with metastatic melanoma, surgical resection is rarely curative and usually entails the combination of multiple other cancer therapeutic approaches. Chemotherapy, targeted therapy, and immunotherapy have all been utilized clinically to target melanoma and prolong patient survival. Therefore, novel therapeutic approaches which target both primary tumors and metastatic lesions are necessary to more effectively treat, and someday to cure melanoma.

Gene therapy appears to be an appealing alternative therapeutic option to enhance the efficacy of currently available melanoma therapeutics, leading to better patient prognosis. In the most general terms, gene therapy is a process that corrects genetic defects by replacing defective genes with healthy gene copies (Naldini, 2015). Several vehicles have been developed and utilized to deliver gene therapies including DNA vectors, viruses, micelles, microbubbles as well as various cells that have been modified to carry gene therapy targets (Nayerossadat, Maedeh, & Ali, 2012). In this review, we will provide a synopsis of the various gene therapy targets for melanoma on the molecular level and discuss their associated therapeutics which are in various stages of preclinical and clinical development. We will then discuss some of the most effective gene therapeutic approaches for melanoma and assess the challenges associated with the utilization of these approaches. Finally, we will discuss approaches that utilize gene therapy in combination with other strategies that might aid in enhancing therapeutic outcomes and provide a path which may ultimately bring us closer to a cure for melanomas clinically.

2. TARGETS FOR GENE THERAPY

Selecting appropriate targets for gene therapy is paramount for optimizing patient survival. The key attribute of an optimal gene therapeutic target entails the ability to have maximum therapeutic effects on cancer cells, with minimum toxicity to normal cells. Several different gene therapy targets have been identified and have proven to be effective against melanoma. Gene therapeutic approaches can be used to directly target melanoma cells or aid the patient's immune system in targeting melanoma (Viola, Rafael, Wagner, Besch, & Ogris, 2013). Several gene therapeutic approaches have been used to target melanoma including the introduction of wild-type tumor suppressor genes, expressing suicide genes that can induce apoptosis or sensitize tumors to drugs or radiation, silencing the expression of oncogenes, targeting stromal cells, or introducing immune response modulators to stimulate tumor recognition and destruction by immune cells (Fig. 1) (as reviewed in Das et al., 2015). Other approaches include introducing

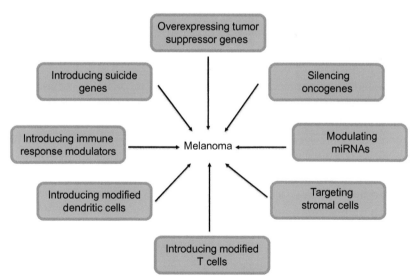

Fig. 1 Gene therapy targets against melanoma. This schematic figure shows various gene therapy targets that have been utilized against melanoma either singly or in combination.

T cells and dendritic cells modified to selectively target cancer cells, enhancing a patient's immune response against melanoma (Boudreau, Bonehill, Thielemans, & Wan, 2011; Kershaw, Westwood, Slaney, & Darcy, 2014). In this section, we will discuss some of the targets used in these approaches.

Genes that regulate key cell survival and cell death pathways are frequently mutated or downregulated in melanoma, making them ideal gene therapy targets. Accordingly, an obvious strategy is to restore the expression of such wild-type genes in tumors lacking these genes. One such gene that is downregulated in melanoma is melanoma differentiation associated gene-7 (MDA-7)/Interleukin-24 (IL-24). MDA-7/IL-24 was initially identified in a differential gene expression screen from terminally differentiating human melanoma cells (Jiang, Lin, Su, Goldstein, & Fisher, 1995). Numerous studies have scrutinized the importance of MDA-7/IL-24 in inhibiting tumor growth in various cancer indications including melanoma (Dash et al., 2010; Menezes et al., 2014). MDA-7/IL-24 induces apoptosis specifically in tumor cells while leaving normal cells unharmed. Infecting melanoma cells with a replication incompetent adenovirus delivering *mda-7* (Ad.*mda-7*) caused a reduction in antiapoptotic proteins Bcl-2 and Bcl-XL and upregulated proapoptotic proteins like Bax and Bak, resulting in the induction of tumor cell death (Lebedeva et al., 2002). In addition, in vivo studies using a cancer terminator virus (*CTV*) that selectively replicates in cancer cells and expresses *mda-7/IL-24* caused a reduction in both primary injected and distant uninjected melanomas in mice (Sarkar et al., 2008) (Fig. 2). Another example is the INK4a/ARF locus that is frequently inactivated during melanoma development (Sharpless & Chin, 2003). The two tumor suppressor proteins, $p16^{INK4a}$ and $p14^{ARF}$, that function via distinct anticancer pathways (the Rb and p53 pathways, respectively) are encoded in this region. In fact, germline mutations in INK4a-ARF have also been identified in familial melanoma (Hayward, 2000; Piepkorn, 2000), underscoring the significance of this gene target for melanoma therapy.

Another strategy to target melanoma involves introducing a suicide gene that can induce apoptosis or sensitize tumors to drugs or radiation. One of the most widely used suicide genes for gene therapy is the herpes simplex virus thymidine kinase (HSVtk) gene. HSVtk in combination with a variety of nucleosides can cause cell death. HSVtk is commonly used in combination with the nucleoside ganciclovir, an acyclic nucleoside analog. HSVtk phosphorylates the prodrug ganciclovir, which is then incorporated into replicating DNA strands, terminating DNA elongation during the

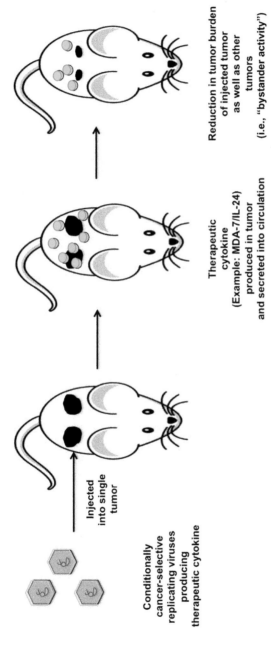

Fig. 2 Conditionally replicating viruses producing therapeutic cytokines can reduce tumor burden in both injected and uninjected tumors. When conditionally replicating adenoviruses carrying therapeutic cytokines (e.g., *mda-7/IL-24*), cancer terminator viruses (*CTVs*), are injected into a single tumor, the therapeutic virus and cytokine produced within the tumor are secreted into the circulation. Although the dissemination of the therapeutic virus may be restricted (due to nonspecific organ trapping, such as in the liver, and clearance by the immune system), the therapeutic cytokine can obtain broader distribution in the body. This secreted therapeutic cytokine can then also target tumors located elsewhere in the body. Moreover, in the context of *mda-7/IL-24*, through receptor engagement, this therapeutic cytokine can stimulate its own production through paracrine/autocrine loop in distant tumor cells. In total, this property of *mda-7/IL-24* promotes a broader therapeutic effect resulting in destruction of the primary injected tumor and distant noninjected tumors.

S-phase of the cell cycle. Therefore, HSVtk positive cells undergoing DNA replication in S-phase of the cell cycle will die by this gene therapeutic mechanism. As a proof of principle study, intratumoral injection with adenoviruses expressing HSVtk followed by intraperitoneal injection with ganciclovir resulted in a 50% reduction in melanoma growth rate in vivo (Bonnekoh et al., 1996).

Silencing genes that are overexpressed in melanoma is another effective strategy to target melanoma using gene therapy. Growth factors that enhance cell growth including basic fibroblast growth factor (bFGF), platelet-derived growth factor (PDGF), epidermal growth factor (EGF), insulin-like growth factors 1 and 2 (IGF-1 and IGF-2), and transforming growth factors α and β (TGFα and TGFβ) have been reported to be overexpressed in melanoma (Polsky & Cordon-Cardo, 2003). Oncogenes comprising the RAS/MAPK signaling pathway are among the most commonly activated in melanoma. These oncogenes respond to signals from their growth factor receptors to mediate cellular responses. RAS oncogenes are frequently activated via mutation in malignant melanoma (Ball et al., 1994). Transgenic or genetically engineered mice that are null for the *ink4a* homolog with melanocyte-specific expression of activated H-*ras*^{G12V} spontaneously develop highly invasive melanoma with short latency (Chin et al., 1997).

Recently, several microRNAs (miRNAs) have been identified for their significance in melanoma development and cancer therapeutic resistance (Fattore et al., 2017), hence modulating miRNA expression may be another strategy to genetically target melanoma. miR-let-7a, that negatively regulates integrin beta 3, is lost in melanoma and this loss of miR-let-7a supported the development and progression of malignant melanoma (Muller & Bosserhoff, 2008). Another tumor-suppressive microRNA, miR-339-3p, is known to inhibit melanoma invasion by downregulating MCL-1 expression (Weber et al., 2016). Furthermore, miR-339-3p overexpression was shown to decrease the ability of melanoma cells to colonize the lungs in an in vivo model (Weber et al., 2016). Another miRNA, miR-579-3p, was determined to be oncosuppressing and involved in melanoma progression as well as resistance to targeted therapy (Fattore et al., 2016). Targeted therapy has been an essential therapeutic option for melanoma patients, however, patients with BRAF activating mutations that were treated with a combination of BRAF and MEK inhibitors inevitably develop drug resistance (Welsh, Rizos, Scolyer, & Long, 2016). miR-579-3p is downregulated as the stage of melanoma increases; furthermore, such decreased expression of miR-579-3p is associated with poor survival (Fattore et al., 2016). miR-579-3p targets BRAF and

MDM2 to act as an oncosuppressor and the overexpression of miR-579-3p impairs the emergence of drug resistance in human melanoma cells (Fattore et al., 2016). Several other miRNAs are also involved in melanoma tumorigenesis as well as the development of resistance to targeted therapy as reviewed by Fattore et al. (2017).

Researchers have also experimented with targeting and/or modulating the stromal tumor microenvironment to treat melanoma (Bhome et al., 2016; Pitt et al., 2016). The stromal tumor microenvironment is comprised of tumor-associated fibroblasts, immune, and inflammatory cells, as well as vascular endothelial cells. Cancer-associated fibroblasts are known to be involved in promoting tumor growth and angiogenesis (Shao et al., 2011), hence modulating these fibroblasts is one strategy to target melanoma. For example, modulating fibroblasts to constitutively activate the Notch1 pathway resulted in the inhibition of melanoma growth and tumor angiogenesis (Shao et al., 2011). Another strategy to target melanoma introduces immune response modulators to stimulate tumor recognition by immune cells. Cancer cells can acquire the ability to evade tumor recognition in several ways including downregulating expression of tumor antigens and costimulatory molecules necessary for immune recognition, expression of surface proteins that cause immune cell inactivation and immune evasion, and modulating the microenvironment to suppress immune response and promote tumor cell proliferation and survival (Restifo et al., 1993; Vinay et al., 2015). Immunostimulatory cytokines such as TNF-α, IFN-γ, granulocyte-macrophage colony-stimulating factors (GM-CSF), as well as costimulatory molecules such as the CD40 ligand, have been overexpressed in tumor cells to augment immune recognition of these tumor cells and enhance tumor antigen presentation and recruitment of tumor-specific T cells and B cells (Driessens, Kline, & Gajewski, 2009). Furthermore, immune checkpoint inhibitors against CTLA4 and inhibitors that block the PD-1/PD-L1 interaction have proven effective in the clinical settings for the treatment of melanoma. Researchers have also developed modified T cells and dendritic cells to enhance a patient's immune response against melanoma (Boudreau et al., 2011; Kershaw et al., 2014).

3. GENE THERAPY IN MELANOMA

Several of the gene therapy approaches discussed in the previous section have been tested clinically; we will now discuss these results here. One of these gene therapy approaches uses oncolytic viruses. Oncolytic viruses specifically replicate in cancerous tissues leading to the lysis of tumors or oncolysis. When infected tumor cells lyse, mature viral particles are released that can further target neighboring tumor cells, allowing for amplification of the killing effect. As discussed previously, MDA-7/IL-24 induces apoptosis in tumor cells. A replication incompetent adenovirus expressing MDA-7/IL-24, Ad.*mda*-7 (INGN-241), was tested in clinical trials in melanoma patients and was well tolerated by patients overall (Cunningham et al., 2005; Fisher et al., 2003, 2007; Tong et al., 2005). The adenovirus induced apoptosis in these treated melanomas and an antitumor response were observed. In one patient with metastatic melanoma, a dramatic antitumor response was observed. The findings from this clinical trial suggested that Ad.*mda*-7 also had a potent "bystander" activity in addition to its immunomodulatory activity (Cunningham et al., 2005; Fisher et al., 2003, 2007; Tong et al., 2005).

Other cancer therapeutic genes such as p53 have also been used in gene therapy applications clinically. The tumor suppressor p53 is frequently genetically altered in human cancers. In vitro and in vivo studies demonstrated that wild-type p53 could restore normal p53 functionality. Hence, Dummer and colleagues performed a phase I dose-escalation study to assess the biological activity and safety of a replication-defective adenoviral vector expressing wild-type p53 in five metastatic melanoma patients with p53 overexpressing tumors (Dummer et al., 2000). Wild-type p53 was observed in tumor tissues and antiadenoviral antibodies were detected in patients indicating biological activity and mild treatment-related effects were observed in patients.

Suicide gene therapy approaches have also been tested clinically with limited success. Klatzmann and colleagues performed a phase I/II dose-escalation study in metastatic melanoma patients to assess retrovirus-mediated herpes simplex virus type I thymidine kinase (HSV-1 TK) suicide gene therapy (Klatzmann et al., 1998). Expression of HSV-1 TK sensitized transduced and bystander cancer cells to ganciclovir toxicity and was the basis of this gene therapy strategy. Cells producing retroviral vector encoding HSV-1 TK were injected multiple times into melanoma cutaneous nodules. Seven days later, ganciclovir was administered for 14 days. Some mild and transient treatment-related effects were observed in patients; however, antitumor effects were limited and disease progression was observed in all patients at long-term follow-up. The authors did observe necrosis in histology sections of some treated tumors, however, they thought poor gene transfer efficiency accounted for these limited therapeutic effects (Klatzmann et al., 1998).

Soiffer and colleagues performed a phase I clinical trial in metastatic melanoma patients to investigate the biologic activity of vaccination with irradiated autologous melanoma cells engineered to secrete human GM-CSF (Soiffer et al., 1998). Intense infiltration of T lymphocytes, dendritic cells, macrophages, and eosinophils were observed at the immunization sites. Additionally, metastatic lesions resected after vaccination were also densely infiltrated with T lymphocytes and plasma cells with extensive tumor destruction, fibrosis, and edema. Furthermore, antimelanoma cytotoxic T cell and antibody responses were associated with tumor destruction. Consequently, this vaccine stimulated potent antitumor immunity in metastatic melanoma patients however due to replacement of necrotic areas by immunological response, clinical tumor regression was not observed (Soiffer et al., 1998).

Duval and colleagues performed a phase I dose-escalation trial using adoptive transfer of allogeneic cytotoxic T lymphocytes equipped with a HLA-A2 restricted MART-1 T-cell receptor (TCR) in patients with metastatic melanoma (Duval et al., 2006). Overall, patients tolerated the treatment well and toxicity ranged from mild to moderate overall, however, only few patients displayed partial responses against primary tumors or the regression of metastases. Robbins and colleagues performed another adoptive immunotherapy clinical trial to test the potential of adoptively transferred autologous T cells transduced with a TCR directed against the NY-ESO-1 cancer/testis antigen that is expressed in about 25% of melanoma patients (Robbins et al., 2011). Encouragingly, responses were observed in 5 of 11 melanoma patients treated with T cells targeting NY-ESO-1; two patients displayed complete regression lasting up to 1 year.

A combination of various gene therapy approaches has also been tested in the clinical setting. Talimogene laherparepvec (T-VEC) is an attenuated herpes simplex virus type 1 that is modified to selectively lyse cancer cells and promote global antitumor immunity via expression of GM-CSF (Killock, 2015). T-VEC is a first-in-class intratumoral oncolytic viral immunotherapy that recently received regulatory approval for the treatment of advanced melanoma in the United States, Europe, and Australia (Rehman, Silk, Kane, & Kaufman, 2016). Results from a recent phase III clinical trial in patients with unresected stages IIIB–IV melanoma demonstrated that T-VEC decreased the size of 64% of injected lesions, 34% of uninjected nonvisceral lesions, and 15% of visceral lesions while complete resolution of lesions was observed in 47% of injected lesions, 22% of uninjected nonvisceral lesions, and 9% of visceral lesions (Andtbacka et al., 2016). Additional clinical trials using this approach are ongoing.

Newer technologies with advanced approaches to target melanoma include using CRISPR technology and mRNA-based therapeutics (Sahin, Kariko, & Tureci, 2014). Wilgnehof and colleagues performed a phase IB study using intravenous dendritic cells electroporated with synthetic mRNA to induce immunotherapeutic effects in patients with advanced melanoma (Wilgenhof et al., 2013). Autologous monocyte-derived dendritic cells were electroporated with synthetic mRNA encoding a CD40 ligand, a constitutively active Toll-like receptor 4 and CD70, along with mRNA encoding a fusion protein of a human leukocyte antigen–class II targeting signal and a melanoma-associated antigen (TriMixDC-MEL). These dendritic cells were administered both intradermally and intravenously and immune responses were assessed. This regimen was tolerated well by all 15 patients. Complete response was observed in two patients and a partial response was observed in another two patients (Wilgenhof et al., 2013). Recently, the first clinical trial proposal using CRISPR technology to modify a patient's T cells received approval in the United States. This phase I trial will consist of 18 patients with several cancer types including melanoma, aimed primarily at determining the safety and potential contraindications of this approach (Cyranoski, 2016). The results from this trial are much awaited as they would open up new avenues to potentially engineer key changes in T cells in order to successfully target melanoma and other cancers.

4. CHALLENGES OF GENE THERAPY

The field of cancer gene therapy is developing rapidly and it is widely believed that this technology will be a significant part of the future of cancer therapeutics (Das et al., 2015). At this juncture, therapeutic gene transfer has had limited success against human disease (Cross & Burmester, 2006; Naldini, 2015; Williams & Baum, 2003). However, recent clinical trials have demonstrated their excellent therapeutic benefits and safety for diseases/disorders with terminal or severely disabling consequences (Naldini, 2015). This progress has been achieved by overcoming hurdles such as improved study designs, vector manufacturing, characterization as well as vectors which can safely deliver therapeutic genes to target cells, a better understanding of the specific biological processes, technologies for editing genes and correcting inherited mutations, the engagement of stem cells to regenerate tissues, improved methods for ex vivo cell handling, improved monitoring and management of patients, and the effective exploitation of powerful immune responses (Naldini, 2015).

Despite all these advances, researchers still face difficulties in quantitative clinical prediction of oncogenic risk/genotoxicity (Naldini, 2015). It is also difficult to predict the progression to malignancy of aberrant clones in individual patients. Since these vectors have the ability to integrate throughout the genome, as dictated by the parental virus strain and particle composition, the risk of gene activation and inactivation at insertion sites cannot be fully abrogated (Naldini, 2015). Currently, improved vector types and designs are also being tested which could potentially solve some of the problems related to unintended oncogene activation however, the safety of these vectors would have to be screened using large-scale trials with 5–10 years observation (Naldini, 2015). Despite these challenges, development of vectors with improved safety profiles, with decreased propensity for insertional "genotoxicity," defining "safe integration sites" in the genome, and the reduction in the total number of vector-exposed cells (and thus vector integrations) are some relevant approaches (Williams & Baum, 2003). Molecular insertion site analysis, which uses the transgene as a tag to identify neighboring cellular sequences, is a powerful tool that would help advance these strategies as well as aid in analyzing integrating capabilities of different vectors into different genomic regions (Williams & Baum, 2003).

Oncoretroviruses, lentiviruses, adenoviruses, adeno-associated viruses, herpes simplex viruses, and their hybrid vectors are the main clinically applicable virus vectors engineered over recent years to deliver genes efficiently and safely (Thomas, Ehrhardt, & Kay, 2003). Each type of vector is characterized by an assemblage of properties, as well as an assortment of problems, that make it suitable for some applications and unsuitable for others (Table 1) (Buchschacher & Wong-Staal, 2000; Evans, 2011; Gong et al., 2016; Kremer & Perricaudet, 1995; Neve, 2012; Warnock et al., 2011). Some viral vectors have shown increased immunogenicity (adenoviruses) while some show increased risk of oncogenesis (retroviruses) (Thomas et al., 2003). Engineering vectors for targeted delivery into specific cells still remains challenging. An assortment of "transductional retargeting" approaches are presently being investigated to develop further the area of viral vector-mediated gene delivery (Thomas et al., 2003). Highlighting the importance of vector–host interactions, the design of new viruses, the development of better site-specific integrating vector systems, the improvement of the efficiency with which vectors infect specific cell types, and an enhanced understanding of how to predict the response of individual patients' posttreatment, will significantly extend the range of therapeutic applications (Thomas et al., 2003).

Table 1 The Relative Advantages and Disadvantages of Various Viral Vectors for Gene Therapy

Viral Vectors	Advantages	Disadvantages	References
Oncoretroviruses	Cancer selective, only expresses their genes in replicating cells. Infections are controllable by well-established antiretroviral drugs	Randomly integrates into the genome, leading to in-consistent therapeutic responses, and the risk of secondary malignancies. Somewhat limited transgene packaging size (9–10 kB)	Evans (2011)
Lentiviruses	Highly efficient gene transfer, stable expression. Well-established vector construction technologies are available	Randomly integrates into the genome, leading to in-consistent therapeutic responses, and the risk of secondary malignancies. Somewhat limited transgene packaging size (9–10 kB)	Buchschacher and Wong-Staal (2000)
Adeno-associated viruses (AAV)	Highly efficient gene transfer, stable gene expression for many years after treatment. Well-established vector construction technologies are available	Preferentially integrates into chromosome 19. Very limited DNA packaging size (~4.7 kB)	Kremer and Perricaudet (1995)
Adenoviruses	Highly efficient gene transfer, highly immunogenic, and a very large packaging capacity (~36 kB). Does not integrate into the genome. Cancer selective mutants are widely available. High titer production is possible (10^{11} pfu/mL). Well-established vector construction technologies are available	Adenoviruses are rapidly inactivated by neutralizing antibodies in circulation. Adenoviruses are also efficiently sequestered by the liver and spleen. Cannot achieve sustained gene expression and repeated administration typically requires the use of other Ad serotypes such as Ad3, Ad11, Ad 35, Ad45, or Ad50	Kremer and Perricaudet (1995)
Herpes simplex viruses (HSV)	Highly efficient gene transfer, highly immunogenic, and a very large transgene packaging capacity (~50 kB). High titer production is possible (10^{10-12} pfu/mL). Does not integrate into the genome. Cancer selective, Tk-deleted, mutants are available. Drugs are available to inhibit HSV replication if vector side effects are observed. Well-established vector construction technologies are available	HSV vectors are rapidly inactivated by neutralizing antibodies in circulation. HSV has a latent infection phase and could recombine with latent HSVs (chicken pox) in patients, reactivating pathogenic HSV strains. HSV vectors despite their infection latency, cannot achieve prolonged transgene expression. HSV vectors also accumulate in the liver and spleen similar to Ads	Neve (2012)

Poxviruses	Highly immunogenic, a very large transgene packaging capacity (~25–30 kB). Fairly high titer production (10^9 pfu/mL). Does not integrate into the genome. Low preexisting immunity. Appears to lyse cancer cells selectively. As a blood borne pathogen, poxviruses offer the best blood stability of any gene vector at this time	The production of poxviruses encoding cancer therapeutic transgenes is complex and is not widely available. Expression of transgenes by poxviruses is transient. Not clear if poxviruses infect cancer cells as efficiently as Ads or HSV vectors	Warnock, Daigre, and Al-Rubeai (2011)
Reoviruses	Reoviruses are not associated with any major human diseases and their replication is shown to be supported by Ras mutations, supporting their cancer selectivity. Favorable packaging size (10 segments, 23.5 kB). Does not integrate into the genome. Does not appear to induce profound side effects in patients clinically	The production of reoviruses encoding cancer therapeutic transgenes is more complex as reoviruses have a dsRNA genome. The expression of transgenes by reoviruses is transient. It is also unclear if reoviruses infect cancer cells as efficiently as Ads or HSV vectors	Gong, Sachdev, Mita, and Mita (2016)

Viral gene therapy is closely dependent on the binding of viral proteins with strain/serotype-specific host receptors. Transductional retargeting, the modification of viral surface proteins to express ligands which specifically bind receptors preferentially or exclusively expressed on tumor cells such as chimeric antigen receptor (CAR), is a common strategy in viral-mediated gene therapy (Haisma et al., 1999). However, technologies utilizing CAR expression need to define a way to control toxicities resulting from CAR-mediated recognition of healthy tissues (Naldini, 2015). Transferring TCR instead of CAR genes could prove more effective in the long term, and additionally, TCR genes are a better fit for cancers that tend not to accumulate passenger mutations and are less responsive to checkpoint blockade drugs (Naldini, 2015). Utilization of epithelial cell adhesion molecule (EpCAM), present in tumor cells, in the design of the vectors may also be helpful (Haisma et al., 1999). The immunogenicity of viral vectors can also be ameliorated by temporary immune suppression, use of Ad capsid chimeras developed from less immunogenic Ads, use of hexon chimeric vectors generated via replacement of immunodominant hypervariable regions, microbubble-mediated delivery, PEG/PLGA polymer coating of vectors, next-generation "gutless" vectors, incorporation of host proteins within membranes, and expression of organ/tissue-specific promoters in the vectors (Viola et al., 2013).

Nonviral delivery systems usually comprise chemical approaches, such as nanoparticles, cationic liposomes and polymers, or physical approaches, such as magnetofection, gene gun, ultrasound, electroporation, or particle bombardment. Nanoparticle-based delivery such as lipid, polymer, PLGA, polyion, peptide, or inorganic-based nanoparticles have also been studied for gene delivery (Liu & Zhang, 2011). The effectiveness of nonviral delivery systems is usually less than that of viral systems, however, they are cheaper, readily available, are significantly less immunologically stimulating and they have no limitation in size of transgenic DNA compared to the viral systems (Nayerossadat et al., 2012).

Considering the greater efficacy of viral gene therapeutics, we will conclude with a discussion of novel mechanisms to restrict the replication of viral therapeutics for melanoma. The Tyrosinase promoter could be used to regulate the expression of the crucial viral replication genes (E1A and E1B), increase target specificity, further limiting toxicity. Target-specific oncolytic viruses have been found to be moderately effective in clinical trials, but the addition of a second therapeutic arm with a therapy agent in the same virus would likely significantly enhance clinical efficacy and decrease the likelihood of resistance emerging. The development of a *CTV* (Azab et al., 2014; Das et al., 2012; Greco et al., 2010; Sarkar et al., 2008; Sarkar, Su, & Fisher, 2006), is one example of this therapeutic strategy. The minimal active region of progression elevated gene-3 (PEG-3) promoter (Su, Shi, & Fisher, 1997) was used in this virus to direct *CTV* oncolysis and simultaneously express cancer therapeutic cytokine genes, e.g., *mda-7/IL-24*, in a cancer-selective manner (Jiang & Fisher, 1993; Jiang et al., 1995). Significant therapeutic potential of the resultant *CTV* was established in melanoma (Sarkar et al., 2008) supporting its potential for targeted therapeutic applications in melanoma.

5. IMMUNOTHERAPY AND COMBINATION THERAPY

Multidrug treatment regimens are rapidly becoming the clinical standard of care, as monotherapy has been found to be relatively ineffective (Nastiuk & Krolewski, 2016). Combinations of anticancer drug therapy regimens are rapidly becoming first-line approaches for the treatment of melanoma (Nastiuk & Krolewski, 2016). For example, the FDA has approved the combination of some new drugs, such as dabrafenib + trametinib, vemurafenib + cobimetinib, and nivolumab + ipilimumab, for the treatment of advanced melanoma (Johnson & Sosman, 2015; Menzies & Long, 2014; Postow et al., 2015). Chemotherapy, hormone therapy, targeted therapy, and immunological checkpoint blockade therapy, are only some of the therapeutic strategies displaying additive or synergistic anticancer efficacy (Nastiuk & Krolewski, 2016). In spite of the projected success of these combinations clinically, many pharmaceutical companies have been hesitant in collaborating in the trials and comarketing a combination of different drugs developed from multiple pharmaceutical companies (Deng, 2015; Nastiuk & Krolewski, 2016). There is also a consensus impetus for the use of targeted therapeutics, as many are of the opinion that standard chemotherapy has reached, or is nearing, its limits of clinical efficacy (Nastiuk & Krolewski, 2016). Conversely, a multimodal delivery system, which specifically targets tumor cells, would be able to enhance chemotherapy effectiveness by minimizing exposure to normal, nonmalignant tissues. A major advantage of gene therapy is the capacity to impact targets which are otherwise undruggable, as well as providing the tools to simultaneously affect multiple biologically relevant genes (Nastiuk & Krolewski, 2016). Tumors undergoing chemotreatment may ultimately develop resistance leading to relapse, and this might be where gene therapy can have a profound effect in combination with chemo/radiotherapy. For example, BRAF-targeted therapy is reported to foster early drug resistance, with median progression-free survival being 6–7 months (Curti, 2014). The drug treatments mostly resulted in grade 2 toxic effects, and 5%–13% of patients stopped treatment due to adverse events (Curti, 2014). The therapeutic strategies also often underestimate the remarkable adaptability of melanoma to therapy (Curti, 2014). Melanoma can develop radiographically detectable progression with BRAF monotherapy and 9–10 months' post-BRAF and MEK inhibition. The inhibitors induce selection pressure on melanoma cells, driving adaptation on a timescale that is massively compacted with relation to Darwinian natural selection (Curti, 2014). Targeted therapy is necessary in patients with aggressive melanoma with BRAF mutations, however, resistance appears in almost all patients (Curti, 2014). This is where combination gene therapy can play an important role in treating patients with melanoma.

Immunotherapy has proven to be an effective therapy for patients with advanced cancers, including melanoma. There are various forms of immunotherapy, including nonspecific immunotherapies, T-cell therapy, cancer vaccines, monoclonal antibodies, and oncolytic virus therapies. Nonspecific immunotherapies, including interleukins and interferons, stimulate the immune system to eradicate tumor cells when combined with other cancer treatments, including radiation, chemotherapy, targeted small molecule therapy, etc., or can kill cancer cells on their own. T-cell therapy exploits our body's own immune cells to attack cancer cells. T cells are isolated from a patient, genetically modified to express targeting ligands on their surface (CAR) for locating and destroying tumor cells, expanded in large numbers ex vivo, and then reintroduced into the patient's body. This approach has proven highly successful in a subset of patients with advanced metastatic melanoma and renal cancer. Cancer vaccines are designed to augment the ability of the body to recognize and destroy cancer cells. These include vaccines that provide treatment strategies for patients and those that provide preventive functions. Monoclonal antibodies provide valuable components in immunotherapy strategies, particularly those that target "checkpoints" that include molecules on immune cells requiring activation (or inactivation) to initiate an immune response. These include antibodies targeting "checkpoint" inhibitors PD-1, such as Pembrolizumab (Keytruda), Nivolumab (Opdivo), PD-L1, such as Atezolizumab (Tecentriq), Avelumab (Bavencio), Durvalumab (Imfinzi), and CTLA-1, such as Ipilimumab (Yervoy). A potential downside of these agents, which have shown profound effects in patients with a number of cancers including melanoma, kidney cancer, bladder cancer, nonsmall cell lung cancer, head and neck cancer, Hodgkin lymphoma, Merkel cell carcinoma and other cancers, involve potential immune system attack on normal organs in the body including the lungs, intestines, liver, kidneys, hormone-producing glands, or other organs. Immunotherapy involving oncolytic virus therapy is receiving increasing scrutiny as a means of attacking cancer cells locally and systemically. Injection of genetically modified oncolytic viruses, such as talimogene laherparepvec or T-VEC, or conditionally cancer-selective oncolytic viruses producing a therapeutic interleukin (mda-7/IL-24), CTV, kills primary tumor cells, and stimulates the immune system to attack noninfected cancer cells. Although we are early in the process, it is clear that various permutations of immunotherapy as described earlier are becoming "first-line" standard therapeutic approaches to treat primary and metastatic cancers.

Combinations of ipilimumab and lambrolizumab treatments have also been studied in patients with advanced melanoma. Even in patients whose disease progressed during ipilimumab treatment, lambrolizumab medication resulted in a high rate of sustained tumor regression, with mainly grade 1 or 2 toxic effects (Hamid et al., 2013). In another example, ipilimumab and nivolumab combinatorial treatment lead to positive responses with significantly higher objective response rates, longer progression-free survival, and higher rates of complete response than ipilimumab monotherapy among patients with both BRAF wild-type and mutant advanced melanoma. However, the frequency of grade 3/4 adverse events was reported to be elevated in combination therapy (Postow et al., 2015). In two large randomized phase III trial studies, involving a total of 1178 patients, it was observed that ipilimumab treatment resulted in a significant survival benefit; however, approximately 20% of patients also reported grades 3 and 4 immune-related adverse effects (Drake, Lipson, & Brahmer, 2014). PD-1 blockade (nivolumab) treatment seems to have induce a lower rate of such grades 3 and 4 immune-related adverse effects than CTLA-4 blockade (ipilimumab), possibly due to the fact that PD-1/PD-ligand (PD-L1) pathway acts more peripherally to immune regulation than the CTLA-4/B7-1 pathway (Drake et al., 2014). The successful combinatorial therapeutic effect of these two-immune checkpoint-based therapies can also be multiplied by adding other immune modulators, such as CD25 or CD40 antibodies, conventional therapies, and effective cancer vaccines (Riley, 2013). The most promising vaccines rely on autologous cells for vaccine production, thereby providing a personalized approach. Another solution could be the creation of allogeneic alternative vaccines such as GM.CD40L, a vaccine composed of autologous tumor cells mixed with allogeneic tumor cells that have been engineered to produce both GM-CSF and CD40L in melanoma (Cross & Burmester, 2006) and other cancers (Creelan et al., 2013; Soliman, Mediavilla-Varela, & Antonia, 2015). Combinatorial effects of multiple strategies may lead to a more potent immune response than a single gene used alone.

Oncolytic viruses such as talimogene laherparepvec (T-vec/Imlygic) is a modified type I herpes simplex virus, have been used in combination with the immune checkpoint inhibitors (ICI): pembrolizumab (anti-PD-1) (NCT02263508) and ipilimumab (anti-CTLA-4) (NCT01740297) for unresectable stage IIIB to IVM1c melanoma. Combination of T-Vec with ipilimumab is in phase Ib clinical trial for the treatment of metastatic melanoma (Fonteneau, 2016; Puzanov et al., 2016). Other oncolytic viruses in phase III clinical trials include JX594, combination with sorafenib for the treatment of advanced hepatocellular carcinoma (NCT02562755), CG0070, for the treatment of BCG therapy-resistant nonmuscle invasive bladder cancer (NCT02365818), Reolysin, in combination with paclitaxel and carboplatin for the treatment of squamous cell carcinoma of the head and neck (NCT01166542) or Toca511, a modified gamma retrovirus and TocaFC (5-fluorocytosine) for the treatment of glioblastoma multiforme after surgical resection (NCT02414165; Fonteneau, 2016; Pol et al., 2014, 2016; Vacchelli et al., 2013). T-Vec, JX-594, and CG0070 oncolytic viruses were each designed to have GM-CSF gene inserted in the viral genome, "arming" oncolytic viruses with transgene(s) to enhance their antitumor functions further. However, studies suggest that interleukin 12, interleukin 18, or soluble B7-1 could also significantly enhance the antitumor efficacy of immunotherpeutics such as GM-CSF via augmenting the antitumor immunity induction. In addition to GM-CSF transgenes offering other antitumor functions, such as antiangiogenesis genes, could also be utilized to arm oncolytic viruses (Fukuhara, Ino, & Todo, 2016). Another effective approach could be a combination of gene therapy with MDA-7/IL-24. As mentioned earlier, MDA-7/IL-24 has both "bystander" and immune-modulatory activity (Menezes et al., 2015; Sarkar et al., 2016; Weber et al., 2016), along with melanoma-selective apoptosis induction (Fisher, 2005; Jiang et al., 1995; Menezes et al., 2014; Sarkar et al., 2007). These combinatorial strategies also hold potential for combinatorial melanoma gene therapy.

6. CONCLUSIONS AND FUTURE DIRECTIONS

Gene therapy for melanoma has been studied extensively, and this approach has been evaluated in clinical trials. Most reported gene therapy approaches have shown positive outcomes, however, the review provides additional context for the future of gene therapy in melanoma. The emergence and extensive use of patient and tumor genomic analysis tools, as well as the exploitation of both the unique and common molecules associated with humoral and cellular immunity, should enable the development of clinically effective personalized gene therapies. Combining diagnostic and therapeutic components in the same viral vector, a "theranostic" system, represents a novel approach with the potential to enhance therapeutic monitoring of the efficacy of viral-based gene therapy used alone or in combination with other agents such as radiation, chemotherapy, immune therapy, and antibody therapy (depicted in Fig. 1). Continued development and progress in the field of safer vector design strategies such as synthetic viruses and nonviral methods, autologous and allogenic adoptive immunotherapy, are expected to increase both the efficacy and safety profile of gene therapy overall. Targeted combination gene therapy with other previously mentioned checkpoint blockers and conventional therapies might prove to be clinically effective with minimal toxicity. Interaction between immune checkpoint inhibitors and oncolytic virotherapy is inherently complex, which requires appropriate selection of viral strain, antibody, and timing (Rojas, Sampath, Hou, & Thorne, 2015). These factors are critical in the experimental design of these combination approaches for synergistic effects (Rojas et al., 2015). The potential exists to target multiple melanoma specific genes simultaneously, which when combined with targeted and immunotherapeutic approaches, has significant potential to enhance melanoma therapy. Advancement in biological research, will lead to the availability of less expensive gene vectors that would make gene therapy accessible to the majority of cancer patients, worldwide. The application of these approaches has the potential to transform the future of cancer therapy, resulting in personalized cancer treatment strategies, which would be fast, effective, relatively safe and cheap, with high cure rates, and maybe even methods for cancer prevention. We are optimistic that with further refinement, the efficacy of gene therapy will increase with the potential to become a "gold standard" for the therapy of melanoma and other cancers.

ACKNOWLEDGMENTS
Support for our laboratories was provided in part by National Institutes of Health Grants R01 CA097318 (P.B.F.), R01 CA168517 (Maurizio Pellecchia and P.B.F.), GM093857 VCU IRACDA (P.B.F. and Joyce A. Lloyd), and P50 CA058326 (Martin G. Pomper and P.B.F.); the Samuel Waxman Cancer Research Foundation (P.B.F. and D.S.); National Foundation for Cancer Research (P.B.F.); NCI Cancer Center Support Grant to VCU Massey Cancer Center P30 CA016059 (P.B.F. and D.S.); and VCU Massey Cancer Center developmental funds (P.B.F.). P.B.F. and D.S. are SWCRF investigators. P.B.F. holds the Thelma Newmeyer Corman Chair in Cancer Research in the VCU Massey Cancer Center. P.B.F. is a NFCR Fellow.

CONFLICT OF INTEREST
P.B.F. is a cofounder and has ownership interest in Cancer Targeting Systems (CTS), Inc. VCU, Johns Hopkins, and Columbia University have ownership interest in CTS.

REFERENCES
Andtbacka, R. H., Ross, M., Puzanov, I., Milhem, M., Collichio, F., Delman, K. A., et al. (2016). Patterns of clinical response with talimogene laherparepvec (T-VEC) in patients with melanoma treated in the OPTiM phase III clinical trial. *Annals of Surgical Oncology*, *23*(13), 4169–4177. https://doi.org/10.1245/s10434-016-5286-0.

Azab, B. M., Dash, R., Das, S. K., Bhutia, S. K., Sarkar, S., Shen, X. N., et al. (2014). Enhanced prostate cancer gene transfer and therapy using a novel serotype chimera cancer terminator virus (Ad.5/3-CTV). *Journal of Cellular Physiology*, *229*(1), 34–43. https://doi.org/10.1002/jcp.24408.

Ball, N. J., Yohn, J. J., Morelli, J. G., Norris, D. A., Golitz, L. E., & Hoeffler, J. P. (1994). Ras mutations in human melanoma: A marker of malignant progression. *The Journal of Investigative Dermatology*, *102*(3), 285–290.

Bhome, R., Al Saihati, H. A., Goh, R. W., Bullock, M. D., Primrose, J. N., Thomas, G. J., et al. (2016). Translational aspects in targeting the stromal tumour microenvironment: From bench to bedside. *New Horizons in Translational Medicine*, *3*(1), 9–21. https://doi.org/10.1016/j.nhtm.2016.03.001.

Bonnekoh, B., Greenhalgh, D. A., Bundman, D. S., Kosai, K., Chen, S. H., Finegold, M. J., et al. (1996). Adenoviral-mediated herpes simplex virus-thymidine kinase gene transfer in vivo for treatment of experimental human melanoma. *The Journal of Investigative Dermatology*, *106*(6), 1163–1168.

Boudreau, J. E., Bonehill, A., Thielemans, K., & Wan, Y. (2011). Engineering dendritic cells to enhance cancer immunotherapy. *Molecular Therapy*, *19*(5), 841–853. https://doi.org/10.1038/mt.2011.57.

Buchschacher, G. L., Jr., & Wong-Staal, F. (2000). Development of lentiviral vectors for gene therapy for human diseases. *Blood*, *95*(8), 2499–2504.

Chin, L., Pomerantz, J., Polsky, D., Jacobson, M., Cohen, C., Cordon-Cardo, C., et al. (1997). Cooperative effects of INK4a and ras in melanoma susceptibility in vivo. *Genes & Development*, *11*(21), 2822–2834.

Creelan, B. C., Antonia, S., Noyes, D., Hunter, T. B., Simon, G. R., Bepler, G., et al. (2013). Phase II trial of a GM-CSF-producing and CD40L-expressing bystander cell line combined with an allogeneic tumor cell-based vaccine for refractory lung adenocarcinoma. *Journal of Immunotherapy*, *36*(8), 442–450. https://doi.org/10.1097/CJI.0b013e3182a80237.

Cross, D., & Burmester, J. K. (2006). Gene therapy for cancer treatment: Past, present and future. *Clinical Medicine & Research, 4*(3), 218–227.

Cunningham, C. C., Chada, S., Merritt, J. A., Tong, A., Senzer, N., Zhang, Y., et al. (2005). Clinical and local biological effects of an intratumoral injection of mda-7 (IL24; INGN 241) in patients with advanced carcinoma: A phase I study. *Molecular Therapy, 11*(1), 149–159. https://doi.org/10.1016/j.ymthe.2004.09.019.

Curti, B. D. (2014). Rapid evolution of combination therapy in melanoma. *The New England Journal of Medicine, 371*(20), 1929–1930. https://doi.org/10.1056/NEJMe1411158.

Cyranoski, D. (2016). CRISPR gene-editing tested in a person for the first time. *Nature, 539*(7630), 479. https://doi.org/10.1038/nature.2016.20988.

Das, S. K., Menezes, M. E., Bhatia, S., Wang, X. Y., Emdad, L., Sarkar, D., et al. (2015). Gene therapies for cancer: Strategies, challenges and successes. *Journal of Cellular Physiology, 230*(2), 259–271. https://doi.org/10.1002/jcp.24791.

Das, S. K., Sarkar, S., Dash, R., Dent, P., Wang, X. Y., Sarkar, D., et al. (2012). Cancer terminator viruses and approaches for enhancing therapeutic outcomes. *Advances in Cancer Research, 115*, 1–38. https://doi.org/10.1016/B978-0-12-398342-8.00001-X.

Dash, R., Bhutia, S. K., Azab, B., Su, Z. Z., Quinn, B. A., Kegelmen, T. P., et al. (2010). Mda-7/IL-24: A unique member of the IL-10 gene family promoting cancer-targeted toxicity. *Cytokine & Growth Factor Reviews, 21*(5), 381–391. https://doi.org/10.1016/j.cytogfr.2010.08.004.

Deng, B. (2015). Approval may embolden industry to combine cancer therapies. *Nature Medicine, 21*(2), 105. https://doi.org/10.1038/nm0215-105.

Drake, C. G., Lipson, E. J., & Brahmer, J. R. (2014). Breathing new life into immunotherapy: Review of melanoma, lung and kidney cancer. *Nature Reviews. Clinical Oncology, 11*(1), 24–37. https://doi.org/10.1038/nrclinonc.2013.208.

Driessens, G., Kline, J., & Gajewski, T. F. (2009). Costimulatory and coinhibitory receptors in anti-tumor immunity. *Immunological Reviews, 229*(1), 126–144. https://doi.org/10.1111/j.1600-065X.2009.00771.x.

Dummer, R., Bergh, J., Karlsson, Y., Horowitz, J. A., Mulder, N. H., Huinink, D. T. B., et al. (2000). Biological activity and safety of adenoviral vector-expressed wild-type p53 after intratumoral injection in melanoma and breast cancer patients with p53-overexpressing tumors. *Cancer Gene Therapy, 7*(7), 1069–1076. https://doi.org/10.1038/sj.cgt.7700214.

Duval, L., Schmidt, H., Kaltoft, K., Fode, K., Jensen, J. J., Sorensen, S. M., et al. (2006). Adoptive transfer of allogeneic cytotoxic T lymphocytes equipped with a HLA-A2 restricted MART-1 T-cell receptor: A phase I trial in metastatic melanoma. *Clinical Cancer Research, 12*(4), 1229–1236. https://doi.org/10.1158/1078-0432.CCR-05-1485.

Evans, C. (2011). Gene therapy for the regeneration of bone. *Injury, 42*(6), 599–604. https://doi.org/10.1016/j.injury.2011.03.032.

Fattore, L., Costantini, S., Malpicci, D., Ruggiero, C. F., Ascierto, P. A., Croce, C. M., et al. (2017). MicroRNAs in melanoma development and resistance to target therapy. *Oncotarget, 8*(13), 22262–22278. https://doi.org/10.18632/oncotarget.14763.

Fattore, L., Mancini, R., Acunzo, M., Romano, G., Lagana, A., Pisanu, M. E., et al. (2016). miR-579-3p controls melanoma progression and resistance to target therapy. *Proceedings of the National Academy of Sciences of the United States of America, 113*(34), E5005–5013. https://doi.org/10.1073/pnas.1607753113.

Fisher, P. B. (2005). Is mda-7/IL-24 a "magic bullet" for cancer? *Cancer Research, 65*(22), 10128–10138. https://doi.org/10.1158/0008-5472.CAN-05-3127.

Fisher, P. B., Gopalkrishnan, R. V., Chada, S., Ramesh, R., Grimm, E. A., Rosenfeld, M. R., et al. (2003). mda-7/IL-24, a novel cancer selective apoptosis inducing cytokine gene: From the laboratory into the clinic. *Cancer Biology & Therapy, 2*(4 Suppl. 1), S23–37.

Fisher, P. B., Sarkar, D., Lebedeva, I. V., Emdad, L., Gupta, P., Sauane, M., et al. (2007). Melanoma differentiation associated gene-7/interleukin-24 (mda-7/IL-24): Novel gene therapeutic for metastatic melanoma. *Toxicology and Applied Pharmacology, 224*(3), 300–307. https://doi.org/10.1016/j.taap.2006.11.021.

Fonteneau, J. F. (2016). Oncolytic viruses and immune checkpoint inhibitors. *Immunotherapy: Open Access, 2*, e105.

Fukuhara, H., Ino, Y., & Todo, T. (2016). Oncolytic virus therapy: A new era of cancer treatment at dawn. *Cancer Science, 107*(10), 1373–1379. https://doi.org/10.1111/cas.13027.

Gong, J., Sachdev, E., Mita, A. C., & Mita, M. M. (2016). Clinical development of reovirus for cancer therapy: An oncolytic virus with immune-mediated antitumor activity. *World Journal of Methodology, 6*(1), 25–42. https://doi.org/10.5662/wjm.v6.i1.25.

Greco, A., Di Benedetto, A., Howard, C. M., Kelly, S., Nande, R., Dementieva, Y., et al. (2010). Eradication of therapy-resistant human prostate tumors using an ultrasound-guided site-specific cancer terminator virus delivery approach. *Molecular Therapy, 18*(2), 295–306. https://doi.org/10.1038/mt.2009.252.

Haisma, H. J., Pinedo, H. M., Rijswijk, A., der Meulen-Muileman, I., Sosnowski, B. A., Ying, W., et al. (1999). Tumor-specific gene transfer via an adenoviral vector targeted to the pan-carcinoma antigen EpCAM. *Gene Therapy, 6*(8), 1469–1474. https://doi.org/10.1038/sj.gt.3300969.

Hamid, O., Robert, C., Daud, A., Hodi, F. S., Hwu, W. J., Kefford, R., et al. (2013). Safety and tumor responses with lambrolizumab (anti-PD-1) in melanoma. *The New England Journal of Medicine, 369*(2), 134–144. https://doi.org/10.1056/NEJMoa1305133.

Hayward, N. (2000). New developments in melanoma genetics. *Current Oncology Reports, 2*(4), 300–306.

Jiang, H., & Fisher, P. B. (1993). Use of a sensitive and efficient subtraction hybridization protocol for the identification of genes differentially regulated during the induction of differentiation in human melanoma cells. *Molecular and Cellular Differentiation, 1*, 285–299.

Jiang, H., Lin, J. J., Su, Z. Z., Goldstein, N. I., & Fisher, P. B. (1995). Subtraction hybridization identifies a novel melanoma differentiation associated gene, mda-7, modulated during human melanoma differentiation, growth and progression. *Oncogene, 11*(12), 2477–2486.

Johnson, D. B., & Sosman, J. A. (2015). Therapeutic advances and treatment options in metastatic melanoma. *JAMA Oncology, 1*(3), 380–386. https://doi.org/10.1001/jamaoncol.2015.0565.

Kershaw, M. H., Westwood, J. A., Slaney, C. Y., & Darcy, P. K. (2014). Clinical application of genetically modified T cells in cancer therapy. *Clinical & Translational Immunology, 3*(5), e16https://doi.org/10.1038/cti.2014.7.

Killock, D. (2015). Skin cancer: T-VEC oncolytic viral therapy shows promise in melanoma. *Nature Reviews Clinical Oncology, 12*(8), 438. https://doi.org/10.1038/nrclinonc.2015.106.

Klatzmann, D., Cherin, P., Bensimon, G., Boyer, O., Coutellier, A., Charlotte, F., et al. (1998). A phase I/II dose-escalation study of herpes simplex virus type 1 thymidine kinase "suicide" gene therapy for metastatic melanoma. Study group on gene therapy of metastatic melanoma. *Human Gene Therapy, 9*(17), 2585–2594. https://doi.org/10.1089/hum.1998.9.17-2585.

Kremer, E. J., & Perricaudet, M. (1995). Adenovirus and adeno-associated virus mediated gene transfer. *British Medical Bulletin, 51*(1), 31–44.

Lebedeva, I. V., Su, Z. Z., Chang, Y., Kitada, S., Reed, J. C., & Fisher, P. B. (2002). The cancer growth suppressing gene mda-7 induces apoptosis selectively in human melanoma cells. *Oncogene, 21*(5), 708–718. https://doi.org/10.1038/sj.onc.1205116.

Liu, C., & Zhang, N. (2011). Nanoparticles in gene therapy principles, prospects, and challenges. *Progress in Molecular Biology and Translational Science, 104,* 509–562. https://doi.org/10.1016/B978-0-12-416020-0.00013-9.
Menezes, M. E., Bhatia, S., Bhoopathi, P., Das, S. K., Emdad, L., Dasgupta, S., et al. (2014). MDA-7/IL-24: Multifunctional cancer killing cytokine. *Advances in Experimental Medicine and Biology, 818,* 127–153. https://doi.org/10.1007/978-1-4471-6458-6_6.
Menezes, M. E., Shen, X. N., Das, S. K., Emdad, L., Guo, C., Yuan, F., et al. (2015). MDA-7/IL-24 functions as a tumor suppressor gene in vivo in transgenic mouse models of breast cancer. *Oncotarget, 6*(35), 36928–36942. https://doi.org/10.18632/oncotarget.6047.
Menzies, A. M., & Long, G. V. (2014). Dabrafenib and trametinib, alone and in combination for BRAF-mutant metastatic melanoma. *Clinical Cancer Research, 20*(8), 2035–2043. https://doi.org/10.1158/1078-0432.CCR-13-2054.
Muller, D. W., & Bosserhoff, A. K. (2008). Integrin beta 3 expression is regulated by let-7a miRNA in malignant melanoma. *Oncogene, 27*(52), 6698–6706. https://doi.org/10.1038/onc.2008.282.
Naldini, L. (2015). Gene therapy returns to centre stage. *Nature, 526*(7573), 351–360. https://doi.org/10.1038/nature15818.
Nastiuk, K. L., & Krolewski, J. J. (2016). Opportunities and challenges in combination gene cancer therapy. *Advanced Drug Delivery Reviews, 98,* 35–40. https://doi.org/10.1016/j.addr.2015.12.005.
Nayerossadat, N., Maedeh, T., & Ali, P. A. (2012). Viral and nonviral delivery systems for gene delivery. *Advanced Biomedical Research, 1,* 27. https://doi.org/10.4103/2277-9175.98152.
Neve, R. L. (2012). Overview of gene delivery into cells using HSV-1-based vectors. *Current Protocols in Neuroscience.* Chapter 4, Unit 4.12. https://doi.org/10.1002/0471142301.ns0412s61.
Piepkorn, M. (2000). Melanoma genetics: An update with focus on the CDKN2A(p16)/ARF tumor suppressors. *Journal of the American Academy of Dermatology, 42*(5 Pt. 1), 705–722 [quiz 723–706].
Pitt, J. M., Marabelle, A., Eggermont, A., Soria, J. C., Kroemer, G., & Zitvogel, L. (2016). Targeting the tumor microenvironment: Removing obstruction to anticancer immune responses and immunotherapy. *Annals of Oncology, 27*(8), 1482–1492. https://doi.org/10.1093/annonc/mdw168.
Pol, J., Bloy, N., Obrist, F., Eggermont, A., Galon, J., Cremer, I., et al. (2014). Trial watch: Oncolytic viruses for cancer therapy. *Oncoimmunology 3,* e28694. https://doi.org/10.4161/onci.28694.
Pol, J., Buque, A., Aranda, F., Bloy, N., Cremer, I., Eggermont, A., et al. (2016). Trial watch-oncolytic viruses and cancer therapy. *Oncoimmunology, 5*(2), e1117740. https://doi.org/10.1080/2162402X.2015.1117740.
Polsky, D., & Cordon-Cardo, C. (2003). Oncogenes in melanoma. *Oncogene, 22*(20), 3087–3091. https://doi.org/10.1038/sj.onc.1206449.
Postow, M. A., Chesney, J., Pavlick, A. C., Robert, C., Grossmann, K., McDermott, D., et al. (2015). Nivolumab and ipilimumab versus ipilimumab in untreated melanoma. *The New England Journal of Medicine, 372*(21), 2006–2017. https://doi.org/10.1056/NEJMoa1414428.
Puzanov, I., Milhem, M. M., Minor, D., Hamid, O., Li, A., Chen, L., et al. (2016). Talimogene laherparepvec in combination with ipilimumab in previously untreated, unresectable stage IIIB-IV melanoma. *Journal of Clinical Oncology, 34*(22), 2619–2626. https://doi.org/10.1200/JCO.2016.67.1529.
Rehman, H., Silk, A. W., Kane, M. P., & Kaufman, H. L. (2016). Into the clinic: Talimogene laherparepvec (T-VEC), a first-in-class intratumoral oncolytic viral therapy. *Journal for ImmunoTherapy of Cancer, 4,* 53. https://doi.org/10.1186/s40425-016-0158-5.

Restifo, N. P., Kawakami, Y., Marincola, F., Shamamian, P., Taggarse, A., Esquivel, F., et al. (1993). Molecular mechanisms used by tumors to escape immune recognition: Immunogenetherapy and the cell biology of major histocompatibility complex class I. *Journal of Immunotherapy with Emphasis on Tumor Immunology, 14*(3), 182–190.

Riley, J. L. (2013). Combination checkpoint blockade—Taking melanoma immunotherapy to the next level. *The New England Journal of Medicine, 369*(2), 187–189. https://doi.org/10.1056/NEJMe1305484.

Robbins, P. F., Morgan, R. A., Feldman, S. A., Yang, J. C., Sherry, R. M., Dudley, M. E., et al. (2011). Tumor regression in patients with metastatic synovial cell sarcoma and melanoma using genetically engineered lymphocytes reactive with NY-ESO-1. *Journal of Clinical Oncology, 29*(7), 917–924. https://doi.org/10.1200/JCO.2010.32.2537.

Rojas, J. J., Sampath, P., Hou, W., & Thorne, S. H. (2015). Defining effective combinations of immune checkpoint blockade and oncolytic virotherapy. *Clinical Cancer Research, 21*(24), 5543–5551. https://doi.org/10.1158/1078-0432.CCR-14-2009.

Sahin, U., Kariko, K., & Tureci, O. (2014). mRNA-based therapeutics—Developing a new class of drugs. *Nature Reviews. Drug Discovery, 13*(10), 759–780. https://doi.org/10.1038/nrd4278.

Sarkar, D., Lebedeva, I. V., Gupta, P., Emdad, L., Sauane, M., Dent, P., et al. (2007). Melanoma differentiation associated gene-7 (mda-7)/IL-24: A 'magic bullet' for cancer therapy? *Expert Opinion on Biological Therapy, 7*(5), 577–586. https://doi.org/10.1517/14712598.7.5.577.

Sarkar, S., Pradhan, A., Das, S. K., Emdad, L., Sarkar, D., Pellecchia, M., et al. (2016). Novel therapy of prostate cancer employing a combination of viral-based immunotherapy and a small molecule BH3 mimetic. *Oncoimmunology, 5*(3), e1078059, https://doi.org/10.1080/2162402X.2015.1078059.

Sarkar, D., Su, Z. Z., & Fisher, P. B. (2006). Unique conditionally replication competent bipartite adenoviruses-cancer terminator viruses (CTV): Efficacious reagents for cancer gene therapy. *Cell Cycle, 5*(14), 1531–1536. https://doi.org/10.4161/cc.5.14.3095.

Sarkar, D., Su, Z. Z., Park, E. S., Vozhilla, N., Dent, P., Curiel, D. T., et al. (2008). A cancer terminator virus eradicates both primary and distant human melanomas. *Cancer Gene Therapy, 15*(5), 293–302. https://doi.org/10.1038/cgt.2008.14.

Shao, H., Cai, L., Grichnik, J. M., Livingstone, A. S., Velazquez, O. C., & Liu, Z. J. (2011). Activation of Notch1 signaling in stromal fibroblasts inhibits melanoma growth by upregulating WISP-1. *Oncogene, 30*(42), 4316–4326. https://doi.org/10.1038/onc.2011.142.

Sharpless, E., & Chin, L. (2003). The INK4a/ARF locus and melanoma. *Oncogene, 22*(20), 3092–3098. https://doi.org/10.1038/sj.onc.1206461.

Soiffer, R., Lynch, T., Mihm, M., Jung, K., Rhuda, C., Schmollinger, J. C., et al. (1998). Vaccination with irradiated autologous melanoma cells engineered to secrete human granulocyte-macrophage colony-stimulating factor generates potent antitumor immunity in patients with metastatic melanoma. *Proceedings of the National Academy of Sciences of the United States of America, 95*(22), 13141–13146.

Soliman, H., Mediavilla-Varela, M., & Antonia, S. J. (2015). A GM-CSF and CD40L bystander vaccine is effective in a murine breast cancer model. *Breast Cancer (Dove Med Press), 7*, 389–397. https://doi.org/10.2147/BCTT.S89563.

Su, Z. Z., Shi, Y., & Fisher, P. B. (1997). Subtraction hybridization identifies a transformation progression-associated gene PEG-3 with sequence homology to a growth arrest and DNA damage-inducible gene. *Proceedings of the National Academy of Sciences of the United States of America, 94*(17), 9125–9130.

Tas, F. (2012). Metastatic behavior in melanoma: Timing, pattern, survival, and influencing factors. *Journal of Oncology, 2012*. 647684. https://doi.org/10.1155/2012/647684.

Thomas, C. E., Ehrhardt, A., & Kay, M. A. (2003). Progress and problems with the use of viral vectors for gene therapy. *Nature Reviews Genetics, 4*(5), 346–358. https://doi.org/10.1038/nrg1066.

Tong, A. W., Nemunaitis, J., Su, D., Zhang, Y., Cunningham, C., Senzer, N., et al. (2005). Intratumoral injection of INGN 241, a nonreplicating adenovector expressing the melanoma-differentiation associated gene-7 (mda-7/IL24): Biologic outcome in advanced cancer patients. *Molecular Therapy, 11*(1), 160–172. https://doi.org/10.1016/j.ymthe.2004.09.021.

Vacchelli, E., Eggermont, A., Sautes-Fridman, C., Galon, J., Zitvogel, L., Kroemer, G., et al. (2013). Trial watch: Oncolytic viruses for cancer therapy. *Oncoimmunology, 2*(6), e24612. https://doi.org/10.4161/onci.24612.

Vinay, D. S., Ryan, E. P., Pawelec, G., Talib, W. H., Stagg, J., Elkord, E., et al. (2015). Immune evasion in cancer: Mechanistic basis and therapeutic strategies. *Seminars in Cancer Biology, 35*(Suppl), S185–S198. https://doi.org/10.1016/j.semcancer.2015.03.004.

Viola, J. R., Rafael, D. F., Wagner, E., Besch, R., & Ogris, M. (2013). Gene therapy for advanced melanoma: Selective targeting and therapeutic nucleic acids. *Journal of Drug Delivery, 2013*. 897348. https://doi.org/10.1155/2013/897348.

Warnock, J. N., Daigre, C., & Al-Rubeai, M. (2011). Introduction to viral vectors. *Methods in Molecular Biology, 737*, 1–25. https://doi.org/10.1007/978-1-61779-095-9_1.

Weber, C. E., Luo, C., Hotz-Wagenblatt, A., Gardyan, A., Kordass, T., Holland-Letz, T., et al. (2016). miR-339-3p is a tumor suppressor in melanoma. *Cancer Research, 76*(12), 3562–3571. https://doi.org/10.1158/0008-5472.CAN-15-2932.

Welsh, S. J., Rizos, H., Scolyer, R. A., & Long, G. V. (2016). Resistance to combination BRAF and MEK inhibition in metastatic melanoma: Where to next? *European Journal of Cancer, 62*, 76–85. https://doi.org/10.1016/j.ejca.2016.04.005.

Wilgenhof, S., Van Nuffel, A. M., Benteyn, D., Corthals, J., Aerts, C., Heirman, C., et al. (2013). A phase IB study on intravenous synthetic mRNA electroporated dendritic cell immunotherapy in pretreated advanced melanoma patients. *Annals of Oncology, 24*(10), 2686–2693. https://doi.org/10.1093/annonc/mdt245.

Williams, D. A., & Baum, C. (2003). Medicine. Gene therapy—New challenges ahead. *Science, 302*(5644), 400–401. https://doi.org/10.1126/science.1091258.